D1559191

Beyond the Gene

Monographs on the History and Philosophy of Biology

RICHARD BURIAN, RICHARD BURKHARDT, JR.,
RICHARD LEWONTIN, JOHN MAYNARD SMITH
EDITORS

The Cuvier-Geoffroy Debate: French Biology in the Decades Before Darwin
TOBY A. APPEL

Beyond the Gene: Cytoplasmic Inheritance and the Struggle for Authority in Genetics
JAN SAPP

Controlling Life: Jacques Loeb and the Engineering Ideal in Biology
PHILIP J. PAULY

Beyond the Gene

Cytoplasmic Inheritance and the Struggle for Authority in Genetics

JAN SAPP

New York Oxford
OXFORD UNIVERSITY PRESS
1987

Oxford University Press

Oxford New York Toronto
Delhi Bombay Calcutta Madras Karachi
Petaling Jaya Singapore Hong Kong Tokyo
Nairobi Dar es Salaam Cape Town
Melbourne Auckland

and associated companies in
Beirut Berlin Ibadan Nicosia

Published by Oxford University Press, Inc.,
200 Madison Avenue, New York, New York 10016

Library of Congress Cataloging-in-Publication Data
Sapp, Jan.
Beyond the gene.
(Monographs on the history and philosophy of biology)
Originally presented as the author's thesis (doctoral—University of Montreal)
Bibliography: p. 1. Cytoplasmic inheritance. I. Title. II. Series.
QH452.S27 1987 575.1'09 86-12845
ISBN 0-19-504206-9

2 4 6 8 9 7 5 3 1

Printed in the United States of America
on acid-free paper

To Carole

Acknowledgments

This book is based on my doctoral dissertation carried out under the supervision of Camille Limoges and submitted to the University of Montreal in 1984. I owe much to Camille and to many others who helped in the creation of this work. I am particularly delighted in thanking Carole McKinnon who accompanied me and gave me unending support throughout all phases of this project. David Nanney and Richard Burian gave encouragement, valuable advice, and generous commentary on each of the chapters of this work.

Mrs. Ruth Sonneborn kindly invited me into her home at Bloomington, Indiana, and allowed me to search through and quote from letters in T. M. Sonneborn's papers. Janine Beisson made me her guest at the *Centre de Génétique Moléculaire du C.N.R.S.* at Gif-sur-Yvette and gave me access to Boris Ephrussi's papers. I am happy to thank my Aunt Olive Cassidy for lending me her beautiful cottage in Woods Hole, which allowed me to spend a summer near the excellent library of the Marine Biological Laboratory. I also appreciate the help of the staffs at the Rockefeller Archive Center in Tarrytown, New York, and at the *Bibliothek und Archiv der Max-Planck-Gesellschaft* at Berlin.

For interviews and correspondence by mail I have great pleasure in thanking André Adoutte, Geoffrey Beale, Jean Brachet, Ruth Dippell, Joseph Frankel, Nicholas Gillham, John Jinks, Joshua Lederberg, André Lwoff, Georg Melchers, John Preer, Marcus Rhoades, Ruth Sager, Piotr Slonimski, Ruth Sonneborn, and Diter von Wettstein. I also thank Guido Pontecorvo and Barbara McClintock for giving me permission to quote passages from their letters and Paul Harvey for those of C. D. Darlington.

Many friends and colleagues have provided intellectual support, and in particular, I would like to mention Randall Albury, Garland Allen, Alberto Cambrosio, Stephen Cross, Richard Gillespie, Yves Gingras, Tim Jordan, Peter Keating, Homer Le Grand, Robert Olby, and David Turnbull.

I would also like to express my gratitude to Bob Lee, who first introduced me to cytoplasmic genetics as a student at Dalhousie University, and to Rita Hutchison, who typed and retyped this manuscript.

This research was aided by a grant from the Social Sciences and Humanities Research Council of Canada in 1981. Financial assistance for typing was received from the Faculty of Arts Research and Development Scheme, University of Melbourne, 1984.

I am grateful to Reidel Publishing Company for giving me permission to use materials from:

Jan Sapp, "The Struggle for Authority in the Field of Heredity, 1900–1932: New Perspectives on the Rise of Genetics," *Journal of the History of Biology* 16, no. 3, fall 1983, pp. 311–342. Copyright © 1983 by Reidel Publishing Company, Dordrecht, Holland.

Jan Sapp, "Inside the Cell: Genetic Methodology and the Case of the Cytoplasm," in J. A. Schuster and R. R. Yeo, eds., *The Politics and Rhetoric of Scientific Method*. 1986, pp. 167–202. Copyright © 1986 by Reidel Publishing Company, Dordrecht, Holland.

Contents

Introduction

When a subject is highly controversial . . . one cannot hope to tell the truth. One can only show how one came to hold whatever opinion one does hold. One can only give one's audience the chance of drawing their own conclusions as they observe the limitations, the prejudices, the idiosyncrasies of the speaker. (Virginia Woolf, 1928, p. 6)

Genetics, more than any other aspect of the life sciences in the twentieth century, has become the object of intensive historical investigation over the past twenty years. Writers have generally focused on three major developments; the rise of Mendelian-chromosome theory (see Allen, 1975, 1978), the emergence of population genetics and the evolutionary synthesis (see Provine, 1971; Mayr and Provine, 1980), and the transformation of the chromosome theory to the nucleic acid (DNA) theory (see Olby, 1974; Judson, 1979). The intense concentration on these special research areas has resulted in a rich body of historical information and commentary. At the same time, however, it has led to a somewhat selective history of genetics. A vital aspect of the history of genetics that has been consistently neglected by writers focusing on these developments is the research and theories of cytoplasmic inheritance. Cytoplasmic inheritance concerns hereditary materials and principles that are located outside the nucleus of the cell. Mendelian inheritance is due to genes (DNA) located in the chromosomes of the nucleus of the cell. Both parents (sperm and egg) contribute equally to Mendelian inheritance. However, cytoplasmic traits are transmitted largely maternally. Egg cells contain far more cytoplasm than sperm cells, which are generally made up of little more than a nucleus and its delivery system.

The history of cytoplasmic inheritance research is a complex one, and its neglect is due largely to two developments. First, only in recent years has the importance of cytoplasmic inheritance come to be generally recognized by geneticists. As a result, historians are generally unfamiliar with this aspect of modern genetics. Today, various kinds of phenomena are discussed under the rubric of non-Mendelian, cytoplasmic inheritance. Some of the principal and generally recognized cytoplasmic genetic factors are cytoplasmic DNA-based genes associated with the energy-generating organelles of the cell: mitochondria and chloroplasts. There are also cases of cytoplasmic inheritance due to the transmission of symbiotic microorganisms. In addition to these mechanisms, there remain two other classes of cytoplasmic inheritance that geneticists do not readily associate with the information carried in DNA. One involves information carried in self-perpetuating metabolic patterns. The other involves the perpetuation of cellular organization and the informational role of supramolecular structure.

Nevertheless, the inheritance of cytoplasmic submicroscopic particles, structures, and patterns has been continuously discussed and investigated by various groups of biologists throughout the history of genetics and has been the subject of intense controversy. During the first half of the century investigations of cytoplasmic inheritance were carried out close to the margins of, and developed in constant tension and conflict with, the research program and doctrines of Mendelian genetics. They became allied with, or appropriated by, those forces which opposed the general genetic synthesis, specifically those advocating the Lamarckian notion of the inheritance of acquired characteristics, macromutations, and, during the late 1940s, Lysenkoism. The significance of some of the more recent experimental results concerning cytoplasmic genes and supramolecular structure remains a subject of controversy in the field.

The second reason for the lack of discussion of cytoplasmic inheritance in the history of genetics is that historical studies have typically been carried out so as to construct a synthesis and convergence of disciplines. That is to say, on the one hand, how biometry, Darwinian evolution and natural history, and the Mendelian-chromosome theory converged to provide a logically consistent theory of the origin of species; and on the other hand, how embryology, biochemistry, cytology, and Mendelian genetics came together to form a unified, cellularly and physiologically oriented view of development. The major developments in these areas of inquiry may be summarized very briefly as follows: In 1900, Mendel's theory for the transmission of hereditary characteristics, written during the mid-1860s, was rediscovered independently by Carl Correns, E. von Tschermak, and Hugo de Vries. After 1910, the advancement of Mendelism was led by the *Drosophila* group headed by T. H. Morgan in the United States, who established a physical basis for genes in chromosomes. By 1915 the Mendelian chromosome theory had met with "general acceptance" by the biological community.

The development of Mendelian genetics was not always smooth, of course. Many biologists had initially objected to Mendelian theory, which appeared to focus attention on morphological units, as opposed to the process by which the units functioned. The early geneticists themselves were divided on significant issues. In England William Bateson, the first major Mendelian publicist, was unable to follow genetics all the way to the chromosome theory. Moreover, the first generation of Mendelian geneticists, who supported discontinuous evolution, were in sharp conflict with the intellectual offspring of Francis Galton, statisticians and Darwinians who supported continuous variation. The American Mendelians were in conflict with American Darwinians and naturalists. The gap between Mendelism and Darwinism began to be bridged during the 1930s and 1940s when statistical foundations were laid for extending the Mendelian theory of individuals to populations. The early statistical work in this area was led by R. A. Fisher and J. B. S. Haldane in England and Sewall Wright in the United States, who laid the groundwork for the synthetic theory of evolution and its development during the 1950s and 1960s.

During the 1940s the biochemical genetic research programs led by George Beadle in the United States established that genes controlled biochemical process in the cell by controlling the specificity of enzymes. By the 1950s diverse studies led to the conclusion that nucleic acids rather than proteins or nucleoproteins were the principal substance of heredity, and in 1953 J. D. Watson and Francis Crick formulated the structure of DNA as a double helix. By the mid-1960s the genetic code had been solved and an outline was constructed of the nature of the gene, how it was expressed in the assembly of proteins, and how its expression was controlled.

Within the theoretical orientation of the synthetic approach to the history of genetics, there is very little room for research on cytoplasmic inheritance. Yet cytoplasmic inheritance, recognized as a significant area of research, emerged at the same time as Mendelism, and as we shall see, it was persistently attacked and criticized by nucleocentric geneticists. Throughout most of the twentieth century, the theoretical limits of this dispute were framed by two extreme positions. On the one hand, many biologists who protested against neo-Darwinian conceptions of heredity and evolution (especially embryologists) maintained that the cytoplasm of the egg cell was primarily responsible for the "fundamental" characteristics of the organism, that is, those characteristics which distinguished higher taxonomic groups. According to this view, Mendelian or nuclear inheritance was largely concerned with relatively trivial differences. Mendelian genes added the finishing touches to the organism, such as eye color, hair color, or tail length. On the other hand, Mendelian geneticists, especially in the United States, and neo-Darwinian evolutionists upheld the predominant, if not exclusive, role of the Mendelian gene in heredity. By the 1920s most American geneticists had come to view genes as discrete physical units located on chromosomes. They claimed that genes were the "governing elements" of the cell, largely immune from the rest of the cell, yet dictating its activities. T. H. Morgan expressed this view succinctly in 1926: "In a word the cytoplasm may be ignored genetically."

This opposition to the dominant role of the nuclear gene in heredity and evolution cannot be taken seriously by historians of the grand synthesis whose writings reflect the old extreme views of many American geneticists. The synthetic approach prevents us from seeing or making sense out of the history of the research and theories concerning cytoplasmic inheritance. Writers who have adopted this approach often begin with the explicit or implicit assumption that the dominant theories of our times are correct or at least represent a closer approximation to the truth than competing theories. They have come to believe in them so strongly that they often fail to recognize the social dimension of that very triumph. When they mention non-Mendelian views at all, they treat them as "obstacles" to the progress of reason. They account for them as "premature" or as the result of "confusion," the intrusion of "nonscientific" considerations into science, or an improper "scientific method." This approach would be acceptable if one reliable scientific method (or set of methods) existed that invariably produced objective knowledge. But the problematic character of such an assumption is recognized by many historians and philosophers, as well as by scientists themselves. Once we fully realize that scientists are constantly negotiating what science is, then it is necessary to consider social interests on both sides of scientific controversy.

In this book the field of heredity in the twentieth century is viewed as the product of a social system of conflict and competition. Instead of investigating the conceptual steps leading to a "synthesis," I focus on the social and cognitive relations of biologists engaged in a struggle for authority. Scientific authority, as partly manifested in prestige and fame, is understood as each participant's socially recognized capactiy to act and speak legitimately in scientific matters. What is at stake in this struggle is the definition of the field itself: What questions are important, what answers are acceptable, what organisms are useful, what techniques are appropriate, and what phenomena are interesting. However, the competition between scientists and scientific controversies cannot be reduced to a "free" competition of ideas where the strength of the "true" idea decides the outcome. Scientists do not acquire authority and transform the nature of science simply through their contributions to scientific knowledge; they are also actively engaged in changing the field socially. They not only receive recognition; they grant it as well through various strategies and tactics ranging from teaching to refereeing papers and reviewing research grants. As we shall see in this book, there is no neutral or value-gree way of engaging in this activity. When scientists attempt to impose a definition of the field, each participant tends to uphold those scientific values which are most closely related to him or her personally or institutionally. This principle illuminates the links between controversies at different levels, such as those between laboratories, institutions, disciplines, and countries. It is developed from the conception of the scientific field as first formulated by Pierre Bourdieu (1975).

The struggle for authority throughout the twentieth century will be highlighted as we follow the development of the research and theories concerning cytoplasmic inheritance. We shall see in Chapter 1 that the idea that the cytoplasm played the predominant role in heredity was first proposed by many embryologists who could not account for the orderly and "directed" nature of the development of the organism as a whole in terms of Mendelian genes. In Chapter 2 I will investigate the discursive and social process by which Mendelian geneticists in the United States who upheld the predominant if not exclusive role of the nuclear genes rose to an authoritative position in the field by the 1920s and 1930s. It will be argued that American geneticists established their authority by forming their own discipline with its own well-defined objectives, techniques, explanatory standards, doctrines, journals, and societies and by restricting their investigations to problems which could be effectively dealt with by Mendelian procedures. In Chapter 3 we shall see that between the two World Wars genetic investigations of cytoplasmic inheritance were carried out primarily in Europe, especially in Germany, where many investigators attempted to provide definitive evidence to challenge what they called the "nuclear monopoly." However, the evidence for cytoplasmic inheritance was attacked and criticized especially by American geneticists and classical neo-Darwinian evolutionists.

The controversy over the relative importance of the cytoplasm and the nucleus intensified in the late 1940s and 1950s with the rise of microbial and biochemical genetics. Nonetheless, investigations of cytoplasmic inheritance were still carried out on a relatively modest level in the United States compared to the extensive work on nuclear inheritance. As will be discussed in Chapter 4, research on and theories of cytoplasmic heredity in the United States developed in constant conflict and tension with the predominant biochemical genetic research programs led by George Beadle and his many followers. Although the microbial genetic evidence for cytoplasmic hereditary materials was continually criticized and trivialized by many American geneticists led by Beadle and H. J. Muller during the 1940s and 1950s, it was well received in France following World War II. In Chapter 5 we shall see that genetic research programs on cytoplasmic inheritance and problems of genetic regulation were developed in France during the 1940s and 1950s as an explicit strategy to compete with American nucleocentric genetics. During the same period the genetic evidence for cytoplasmic inheritance became caught up in a Cold War dispute when, as we shall see in Chapter 6, it was used by Lysenkoists to dismiss the Mendelian chromosome theory altogether. In Chapter 7 we shall see that cytoplasmic inheritance declined from theoretical discussions with the rise of molecular biology during the late 1950s and 1960s, when new explanatory standards, concepts, and doctrines emerged in the field. However, genetic investigations of the nature of cytoplasmic genes and supramolecular structure carried out from the 1960s through the 1980s continued to challenge genetic and evolutionary orthodoxy.

When detailing this history special attention will be given to methodological discourse when the role of the cytoplasm in heredity was being evaluated by disputants. This includes not only formal methodological doctrines (inductivism, hypothetico-deductivism, falsificationism, etc.) but also social accounts. As discussed most prominently by Mulkay and Gilbert (1982), scientists use a variety of social accounts when explaining the conflicting beliefs of their competitors, such as "a defensive attitude," "prejudice," "dislike," "failure to put enough effort," and "being trained in terms of a false theory." As we shall see, these kinds of accounts were used by biologists on both sides of the nucleo-cytoplasmic controversy. Proponents of cytoplasmic inheritance also pointed to the reward system in the field and the motivations of their competitors in an attempt to delegitimate nucleocentric theory by attempting to show that its dominance was due to the social power of its adherents rather than to its inherent cognitive superiority. At the same time, geneticists in the controversy employed a variety of politicoeconomic metaphors when discussing modes of genetic control in the cell, such as "nuclear monopoly," "dictatorial agents," "democratic organization," "the cell as an empire," "republic of chromosomes," etc. We shall see that these metaphors not only reflected the larger cultural ideology of the disputant, but served as rhetoric to reflect perceived power relations among the competitors inside the field itself. Proponents of the importance of cytoplasmic inheritance also frequently called upon what I term the "technique-ladenness of observations" in their attempts to advance alternative theories and to account for the beliefs of those geneticists who upheld the predominant or exclusive role of nuclear genes in heredity. By "technique-ladenness of observations" I refer to the means by which scientific concepts become bound to phenomena studied by certain techniques. It is meant to embrace the scientific equipment, procedures, materials, tools, and skills that are involved in the production of scientific results.

National dichotomies due to differences in institutionalized modes of social control (such as job control, the referee system of scientific journals, and means of gaining financial support from private and public funding agencies), established scientific traditions, and international competition have to be taken into consideration when accounting for the different nature of genetic research and orthodoxy in various countries throughout the century. The issue of specialization and disciplinary formation will also be discussed when accounting for the rise of cytoplasmic genetic research in Germany between the two World Wars. It will also be discussed when investigating the institutionalization of genetics in France after World War II. It will be shown that the centralized control of the university system in those countries helped to greatly impede the institutional development of genetics.

In most cases discussed in this book scientific concepts have been formed primarily out of the internal or domestic politics of science. However, the broader political and cultural context of science also has to be taken into consideration when attempting to evaluate the significance of cytoplasmic inheritance and the problems it posed for its researchers. These issues become especially poignant when investigating concepts of heredity at the outset of the Cold War inside and outside of biology.

Beyond the Gene

CHAPTER 1

Defining the Organism

Inheritance must be looked at as merely a form of growth. . . . (Charles Darwin, 1868, p. 404)

With the rise of Mendelian-chromosome theory as formulated by T. H. Morgan and his *Drosophila* school, many embryological investigators voiced opposition to the exclusive role of the nuclear gene in heredity. Many went so far as to claim that Mendelian genetics was concerned only with characteristics that did not exceed the framework of the species. In effect, they asserted that chromosomal genes determined trivial characteristics with little evolutionary significance, and in direct conflict with the views of Mendelian geneticists in the United States, they claimed that the cytoplasm determined the fundamental constitution of plants and animals. Among the particularly outspoken early supporters of the importance of the cytoplasm in heredity were a number of leading experimental biologists, including E. G. Conklin, J. W. Jenkinson, F. R. Lillie, Albert Brachet, and Jacques Loeb. These biologists and many others viewed the Mendelian-chromosome theory to be incomplete and of little value for understanding the processes of development and evolution.

The idea that the cytoplasm played a predominant role in heredity was not new during the first decades of the present century. The belief in the determinative action of the cytoplasm in heredity emerged from observations and theoretical arguments of experimental embryologists in relation to the nuclear theory of heredity which rose to prominence during the 1880s and 1890s. To a great extent the major controversy which raged over the roles of the nucleus and cytoplasm in heredity throughout most of the twentieth century grew out of the historical context of late nineteenth-century embryology and cytology and cannot be understood without its consideration.

The Organism as a Chinese Box

The nucleus cannot operate without a cytoplasmic field in which its peculiar powers may come into play but this field is created and molded by itself. Both are necessary

> to *development;* the nucleus alone suffices for the *inheritance* of specific possibilities of development. (E. B. Wilson, 1896, p. 327)

Between 1875 and 1890 biologists' conception of the cell was undergoing a radical transformation. During this period cytologists attempted to bring cell theory and evolutionary theory into organic connection. Their theoretical construction was based upon a distinction between the roles played by the constituents of the cell. With the aid of the light microscope coupled with the use of chemical substances which selectively stained cellular structures, little by little the nucleus emerged in the cell (Coleman, 1965). Within it, the darkly staining chromosomes began to assume the chief role in providing the physical link between the cell and evolution. Their physical continuity through the cell cycle, the constancy of their numbers, the accuracy of their movements, and the longitudinal splitting of the chromatin threads, along with the fusion of male and female pronuclei in fertilization, all combined to give them an exceptional position in cytological discourse. Like the cell, the nucleus exhibited physical continuity. It was never formed *de novo,* but always arose by the division of a preexisting nucleus and thus satisifed the cytological criteria as a "bearer of heredity."

The nucleus was sharply distinguished by Oscar Hertwig, E. Strasburger, Hugo de Vries, and August Weismann and their many followers from the "cytoplasm" of the cell. They claimed that the quality of the nuclear substance dictated whether the unborn organism would become a dog or a cat, whether it would be large or small, male or female. They regarded the nucleus as the "controlling center of cell activity" and hence a primary factor in growth, development, and the transmission of specific qualities from cell to cell, and from one generation to another. The nucleus was thought to at once ensure inheritance and allow variation and direct ontogeny, and what is commonly known as the nuclear theory of heredity was formulated.

This merger occurred at a time when Darwinian theory, which previously included the notion of the inheritance of acquired characteristics, was beginning to be identified solely with natural selection. During the 1880s and 1890s leading biologists, including August Weismann in Germany and the British biologists Ray Lankaster and Alfred Russell Wallace, came to advocate the all-sufficiency of natural selection to account for evolution. Weismann led the theorizing on the relationship between the nucleus, heredity, and evolution (Churchill, 1968).

During the 1880s Weismann set out to purge evolutionary theory of any recourse to the idea of the inheritance of acquired characteristics. He postulated that natural selection worked on dispositions which lay hidden in the "germ plasm" of the germ (Weismann, 1893). In their hereditary potential, Weismann envisaged germ cells to be autonomous of the somatic cells. He claimed that they were derived directly from germ cells of the preceding generation and were not a product of the organism. Bypassing the somatic cells, the hereditary potentialities of the germ plasm were sheltered from the modifying influence of the environment of the organism. All changes due to outside influences would be temporary, and disappear with the next generation. Changes resulting from environmental conditions such as injuries, use and disuse of parts, temperature, nutrition, or any other influence of the environment on the body could not be transmitted to the germ cells and therefore were not hereditary variations. The proposed stable structure of the germ plasm in the nucleus, together with pure selection theory, would account for the origin of species.

To account for morphogenesis, Weismann (1893), along with Wilhelm Roux, proposed the idea that the highly complex structure of the nucleus was composed of self-duplicating determinants. Each determinant would represent or determine some character of the organism. To explain the existence of organismic integration, Weismann assumed that the determinants in the nucleus were integrated into groups of higher orders of magnitude corresponding to cells or cell groups. Roux advanced the hypothesis of qualitative nuclear divisions as the basis for the orderly harmonious differentiation of the organism and Weismann adopted it. Embryonic development would consist largely of qualitative nuclear divisions which segregate chromatin material into increasingly divergent and differentiated units. So far as normal development was concerned, environmental factors were confined to an external role; they were important only in that certain factors were essential to the continuance of life and the program of predetermined development.

During the second half of the nineteenth century, biology was dominated by similar particulate theories of heredity which postulated some sort of material corpuscle as the ultimate basis of life (see, for example, Delage, 1903). Recognized as the "bearers of heredity," a large number of such corpuscles were thought to, somehow, build up the individual organism. The "gemmules" of Charles Darwin, the "pangenes" of Hugo de Vries, the "physiological units" of Herbert Spencer, the "granules" of Richard Altmann, and other hypothetical entities were highly considered by biologists for almost half a century. All these theories possessed certain features in common. The determinants, whatever they were called, were all endowed with certain peculiarities which permitted them to play a specific part in the development of the individual organism. From this notion, of what some embryologists would later refer to as the "elementary organism" (see Child, 1924, pp. 18–32), developed the belief held by many biologists, that all problems of the organism must be understood in terms of determinants, granules, particles, cells, etc.

This corpuscular conception was habitually applied to every grade of organization during the second half of the nineteenth century. In effect, the fundamental internal workings of an organism were regarded as a series of Russian dolls or Chinese boxes. The higher organism was regarded as a colony of cells; the cell as a colony of simpler units, nucleus, centrosome, and so on; the nucleus as a colony of chromosomes; the chromosome, according to Weismann's terminology, as a colony of "ids"; the id as a colony of "determinants"; the determinant as a colony of "biophores"; and the biophore as a colony of molecules.

The speculative nature of the Weismannian theory quickly excited considerable adverse criticism on the part of experimentalists during the 1890s. Nonetheless, the theory of the germ plasm played an important role in the development of biological research, for it framed a set of ideas in a manner sufficiently logical to serve as a working hypothesis or basis of attack. It was in relation to the nuclear theory of heredity that many experimental embryologists, turned to the cytoplasm of the cell for an explanation of the principal causes of development, heredity, and evolution.

"The Organism as a Whole"

> The nucleus contains the physical basis of inheritance . . . and chromatin, its essential constituent is the idioplasm. . . . This conclusion is now widely accepted and rests upon a basis so firm that it must be regarded as a working hypothesis of high value. To accept it is, however, to reject the theory of germinal localization in so far as it assumes a prelocalization of the egg-cytoplasm as a fundamental character of the egg. (E. B. Wilson, 1900, p. 403)

The preoccupation of embryologists with the cytoplasm became intense following the celebrated experiment of Hans Driesch (1891), in which he isolated the first two blastomeres of the sea urchin egg. If, during cell division, changes in the hereditary content of the nucleus had taken place, as Weismann had proposed, one would expect the separated blastomeres to develop abnormally. Instead, however, two complete larvae resulted. The hereditary equivalence of the blastomeres indicated that progressive differentiation could not result from qualitative divisions of nuclear material as had been imagined. Driesch's findings were quickly supported by an abundance of evidence which was held up in opposition to the predominant role of the nucleus in the control of development and heredity. It was clear from cytological investigations of the dividing cell that the nuclear material was precisely divided at each mitotic division and evenly distributed to daughter cells. There seemed to be no fundamental nuclear differentiations as Weismann had envisaged, except at maturation divisions. Each cell, then, inherited the sum total of hereditary qualities of the nucleus.

Many embryologists quickly confirmed Driesch's initial results. The methods most generally used by experimental embryology up until the 1930s consisted in the amputation or incomplete separation of parts of the embryo, or in their rearrangement by compression, transplantation, or centrifuging. These methods, of course, did not touch upon the actual nature of the physical and chemical changes underlying development. Nonetheless, from the large mass of work done by these methods, one important generalization emerged: in practically all eggs, a part had the power to give rise to more than it would if left in its normal surroundings, and in many cases a part could give rise to a whole. These results demonstrated for many embryologists the truth of the principle that "the organism as a whole controls the formative processes going on in each part." In the well-known aphorism of Driesch, the whole problem of development seemed to be brought into focus:

> *The relative position of a blastomere in the whole determines in general what develops from it; if its position be changed, it gives rise to something different.* In other words, *its prospective value is a function of its position.* (Quoted in Wilson, 1925, p. 1056)

The principle of the "organism as a whole" became a central tenet of embryology; it was its capital problem to find a physicochemical basis for it.

Both points—the regulative qualities of the egg and the functional equivalence of the blastomere nuclei—led many embryologists between 1891 and 1910 to localize the primary seat of differentiation in the cell cytoplasm. Many embryologists, including C. O. Whitman, T. H. Morgan, William Bateson, Yves Delage, F. R. Lillie, and E. G. Conklin, rejected the conception of the complex organism in terms of preformed elements or "elementary organisms" in the Weismannian sense. They could not accept the view that the organism arose as a result of a secondary adaptation of independent units whether cells, chromosomes, or molecules.

The notion that the whole organism subsisted only by means of reciprocal action of the single elementary parts was for them inadequate to explain the harmonious whole manifested by the organism. The fact that each of the parts of the egg was capable of developing into a complete organism, and yet did not do so when left in its natural position, proved that the developing germ, the embryo, was an integrated unit. It had the properties of a "supracellular continuum"; cell boundaries appeared to be no obstacle to the all-pervading integrative forces of the organism as a whole. They recognize the cell to be the lowest biological unit endowed with the integrative properties of an organism.

Epigenetic theories of development were adopted in direct conflict with the particulate, preformationist understanding of the organism (Fischer, 1976; Gilbert, 1978; Fantini, 1985). In epigenetic theory the role of the "elementary organism" or "unit" was very different from that proposed by Weismann and others. The individual elementary units comprising a complex organism, e.g., the cells of a blastula, were not necessarily predetermined as different parts but could be primarily all alike in constitution. The differences which arose during development were thought to be determined by the action of environmental factors upon whole groups of cells and upon each member of them.

In brief, epigenetic theory conceived the organism as a product of the reaction between a particular kind of protoplasm, whether in the form of a single cell or of many cells, and environmental factors. The cells represented the product of such reactions, and their grouping with others to form complex organisms involved further reactions of the same sort, and so on. From the epigenetic viewpoint, then, a particular organism, whether single-celled or complex, represented the behavior of a particular protoplasm in a particular environment.

Before becoming lost in admiration of the "purposive" or regulatory processes of the embryo and becoming an avowed vitalist (Churchill, 1969), Driesch formulated a theory of development based on nucleo-cytoplasmic interactions. He suggested that the nucleus would effect chemical changes in the cytoplasm by a "fermentative action" on "organogenic materials." "The determining, organogenic materials," he argued, "are not produced directly by the nucleus, but arise in the cytoplasm only under the direction of the nucleus" (Driesch, 1884, translated by Oppenheimer, 1965, p. 217). Driesch contended that the cytoplasm contained a specific plasma-body which responded to stimuli external to the cell, and which in turn influenced the nucleus to produce ferments which, in turn, affected the cytoplasm at different places and at different times.

Epigenetic theory presented difficulties as well. Although the effects of the environment mediated through metabolic reactions in the cell cytoplasm could account for modifications in individual development, it was clear to many embryologists that the process of epigenesis, alone, could not be responsible for the typical and specific *form* of the organism. Morphogenesis was clearly an orderly and purposeful process. Some sort of organization or plan was required to direct the epigenetic process. Experimental embryologists clearly perceived one of the requirements imposed by any system, through which, at each generation, the form of the parents is reproduced in the offspring by the combination of basic constituents: the intervention of a "spatial principle" which would bring parts into place so that they combined in the proper way and at the proper time.

Since nuclear division was not the basis of organogenesis, yet the origin of form was not due to the action of the environment, it seemed to many embryologists that there must be somewhere in the cytoplasm of the egg a definite primordial directive structured organization which would be responsible, at least, for the general orientation and symmetry of the organism. This argument led to numerous extensive cytological studies of egg structure and cytoplasmic prelocalization of organs and experimental demonstrations of the importance of cytoplasmic differences for progressive differentiation.

The existence of some sort of inherent submicroscopic structure, or organization, in the cytoplasm of the cell found embryological support from observations of differentiation without cleavage in ciliated protozoa and was sometimes indicated by the visible spatial distribution of cytoplasmic materials of certain egg cells. Indeed, the observations of American embryologists who conducted cell lineage studies in the 1890s, such as C. O. Whitman, F. R. Lillie, E. G. Conklin, and others, led to the conclusion that the cytoplasm of the egg was not isotropic as O. Hertwig and other cytologists who investigated cell fertilization had imagined. Cell lineage research attempted to trace the organs, tissues, and even the germ layers of the embryo, cell by cell, back to their earliest appearance (Maienschein, 1978). These studies were concerned with explaining the development of the individual organism in physicochemical terms and with elucidating evolutionary relationships among organisms. Besides the existence of an intimate organization of the cytoplasm, spoken of as its polarity and bilaterality, there were even cases in which a real specification of special parts of the germs existed, a relation of these special parts to special organs, and this sort of specification was also shown to exist in the cytoplasm (Conklin, 1905a,b, 1908, 1915; see also Weiss, 1939, pp. 222–256).

In brief, the situation was as follows: In many animal eggs there was a polarity, a symmetry, and a stratification of the cytoplasm that were closely correlated with the early stages of embryonic development and thus with the final stages of differentiation as represented by the adult. In other words, the early cleavage stages of the oosperm, while apparently allotting to each pair of daughter cells equivalent portions of chromatin, did not necessarily allot to the daughter cells equivalent amounts or equivalent types of cytoplasm. The cytoplasm in different parts of the egg exhibited differences in pigmentation, viscosity, and other properties, which could be followed visually during the early stages of cleavage.

From such observations it appeared that different types of cytoplasm existed that directly influenced development. The cytoplasmic materials in the vicinity of the animal pole gave rise to ectoderm or the outer cell layer of the embryo, while those at the vegetative pole gave rise to the endoderm or the inner cell layer; the polar axis of the egg remained the polar axis of the embryo. In fact, the localization of the cytoplasmic materials was so definite and so constant in the egg that characteristic organizational patterns could be recognized for different phyla, and seemingly identifiable portions of these patterns were traceable in the embryo after organ differentiation occurred.

These observations were supported by experimental reports which claimed that (1) when the localization patterns of these fertilized eggs were disturbed experimentally, development exhibited correlated disturbances; (2) when eggs developed without fertilization, by either natural or artificial parthenogenesis, the characteristic polarity, symmetry, and pattern of the adult were found in the cytoplasm just as if the egg had been fertilized; (3) when eggs and other cells were centrifuged, no displacement of the polar axis itself occurred; (4) even the nucleus could be displaced (as shown by the earlier experiments of Driesch), either by centrifuging or by mechanical pressure, and could move extensively through the cytoplasm under normal conditions, without changing the cell polarity.

All this led many embryologists to the conclusion that the fundamental basis of polarity must be sought in the "ground substance" of the cytoplasm and must depend upon some configuration of heterogeneous physical or chemical properties. The theoretical directions investigators took in their advocacy of a determinative for the cytoplasm varied somewhat. In order to understand the particular character of the views of Whitman, Lillie, and Conklin, each will be treated in turn.

C. O. Whitman (1864–1910) was the first director of the Marine Biological Laboratory at Woods Hole and founder of the *Journal of Morphology,* both established in 1888. Whitman was the first biologist to conduct cell lineage studies and first declared in 1888 that germ layers could be traced to special blastomeres. In a theoretical paper written in 1893 and highly acclaimed by American embryologists, entitled "The Inadequacy of the Cell-Theory of Development," Whitman launched one of the first published protests against what he saw as the "anthropomorphic" conception of embryonic development. In its essence, the paper was a plea for the recognition of a microstructure in the cytoplasm of the egg as the basis of the organization of the individual.

Whitman could not accept the view that the organism was simply a community of individual cells bound together by interaction and mutual dependence. For him, the unit of the organism was not the result of a "physiological division of labor," whereby with an "exchange of services" and with "the struggle for existence" in operation here as elsewhere, the units became more and more intimately associated. The properties of the organism represented more than the sum of its parts. In contrast to viewing the unity of the organism in terms of an economic division of labor, he sought a structural foundation for it. He wrote:

> It is not division of labor and mutual dependence that control the union of the blastomeres. It is neither functional *economy* nor social instinct that binds the two halves

> of an egg together, but the constitutional bond of *individual organization*. It is not simple adhesion of independent cells, but integral structural cohesion. (Whitman, 1893, p. 649)

Whether or not the organization of the organism evolved from symbiotic advantages resulting from the struggle for existence was not an immediate concern for Whitman. "It is enough for the present purposes," he argued, "to know that organization exists and that organic unity depends on *intrinsic properties* no less than does molecular unity" (Whitman, 1893, p. 649). In Whitman's view, after the discovery of cell division as the law of cell formation and after the scheme of the cell set up by Schleiden and Schwann had been revised by Leydig, Max Schultz, and others, the next step forward in the cell theory had to be credited to Ernst Brücke. Recognizing that life could not be ascribed to a structureless substance, Brücke wrote:

> We must therefore, ascribe to living cells, in addition to the molecular structure of the organic compounds that they contain, still another structure of different type of complication; and it is this which we call by the name of organization. (Brücke, 1861, p. 368, translated in Wilson, 1925, p. 670)

There exists a "pre-organization," Whitman claimed, "a grade of organization as the result of heredity." This "organization," or "*structural foundation*" was the starting point of each organism. It preceded cell formation and regulated it. Whitman and many other embryologists rejected the Weismannian distinction between germ plasm and somatoplasm. In principle it was possible that body tissue could give rise to a complete organism and that the gonads were not completely insulated from the environmental forces which effect changes in the soma. Whitman and other embryologists who followed him supported the "continuity of organization" by calling upon the experimental work of Driesch:

> The organization of the egg is carried forward to the adult as an unbroken physiological unity, or individuality, through all modifications and transformations. The remarkable inversions of embryonic material in many eggs, all of which are orderly arranged in advance of cleavage, and the interesting pressure experiments of Driesch by which a new distribution of nuclei is *forced* upon the egg, without any sensible modification of the embryo, furnish, I believe, decisive proof of a definite organization in the egg, prior to any cell formation. (Whitman, 1893, p. 657)

Although Whitman's views on the formative influence of the "organism as a whole" were well received by embryologists, his claim of a primordial structural organization in the egg cell remained controversial and was attacked by E. B. Wilson and T. H. Morgan. In Wilson's view only the nucleus contained hereditary potentialities. During the 1890s he remained greatly impressed by the physical continuity and coordinated behavior of the chromosomes during fertilization and cell division. In fact, although he later changed his views, Wilson (1896, pp. 326–327) claimed that all cytoplasmic differences arose epigenetically and could be traced back to "nuclear domination" during the early history of the egg.

To support his nucleocentric views, Wilson embraced the theory of pangenesis as developed by the Dutch plant physiologist Hugo de Vries. In 1889 de Vries proposed a model of cellular differentiation in terms of nuclear control which in

Wilson's view lacked some of the major difficulties of that proposed by Weismann and Roux. De Vries claimed that innumerable self-propagating submicroscopic particles called "pangenes" existed in the nucleus, each one of which predetermined the formation of one of the adult cells. The nuclear pangenes migrated to the cytoplasm step by step thereby determining the successive stages in development and the differentiation of the cytoplasm. This view was afterwards endorsed by Weismann, and Wilson himself saw it as a neo-Darwinian modification of Darwin's celebrated hypothesis of pangenesis which he proposed in 1868. In order to explain the inheritance of acquired characteristics, Darwin postulated that pangenes arose in the body from individual tissue cells and were transported to the germ cells where they accumulated. De Vries, however, denied such transportation of particles from cell to cell and claimed that his pangenes originated in the germ cells, not the somatic cells (see Wilson, 1900, pp. 403–404).

Morgan, on the other hand, continued to maintain a strong epigenetic outlook. He could accept no preformationist beliefs, whether they were nuclear or cytoplasmic. He made his opinion of Whitman's position explicit in the following passage:

> His view of the organization of the cell is by no means clear to me and so far as I understand his meaning I do not agree with his view. On the other hand, most of the statements made in respect to the value of the cell in ontogeny seem to me to carry much truth with them. (Morgan, 1895a, p. 124)

Whitman's views and arguments for a primordial organization in the cell cytoplasm which must be regarded as a leading factor in ontogeny did, however, find vigorous support from other leading American embryologists, including Lillie and Conklin.

F. R. Lillie (1870–1947), a student of Whitman, received his doctoral degree from the University of Chicago in 1894 and later succeeded Whitman as director of the Marine Biological Laboratory (MBL) at Woods Hole. During his sixteen years as director of the laboratory, with the financial assistance of his wife's favorite brother, he would bring the MBL into prominence as an international center of embryological research. Lillie's doctoral research was on the cell lineage of the marine annelid *Chaetopterus*. His emphasis was on how the special features of cleavage in each species were adapted to the needs of the future larva. He carried out intensive cytological studies on eggs with rigid mosaic cleavage (with developmental fates of cleavage cells already determined in early cleavage stages) in an attempt to find some material basis for cleavage patterns, polarity, bilateral symmetry, and so on. His light microscopic studies on the organization of egg cytoplasm in normal and centrifuged eggs, coupled with studies on the extent to which differentiation without cleavage is possible in activated *Chaetopterus* eggs, led him to the conclusion that the control of early development must reside in the architecture of the "ground substance" of the egg.

Following Whitman, Lillie protested against the view that embryonic development arises as a result of a secondary adaptation of cells. Cell division was not a cause of progressive differentiation of cells, but was only a means. As Lillie (1906, p. 252) put it:

> The organism is primary, not secondary, it is an individual, not by virtue of the cooperation of countless lesser individualities but an individual that produces these lesser individualities on which its full expression depends. The persistence of organization is a primary law of embryonic development.

What Whitman called "organization" Lillie (1906, p. 251) termed "action of the organism as a whole":

> *There are certain properties of the whole constituting a principle of unit of organization, that are part of the original inheritance, and thus continuous through the cycles of the generations, and do not arise anew in each.*

In embryonic development the "principle of unity" revealed itself first by axial polarization, second by bilateral polarization and determination of the localization pattern, and third by adaptation in cleavage. In Lillie's view, the "property of direction and localization" resided in the "homogeneous, transparent, semi-fluid matrix that suspends all the visible particles of the protoplasm of the egg" (Lillie, 1909). He wrote about the persistence of organization in 1906 as follows:

> I believe that this conclusion is strongly reinforced by my observations on differentiation without cleavage; for here we see the various substances of the ovum marshalled in order, disposed in a bilateral arrangement and fashioned in the form of a larva; and we see the cilia and other cell-constituents arise in the appropriate locations—and all this without the need of even a single nuclear division. (Lillie, 1906, p. 252)

In direct conflict with Wilson, who claimed that all promorphological characters of the cytoplasm were impressed upon it by original preformations in the nucleus, Lillie saw no reason to "reverse the order and assume that the cytoplasmic diversity may be a cause of new nuclear diversity" (Lillie, 1906, p. 260). Arguing that there was no room in the known laws of chemistry for species specificity to be preformed in chromatin, he maintained a position typical of many early experimental embryologists when he wrote

> It seems to me that all *a priori* considerations should be ruled out of court, unless we are willing to transform biology into a branch of metaphysics dealing with potencies and latencies. (Lillie, 1906, p. 260)

Edwin Conklin (1863–1952) was one of the most vigorous early supporters of the determinative effect of the cytoplasm in morphogenesis. As he remarked in 1933, "Throughout my scientific life I have been waging a fight for the recognition of the importance of the cytoplasm of the egg" (quoted in Plough, 1954, p. 2). Conklin began his doctoral work, like many American embryologists including Morgan, Wilson, Ross Harrison, and others, under the direction of W. K. Brooks at Johns Hopkins University in 1891. Like Lillie's, his doctoral work was concerned with cell lineage studies. As Conklin viewed his situation in 1905:

> From all sides the evidence has accumulated that the chromosomes are the principal seat of the inheritance material; until now this theory practically amounts to a demonstration. On the other hand all persons who have much studied cell-lineage have been impressed with the fact that polarity, symmetry, differentiation and localization are first visible in the cytoplasm and that the positions and proportions of embryonic

> parts are dependent upon the location and size of certain blastomeres or cytoplasmic areas. (Conklin, 1905a, p. 220)

Conklin's approach to cell lineage studies differed somewhat from that of Lillie. Lillie was primarily concerned with the organization of the egg in relation to the adaptive needs of the future larva. Conklin, however, was concerned primarily with the features of the organization of the egg which characterized different phyla, and in relating these features to the problem of macroevolution. Conklin claimed that bilateral animals could be characterized by fundamental similarities in the polarity and symmetry of the unsegmented egg. Moreover, as mentioned earlier, in different phyla there were marked differences in the localization of cytoplasmic substances ("organ-forming substances") corresponding to differences in the location of the organs in the embryo or larva. Many different phyla, therefore, could be distinguished by the type of ooplasmic localizations they showed. In its general features, Conklin reasoned, "the characteristics of the phylum are present in the cytoplasm of the egg cell" (Conklin, 1908, p. 98).

One of the principal difficulties in explaining the origin of different phyla on evolutionary grounds, according to Conklin, had been the dissimilar locations of corresponding organs or parts. For example: how could vertebrates be derived from annelids or from any other invertebrate type? As Conklin saw it, if evolution takes place through the transformation of the egg cell, the problem can be explained. He claimed that relatively slight modifications in the localization of the "formative substances" of the egg could produce profound modifications of the adult, such that changes in the relative positions of the parts may be readily accomplished in the unsegmented egg (Conklin, 1905b, 1908).

The possibility of such changes in the unsegmented egg, Conklin argued, was well illustrated by the case of inverse symmetry. In many groups of animals certain species or individuals existed in which there was a total inversion of all organs and parts with respect to the plane of symmetry. Cases of inverse symmetry had been observed in humans where all the viscera were transposed with regard to the median plane. The heart and great arch of the aorta were found on the right side instead of the left. In fact, all the organs presented a mirror image of the usual condition. Cases were found in invertebrates as well as in vertebrates. Among the former, the best known cases were those presented by sinistral gastropods in which the shell was wound in a left spiral instead of a right one and all the organs were transposed with respect to the plane of symmetry. Conklin (1903) claimed that the causes of inverse symmetry could be traced back step by step through development to the inverse organization of the cytoplasm of the egg.

In 1908 Conklin brought together the diverse embryological evidence for the hereditary role of the cytoplasm, and argued for the cell as a whole as the ultimate unit of structure and function. Similarity of differentiation in successive generations, or what Conklin viewed as "heredity," depended upon similarity of both "intrinsic" and "extrinsic" factors. Like Lillie, Morgan, and others, he argued that divisions of chromosomes were almost always equal both qualitatively and quantitatively and that if daughter chromosomes and nuclei ever became unlike, it was probably due to the action of different kinds of cytoplasm upon the nuclei.

As evidence that the cytoplasm possessed "fundamental differentiations," Conklin summarized the evidence based on observations which showed that the cytoplasm was not composed of "simple undifferentiated protoplasm" as had been assumed by upholders of the nucleus as the sole vehicle of inheritance. To support the view that inheritance took place through the cytoplasm of the egg, Conklin referred to the merogony experiments of Theodor Boveri and others.

Although Boveri is best known for his work concerning chromosomal individuality (Sturtevant, 1965; Baltzer, 1967), he also gave much attention to the role of the cytoplasm in development and heredity. In fact, the evidence Boveri accumulated on the role of the cytoplasm in "determination," the methods he established for investigating it, and his theoretical views on the subject were of the greatest interest to embryologists.

Boveri (1901), on the basis of his studies of sea urchin eggs, was one of the first embryologists to propose and confirm the existence of a polarity in the egg whereby between unequal poles there existed a gradient of some sort. Some years later Boveri (1910b) claimed that in *Ascaris* eggs differentiation of blastomeres was determined first by the cytoplasm and later by the nucleus. He also provided some of the most celebrated early evidence that cytoplasmic gradients could cause the well-known "diminution"—elimination of chromatin at ends of the chromosomes. Although no biologists denied the existence of gradients and polarity in the cell, their exact nature and cause remained in doubt.

The idea that polarity and symmetry and organismic form were fundamentally similar to the spatial pattern of the crystal was repeatedly discussed throughout the first two decades of the century (see Haraway, 1976). Crystals possessed characteristic form, were able to grow in a proper solution, and could regenerate their form in such solutions when broken or injured. However, the crystal analogy was attacked by the American biologist C. M. Child (1924), who developed a gradient theory in terms of metabolic action of different intensities and held this to be responsible for cell polarity and organization. The particular gradient theory developed by Child was extended throughout the century and was adopted by Julian Huxley and Gavin de Beer, who developed the theory of gradient fields. Field theory in turn had been brought into biological discourse most prominently by the celebrated Vienna-born embryologist Paul Weiss (1939, 1947). Conceived in its most general terms, a field is the sum of the reactions which an entire protoplasmic system makes with its external and internal environment. The field concept with its noncentralized causality provided a useful way of understanding the problem set up by Driesch: the harmonious equipotential system of the organism as a whole.

At the same time, many embryologists, including Whitman, Lillie, Driesch, Boveri, Harrison, and Hans Spemann (see Chapter 3), opposed Child's gradient in terms of metabolic function (Haraway, 1976, p. 89). They argued that cell organization and morphogenetic movements were too specific and characteristic to be ascribed only to metabolic gradient fields. Instead, they continued to claim that polarity itself had a structural basis. As Harrison (1921, p. 89) put it, "Such gradients may well be an expression of polarity rather than its cause." In this sense the organization field or polarity was thought by many embryologists to be similar to the magnetic field in physics. It existed independently of the cellular substratum upon which it acted. Again, this directing organization would dictate when and where metabolic products became located and what structures they gave rise to. Many embryologists, including Harrison, continued to claim that the organization of regions of embryonic tissue into organ rudiments could be explained by forces arising from some orientated protein microstructure similar to that of a liquid crystal located in the "ground substance" of the cytoplasm (see Driesch, 1908, p. 65). They considered this microstructure to be on a different level of organization to that of ordinary molecules in solution.

The question of whether cytoplasmic organizational fields and the cytoplasmic substance that responded to them were ultimately traceable to the action of the nucleus, or whether they constituted part of the orginal germ, remained a subject of controversy throughout the century. In the meantime, Boveri had constructed an experimental procedure which was designed to test the relative roles of the nucleus and the cytoplasm in heredity.

The experiment that Boveri constructed, which to many embryologists conclusively demonstrated that cytoplasmic inheritance did occur, became famous as "merogonic hybridization" (see Delage, 1899). Ironically, merogony experiments were first carried out by Boveri in 1889 to try to provide experimental proof that the *nucleus* was the sole bearer of inheritance. Essentially, they consisted of fertilizing sea urchin egg fragments lacking a nucleus with the sperm of a different species of echinoderm, having characteristically different larvae. If the embryo showed only characteristics of the paternal species brought in by the sperm, the nucleus would be considered to be the sole "bearer of inheritance." However, if characteristics of the maternal type appeared as well, a cytoplasmic influence must be operative.

In his first paper, Boveri (1889) reported that the larvae resulting from the fertilization of a presumably enucleated egg of one kind of echinoderm with the sperm of another possessed larval features of the male parent only. His conclusion was clear enough: "Herewith is demonstrated the law that the nucleus alone is the bearer of hereditary qualities" (translated by Morgan, 1893, p. 232). Although Boveri's first conclusion was precise, his supporting results were not. Indeed, they were quite ambiguous and were questioned by Morgan (1895a) and others. Merogony experiments emerged as a whole new avenue of research and represented some of the first experimental techniques for investigating the relative roles of the nucleus and cytoplasm in heredity.

Merogony experiments were undertaken by many embryologists between 1891 and 1915, including Morgan, Delage, Jacques Loeb, and Driesch (see also Chapter 3). To Boveri and many other embryologists, the results indicated that the general type of blastula, the number of primary mesenchyme cells, the pattern of the pigmentation, and even the form of the young pluteus larva were due to maternal, cytoplasmic influences, the influence of the sperm (nucleus) first being shown in the character of the skeleton (see review by East, 1934).

Boveri (1903) attempted to formulate a compromise between the roles assigned to the nucleus and to the cytoplasm in heredity. He made a distinction between "preformed" and "epigenetic" ontogenetic characters. The former were to a certain extent "blocked out" or "prelocalized" in the organization of the egg cytoplasm, independently of the nucleus. The epigenetic characters of the nucleus, on the other hand, were thought to be progressively developed by interactions between nucleus and cytoplasm and by reciprocal interactions among the parts of the embryo.

The characters which he attributed to the cytoplasm were general characters of the embryo, including the plasma structure of the embryonic cells, the form of cleavage, the primary axial relations of the embryo, and, up to a certain point, the size of the embryo. He claimed that the adult characters were controlled by the nucleus and were superimposed on the cytoplasmically controlled characters. As development proceeded, the cytoplasmically controlled characters could become obscured by the nuclear characters, or even altogether lost to view. Boveri (1903, p. 362) included in the epigenetic characters "all the essential characteristics of the individual and of the species." As Wilson (1925, pp. 1102–1108) noted, by implication, at least, he placed among the cytoplasmic qualities those that were common to different species and hence characteristic of higher groups.

Compromising the Chromosomes

> When . . . the biologist is confronted with the fact that in the organism the parts are so adapted to each other as to give rise to a harmonious whole; and that organisms are endowed with structures and instincts calculated to prolong their life and perpetuate their race, doubts as to the adequacy of a purely physico-chemical viewpoint in biology may arise. The difficulties besetting the biologist in this problem have been rather increased than diminished by the discovery of Mendelian heredity, according to which each character is transmitted independently of any other character. Since the number of Mendelian characters in each organism is large, the possibility must be faced that the organism is merely a mosaic of independent hereditary characters. If this be the case the question arises: What moulds these independent characters into a harmonious whole?
>
> The vitalist settles this question by assuming the existence of a pre-established design for each organism and of a guiding "force" or "principle" which directs the working out of this design. Such assumptions remove the problem of accounting for the harmonious character of the organism from the field of physics and chemistry. The theory of natural selection invokes neither design nor purpose, but it is incomplete since it disregards the physico-chemical constitution of living matter about which little was known until recently. (Jacques Loeb, 1916, p. v–vi)

With the rise of the Mendelian-chromosome theory, many embryologists in the United States and Europe attempted to formulate a compromise between the hereditary roles of the cytoplasm and the nucleus. Based on the embryological considerations mentioned above, they claimed that Mendelian genetics was concerned only with characteristics which did not exceed the framework of the species and that the cytoplasm was concerned with the "fundamental" characteristics of the organism. Genes were excluded from playing an important part in morphogenesis (gastrulation, cleavage, and organ initiation).

Michael Guyer (1907, 1909, 1911) at the University of Cincinnati was an early proponent of this view. Guyer (1911, p. 302) stated:

> We must restrict our assertion of equal inheritance to the sexual and specific differences which top off, as it were the more fundamental organismal features.

At Oxford the embryologist J. W. Jenkinson (1913, pp. 92–93) wrote:

> The characters, the determinants of which reside in the cytoplasm, are the large characters which put the animal in its proper phylum, class and order, which make it an Echinoderm and not a Mollusc, a sea-urchin and not a Starfish; and these large characters are transmitted through the cytoplasm and therefore through the female alone. The smaller characters—generic, specific, varietal, individual—are equally transmitted by both germ-cells and the determinants of these are in the chromosomes of their nuclei.

A similar distinction between the determinative roles of the nucleus and cytoplasm was maintained by the influential Belgian embryologist Albert Brachet. Brachet (1917, pp. 176–179) distinguished between what he called *l'hérédité générale,* or that of the species, which had its seat, if not exclusively, at least principally, in the cytoplasm, and that of the individual, due to the chromosomes in the nucleus. In his influential text *Heredity and Environment in the Development of Men,* Conklin (1915, p. 176) articulated his compromise theory as follows:

> We are vertebrates because our mothers were vertebrates and produced eggs of the vertebrate pattern; but the color of our skin and hair and eyes, our sex, stature and mental peculiarities were determined by the sperm as well as by the egg from which we came. There is evidence that the chromosomes of the egg and sperm are the seat of the differential factors or determiners for Mendelian characters while the general polarity, symmetry and pattern of the embryo are determined by the cytoplasm of the egg.

There were some major theoretical problems underlying the reluctance of embryologists to accept the chromosome theory as a complete theory of heredity. From the point of view of development, the chromosome theory presented similar difficulties to those offered by the nuclear theory of heredity, which had been rejected by embryologists. The Mendelian-chromosome theory shared two major characteristics with the Weismannian theory. First, it was a particulate theory. That is to say, it rested on the notion that the germ plasm in the nucleus contained a host of determinants that were more or less independent of each other and could perpetuate themselves unchanged. Secondly, the chromosome theory maintained the paradox of nuclear equivalence during cellular differentiation. In addition to these apparent faults, the first generation of Mendelian geneticists failed in its promise to actually demonstrate how Mendelism could account for the origin of species. These issues, considered as a whole or separately, led many biologists to embrace the cytoplasm as the principle agent of heredity. It is necessary to pause here and take a brief glance at some of the problems of ontogenetic development and evolution during the rise of the Mendelian-chromosome theory—problems which were generally ignored by geneticists.

As discussed above, with the advent of the nuclear theory, many embryologists had rejected the notion that the organism was nothing but a mosaic of independent self-replicating units which somehow forced each other into a harmonious whole. As E. B. Wilson viewed the situations in 1923, "Good biological society has of late looked decidedly askance upon all corpuscular or micromeristic conceptions of the cell" (Wilson, 1923, p. 283). During the first decades of the twentieth century many embryologists continued to maintain the theoretical necessity for the existence of some sort of ordered structure, or "organization," in the cytoplasm of the egg which would bring materials together in time and space and account for the general symmetry and orientation of the developing embryo. Moreover, throughout the classical genetics period, embryologists defended the integrity of the organism and felt it necessary to extend Whitman's argument on "the inadequacy of the cell theory of development" to "the inadequacy of the independent gene theory of heredity" (Harrison, 1940). The constitution of the single-celled or complex organism could not be explained "as a symbiotic swarm of the smallest living things." "Life," wrote Conklin (1940, p. 18), "is not found in atoms or molecules or genes as such, but in organization; not in symbiosis, but in synthesis."

From a purely physicochemical point of view, many investigators acknowledged that the cell was an organized system and that an analysis of its activities by chemical composition alone was inadequate. Like machines, organisms could not be expected to be built up by individual self-assembling parts. Some sort of spatial principle or directive structure as the basis of organization was required in order to integrate and regulate the constituents of the cell. This was a time when great advances were being made in the understanding of the behavior of substances in solution and of semipermeable membranes. It was a time when embryologists became primarily occupied with theories of "colloid chemistry." Colloids were thought to be substances that constantly changed their shape and could not be chemically identified, but rather formed gels. A corollary was that the cell, regarded as a whole, was "a complex of innumerable chemical reactions in the substance of the cell system" (Wilson, 1923, p. 283). Indeed, the conception of the cell as a colloid chemical system seemed to indicate to cell physiologists the necessity of the "organization" of the cell.

"One cannot help assuming," wrote the physiologist L. Jost in 1907, "that the mode of arrangement of the ultimate parts of the organism is of greater importance than the chemical nature of these parts" (see Wilson, 1925, p. 670). Consistent with this belief, J. G. Hopkins, founder of biochemistry at Cambridge and mentor of Joseph Needham, who would later develop what he called "chemical embryology," stressed the need to appreciate the structural geography of the cell. "It is clear," Hopkins wrote in 1913, "that the living cell as we know it, is not a mass of matter composed of living molecules, but a highly differentiated system" (Wilson, 1925, p. 670). Similarly, A. P. Mathews in 1915 emphasized the enormous contrast between living protoplasm and the same protoplasm after it had been ground up in a mortar without altering its chemical and molecular properties: "The orderliness of the chemical reactions is due to the cell-structure, and for the phenomena of life to persist in their entirety that structure must be preserved" (Mathews, 1915, p. 11).

The whole *materialist interpretation of life* rested upon the assumption that the specific character of cells, and particularly their orderly localization in the system, must somehow depend on what was called their "organization." Jacques Loeb, to cite still another biologist who characterized the living organism in physico-chemical terms, also implied the existence of such a configuration. Loeb (1916, p. 39) specifically maintained that "without a structure in the egg to begin with, no formation of a complicated organism is imaginable." This same implication lurked behind every attempt to formulate the unity and order of the individual in materialist terms, that is, by ascribing to it definite relations in both space and time among the reactions occurring in protoplasm.

In his well-known book *The Organism as a Whole* (1916), Loeb attempted to synthesize the work of experimental embryologists and probe, in a systematic way, some of the major problems of making organisms from eggs. He described the object of his text concisely as follows:

> In this book an attempt is made to show that the unity of the organism is due to the fact that the egg (or rather its cytoplasm) is the future embryo upon which the Mendelian factors in the chromosomes can impress only individual characteristics, probably by giving rise to special hormones or enzymes. (Loeb, 1916, vi)

A brief consideration of Loeb's theoretical treatment will give a clearer idea of some of the inadequacies many biologists found in the atomistic chromosome theory. As we shall see in the chapters to follow, many of the issues raised by Loeb would be considered by investigators of cytoplasmic inheritance throughout the twentieth century.

In Loeb's view, the order and control of hereditary potentialities in the individual represented a stumbling block for particulate theories of heredity. Such theories seemed to have to deny or ignore the problem of order and control, or postulate a "supergene" or vitalistic "directive force" which would control and order all the individual genes. By the first decade of the century Driesch came to assume that there was an Aristotelian "entelechy" acting as a directing guide in each organism (Churchill, 1969). Another German biologist, J. von Uexküll (1913, p. 216), suggested a kind of Platonic "idea" or "supergene" as a character of life which would account for the purposeful character of the developing organism. Loeb argued that the attitude of Driesch and Uexküll was not very different from that of the famous French physiologist Claude Bernard. Neither Bernard, Driesch, nor Uexküll would think of treating the processes of digestion, metabolism, production of heat, etc., in any other way than as purely chemical or physicochemical processes. On the other hand, when the actions of the organism as a whole were concerned, one finds a totally different situation. Thus, Bernard declared that the making of a harmonious organism from an egg could be explained only on the assumption of a "directive force":

> There is so to speak a pre-established design of each organ of such a kind that each phenomenon by itself depends upon the general forces of nature, but when taken in connection with others it seems directed by some invisible guide on the road it follows and led to the places it occupies. . . .
>
> We admit that the life phenomena are attached to physico-chemical manifesta-

> tions, but it is true that the essential is not explained thereby; for no fortuitous coming together of physicochemical phenomena constructs each organism after a plan and a fixed design (which are foreseen in advance) and arouses the admirable subordination and harmonious agreement of the acts of life. . . .
>
> We can only know the material conditions and not the intimate nature of the life phenomena. We have therefore only to deal with matter and not with the first causes or the vital force derived therefrom. These causes are inaccessible to us, and if we believe anything else we commit an error and become the dupes of metaphors and take figurative language as real. . . . Determinism can never be but physicochemical determinism. The vital force and life belong to the metaphysical world. (Bernard, 1885, translated in Loeb, 1916, p. 3)

Loeb also found it difficult to understand how a harmonious whole could be made if the organism was nothing but a mosaic of Mendelian characters. Even if one made allowances for "the law of chance," he could not see how genes could force each other into a harmonious whole (Loeb, 1916, p. 7). On the other hand, he could not accept the vitalists' assumptions of "supergenes" or a "directive force" which removed the problem of the working out of this design from the field of physics and chemistry. To Loeb, what determined the position of organic molecules in the developing organism was not a metaphysical quality common to all organic matter, but a special structure. The particles could not assume the parents' shape without a pattern to guide them or mold to shape them.

In his view, the problem of explaining the emergence of organisms from eggs by the "law of chance" was similar to that of accounting for the synthesis of living from dead matter. As Loeb stated the problem:

> It is at least not inconceivable that in an earlier period of the earth's history radioactivity, electrical discharges, and possibly also the action of volcanoes might have furnished the combination of circumstances under which living matter might have formed. The staggering difficulties in imagining such a possibility are not merely on the chemical side—e.g., the production of proteins from CO_2 and N—but also on the physical side if the necessity of a definite cell structure is considered. (Loeb, 1916, p. 39)

Even with synthetic enzymes as a starting point, which might be capable of forming molecules of their own kind from a single nutritive solution, the task of creating cells capable of growth and division was a difficult one. Like many other biologists, Loeb (1916, p. 23) viewed this synthetic power of transforming small "building stones" into the complicated compounds specific for each organism as "the secret of life" or at least one of the secrets of life. He could not imagine how enzymes alone, which were known to be concerned at least with the velocity of chemical reactions in a living organism, could build up an organism. In his view, even the simple structure of a bacterium was as essential for its existence as were its enzymes (Loeb, 1916, p 39). Extrapolating this problem to higher organisms, he maintained that Mendelian factors of heredity must have the rough embryo to work on and that the organism could not be considered as a mere mosaic of Mendelian factors which, through chance, force each other into a harmonious whole.

Loeb contended that the cytoplasm of the egg was not undifferentiated, but contained a simple yet definite "physico-chemical structure" which sufficed to determine the first steps in the differentiation of the organism. Thus he wrote:

> The facts of experimental embryology strongly indicate the possibility that the cytoplasm of the egg is the future embryo (in the rough) and that the Mendelian factors only impress the individual (and variety) characters upon this rough block. . . .
>
> In any case, we can state today that the cytoplasm contains the rough preformation of the future embryo. This would show then that the idea of the organism being a mosaic of Mendelian characters which have to be put into place by "supergenes" is unnecessary. (Loeb, 1916, p. 8)

Like other developmental physiologists, Loeb supported this claim with the results of merogony experiments which indicated that the first development in the sea urchin to the gastrula stage was independent of the nucleus, which was the bearer of the Mendelian factors. If this was true, then it was conceivable that the generic and possibly also the specific characters of the organism were determined by the cytoplasm of the egg and not by the Mendelian factors. Loeb not only recognized a definite, yet chemically undefined, structure in the egg cytoplasm, as indicated by observations of polarity and symmetry and visible stratification of the cytoplasm. He also insisted that the egg must contain specific substances. These substances, he claimed, would determine the "species" and specificity in general, and were in all probability *proteins* (Loeb, 1916, p. 40).

Loeb (1916, p. vii) maintained that Mendelian characters may be determined by hormones which need be neither proteins nor specific, or by enzymes which, he argued, also need not be specific for the species or genus. Within this scheme, Mendelian factors would give rise to specific substances which go into circulation and simply start or accelerate different chemical reactions in different parts of the embryo and thereby determine the finer details of the organism characteristic of the variety and the individual. This conception played a double role. As Loeb repeatedly emphasized, it removed difficulties which the contemporary work on Mendelian inheritance created not only for the problem of the harmonious character of the organism as a whole, but also for the problem of evolution.

Herein lies a second major difficulty many biologists who embraced the cytoplasm found with Mendelian genetics. Geneticists could not make new species. By the turn of the century many biologists voiced opposition to the idea that natural selection acting gradually on minute and random hereditary changes could account for the origin of species (Pfeifer, 1965; Allen, 1968; Mayr and Provine, 1980; Bowler, 1983). These protests transcended both national and disciplinary boundaries. The first generation of geneticists believed that selection had little to do with the origin of species. Many claimed that new species occurred through large discrete changes. Still, other biologists upheld the Lamarckian principle of environmentally induced adaptive hereditary changes to account for the origin of species. As will be discussed further in the next chapter, the geneticists' strategy was to take evolution out of the hands of armchair theorists, field naturalists, statisticians, and the general public and to place it firmly within the laboratory in the hope of directly observing and possibly controlling evolutionary processes. However, they failed in their central objective. This failure was recognized by many of those who helped set up the new discipline of genetics, including William

Bateson, who had turned from embryology to cross-breeding, Wilhelm Johannsen in Denmark, Carl Correns in Berlin (see Chapter 3), and Hugo de Vries in the Netherlands, all of whom had turned from plant physiology to develop cross-breeding analysis, and H. S. Jennings in the United States (see Chapter 4).

De Vries, Bateson, and Johannsen; along with Edwin Conklin and several other leading embryologists, believed that it was possible to construct new species all at once through the sudden mutation of a single hereditary unit. This view was well articulated in de Vries' first volume on *The Mutation Theory* (1901), in which he developed the idea that evolution occurred through discrete saltationist stages rather than by gradual changes accumulated by selection. De Vries recognized two processes: the addition of a new hereditary element that could give rise to a new species, and the inactivation of a hereditary unit already present. He and his followers believed they had constructed new species from the evening primrose *Oenothera lamarckiana*. By the mid-1920s, it became clear to many that the interpretation of *Oenothera* was faulty; the new forms were not new species emerging from a mutation. In the meantime, de Vries's work encouraged others to turn to experimental evolution, and Morgan was among the seduced. In 1910 he left embryology to experiment with the fruit fly *Drosophila* in the hope of discovering mutations that would transform one species into another.

However, Morgan and his school also failed to produce new species. Many of the mutations in *Drosophila* were small changes that made a part a little longer or a little smaller, for example. When the mutations were larger, they seemed only large enough to disturb the integrity of the organism or throw it out of harmony with its environment. In general, the known gene mutations did not fit well with the requirements embryologists expected of controllers of elements of spatial pattern. At the same time many experimentalists came to believe by the early 1920s that the Mendelian gene mutations that Morgan and his school produced in *Drosophila,* and perhaps all Mendelian gene mutations, only produced organisms less capable of surviving outside the laboratory than wild-type organisms.

Geneticists studied primarily inherited defects; they mated purebred defective individuals to other purebred defective individuals. The purebred strains in Morgan's laboratory included *Drosophila* that were eyeless, had abnormal abdomens, had vestigial wings, were bar-eyed, etc. Gene mutations did not seem to be what new species were made of. Mendelian geneticists, in Johannsen's (1923, p. 137) words, were "mostly operating with 'characters' which are rather superficial, in comparison with the fundamental Specific or Generic nature of the organism." Mendelism was concerned only with trivial "differences" between individuals or varieties of a species. As Johannsen (1923, p. 137) phrased it:

> The pomace-flies in Morgan's splendid experiments continue to be pomace-flies even if they lose all "good" genes necessary for a normal fly-life, or if they be possessed with all the "bad" genes, detrimental to the welfare of this little friend of the geneticists.

Johannsen (1923, p. 137) himself came to believe "in a great central 'something' as yet not divisible into separate factors" located in the cytoplasm. Bateson, who had opposed the idea that Mendelian genes were located as discrete parts of the chromosome like "beads on a string" (Coleman, 1970) tended to avoid the localization issue. However, when he acknowledged the chromosome theory by 1922, he also tended to support the embryologists' claims of primary cytoplasmic control over development and heredity. Faced with the chromosome theory, he stated:

> Throughout all this work, with ever-increasing certainty, the conviction has grown that the problem of heredity and variation is intimately connected with that of somatic differentiation, and that in analysis of these two manifestations of cellular diversity lies the best prospect of success. Pending that analysis, the chromosome theory, though providing much that is certainly true and of immense value, has fallen short of the essential discovery. (Bateson, 1926, p. 235)

Spontaneous macromutations were not the only possibility for cytoplasmic transformations. Although Johannsen and Bateson opposed the belief in the inheritance of acquired characteristics, other experimentalists, including Correns, believed that characteristics acquired by the soma could be transmitted to the germ cells through the cytoplasm. This view found wide support from neo-Lamarckian evolutionists who believed that selection could not account for the origin of species. As Ernst Mayr (1980, p. 16) recalls, it was commonly held by many German paleontologists and systematists during the 1920s and 1930s.

Indeed, naturalists repeatedly claimed that they could see no connection between the gene mutations reported by Mendelian geneticists and the evolutionary events at the hierarchical levels of species and higher taxonomic groups. This view was also maintained by Félix Le Dantec, Maurice Caullery, and other leading neo-Lamarckian evolutionists in France during the 1920s and 1930s (see Chapter 5). The belief in the inheritance of acquired characteristics converged with the embryologists' belief that the cytoplasm was largely responsible for development. The complete autonomy and the randomness and rarity of gene mutations seemed to exclude genes from playing a direct role in the orderly process of epigenetic development. Since the belief in the inheritance of acquired characteristics necessarily denied the complete constancy of genetic material, many neo-Lamarckian evolutionsits embraced the pliable cytoplasm, which was thought to contain diffuse species-specific substances as theorized by Loeb and others.

Loeb (1916, 1917) recognized that the question of whether or not species characters of the organism were determined by the cytoplasm was of fundamental importance for the problem of evolution, and he sought to devise a definitive experiment. In 1917 he suggested that a decision could be made by comparing the species specificity (proteins) of an F_1 hybrid with that of the two parent forms. Loeb reasoned that if it could be shown that the species specificity of an F_1 hybrid was identical with that of only one of the two parents, no matter if this parent were the paternal or maternal species, it would be an indication that species specificity is Mendelian. However, if the species specificity was always identical with that of the maternal species, no matter from which of the two parent species the mother was selected, it might indicate that the cytoplasm of the egg determines the inheritance of species specificity. (Crosses between different species, genera, and higher taxonomic groups were carried out by many German botanists between the two World Wars in order to test Loeb's theory) (see Chapter 3.)

A third persistent fault in the Mendelian-chromosome theory concerned the problem of somatic cell differentiation. Experiments on tissue cultures showed that at least some of the differences among cells of one organism persisted when they were taken out of the body. On the other hand, the accepted view, maintained on the basis of chromosomal behavior in cell division as well as experimental embryology, was that no sorting out or differention of nuclear genes or factors occurred during development. One of the best-known experimental reports indicating that early development was under cytoplasmic control and that nuclear differentiation did not occur resulted from the famous "constriction experiments" on newt eggs carried out by the celebrated German embryologist Hans Spemann (1914) (see also Chapter 3). This study remained influential among embryologists well into the 1950s (see Chapter 6).

Spemann constricted fertilized newt eggs with a ligature, thereby separating the cytoplasm into two portions, one with, and one without, a nucleus. After a series of nuclear divisions one of the daughter nuclei escaped into the nonnucleated cytoplasm and there continued its divisions. If the nuclei had undergone any irreversible differentiation in hereditary capacities during these early divisions, abnormal development might be expected in the initially nonnucleated portion of the egg. However a normal—if somewhat retarded—twin developed.

The inheritance of cell differences in the face of equivalent nuclei reinforced the claim that somatic cell variation was primarily under cytoplasmic control. Conklin (1920, p. 403), for example, stated that if the genes or Mendelian factors of the nucleus were the only differential factors of development, then

> these genes would of necessity have to undergo differential division and distribution to the cleavage cells; since this is not true, it must be that some of the differential factors of development lie outside of the nucleus and if they are inherited as most of these early orientations are, they must lie in the cytoplasm.

Similarly, F. R. Lillie wrote (1927, p. 367):

> Those who desire to make genetics the basis of physiology of development will have to explain how an unchanging complex can direct the course of an ordered developmental stream.

Supported by similar reasoning, Ross Harrison (1936, p. 220) stated:

> We must seek also in the cytoplasm, which effects the differentiations, a basis for the characters of the organism. This must be assumed to be some kind of "repeat" configuration, in each unit of which the qualities of the whole, including its symmetry and polarity, are in some way implied.

Once Mendelian genes were placed in the chromosomes, it seemed that there was no escape from the conclusion that somatic cell variation and gene variation were separate phenomena. Experimental embryologists found themselves in general agreement with William Bateson who in his last publication wrote:

> Cytology is providing some knowledge, however scanty, of the material composition of the cell, but of the nature of the control by which a series of orderly differentiations is governed we have no suggestion. (Bateson, 1926, p. 234)

More Parts to the Whole

> To some extent perhaps, our conclusions concerning the chromosomes have thus far been more definitive because we are able to follow their history more readily. (E. B. Wilson, 1928, p. 17)

The idea that the cytoplasm was organized and played a direct role in development and heredity began to gain support from an additional point of view during the first two decades of the twentieth century. Cytoplasmic bodies called centrioles seemed to direct the migration of sister chromosomes to opposite poles of the cell during mitosis. Other cytoplasmic granules called "plastids" and "mitchondria" or "chondriosomes" began to receive a great deal of attention because of their possible significance in cell differentiation and heredity. Mitochondria attracted the attention of botanists, zoologists, anatomists, physiologists, pathologists, and clinicians, who all studied them from varied points of view. Hundreds of papers on mitochondria appeared between 1910 and 1920, widely scattered in the journals of many countries. For many investigators, the origin of these granules could be traced back to the writings of R. Altmann (1890). As E. B. Wilson stated in 1928, the idea itself was not new. What was new was the impetus for its further investigation.

Intense cytological studies indicated that mitochondria consisted of specific material having definite cytological and chemical characteristics. Unlike the mechanical behavior of chromosomes, however, mitochondria seemed to be morphologically highly plastic, so that they could appear under different forms. They were identified in both eggs and sperm, and in blastomeres of the segmenting egg, and E. B. Wilson (1916) claimed that they were distributed with approximate equality to daughter cells. Many other leading cytologists, including C. Benda (1901) and F. Meves (1908) in Germany, J. Duesberg (1913, 1919) in Belgium, E. Fauré-Fremiet (1908) and A. Guilliermond (1913) in France, and others, ascribed to mitochondria the power of independent growth and division. They considered them to be a mechanism of cytoplasmic heredity comparable in importance with chromosomes. They were believed to be fundamentally important for the chemical activities of the cell and for the processes of tissue development and differentiation, forming the source from which arise many of the more specific cell components, including plastids.

By the 1920s, following the morphogenetic and cytological investigations of the cytoplasm and his own cytological studies of mitochondria, E. B. Wilson, the so-called "invaluable ally" of the chromosome theory, conceded the possibility of chondriosomal inheritance and the hereditary nature of cell organization. After summarizing the cytological evidence in favor of self-perpetuating cytoplasmic bodies, Wilson (1923, p. 283) wrote:

> For my part, I am disposed to accept the probability that many of these particles, as if they were submicroscopical plastids, may have a persistent identity, perpetuating themselves by growth and multiplication without loss of their specific individual type. . . . There are many facts made known especially by experimental embryology, which indicate that it is in the apparently structureless hyaloplasm (ground substance) that the real problem of the cytoplasmic organization lies; and the same facts drive us to the conclusion that the submicroscopical components of the hyaloplasm are segregated and distributed according to an ordered system.

In addition to the embryological and cytological arguments that had been put forth in support of cytoplasmic inheritance, in 1909, during the rapid rise of Mendelism, new genetic evidence was reported that seemed to indicate a mode of cytoplasmic inheritance through plastids in plants. Two cases were reported, one by Carl Correns at Münster and another by Erwin Baur at the University of Berlin. Both cases involved chlorophyll variegation in plants (where "normal" green foliage is spotted with white or light green). In the case investigated by Correns the chlorophyll characteristics were inherited strictly maternally in the Four O'Clock, *Mirabilia jalapa*. In Baur's case in *Pelargonium* the sexual transmission was biparental, but not with segregation patterns that could be explained by any known Mendelian mechanism.

Both Correns and Baur interpreted their results in terms of non-Mendelian mechanisms. However, Correns resisted a particulate mechanism operating through plastids themselves. Instead, he claimed his results were due to a sick (*krank*) cytoplasm affecting plastid development (Correns, 1909, p. 321). Baur, on the other hand, claimed that his results were due to the independent inheritance of chloroplasts (chromatophores) themselves. In order to explain the somatic cell differentiation revealed by different chlorophyll colorations in the leaves, Baur reasoned that during cell divisions there was a sorting out of different types of plastids from cell to cell. Cell divisions led to some somatic cells containing only white or abnormal plastids, others containing a mixture of both white and green, and still others with only green (Baur, 1909, pp. 349–350).

Defending the Chromosomes

> Is the whole of Mendelism perhaps nothing but an establishment of very many chromosomical irregularities, disturbances or diseases of enormously practical and theoretical importance but without deeper value for an understanding of the "normal" constitution of natural biotypes? The Problem of Species, Evolution, does not seem to be approached seriously through Mendelism nor through the related modern experiences in mutations. . . .
>
> *Chromosomes* are doubtless vehicles for "Mendelian inheritance" but *Cytoplasm* has its importance too. I cannot here enter into this problem from which in the near future we shall certainly have important news. (W. Johannsen, 1923, p. 140)

When Morgan turned to cross-bredding analysis and played a leading role in placing genes on chromosomes, his previous criticisms (Morgan, 1910) of the limited role of the chromosomes in heredity quickly receded. He rapidly became one of the most vigorous defenders of their exclusive role in heredity and evolution. In their epoch-making text *The Mechanism of Mendelian Heredity,* in a short and discrete section, Morgan, A. H. Sturtevant, H. J. Muller, and C. B. Bridges (1915, pp. 135–139) reviewed the evidence for "cytoplasmic inheritance." It was their contention that although the cytoplasmic materials were essential for the *development* of the organism, they were not hereditary. As they saw it, cytological evidence of physical continuity was an essential condition for any hereditary determinants.

They did acknowledge, however, that there were cytoplasmic bodies such as plastids and possibly also chondriosomes (mitochondria) which, like the chromosomes, were able to grow and divide, and therefore might have the power to perpetuate themselves unchanged indefinitely. In fact, they suggested that such bodies might produce active enzymes which, interacting with other products of development, might determine the characteristics of the "race" (Morgan *et al.,* 1915, p. 136). Nonetheless, there was no genetic evidence for mitochondrial inheritance, and they considered the genetic evidence for plastid inheritance to be inconclusive. Correns's case, they maintained, could be due to a "sort of disease that is carried by the cytoplasm," while for that of Baur, "the interpretation must be uncertain."

The Morgan school upheld the dominance of the chromosomal genes and rejected the theory that the cytoplasm determined the characters of the larger phyletic groups while the nucleus or Mendelian genes determined only minor differences. One of their principal polemical tools concerned two cases of so-called maternal inheritance. In 1913 K. Toyama at the College of Agriculture, Tokyo Imperial University, reported evidence which indicated that egg characters which at first appeared to be "non-Mendelian" ultimately could be attributed to Mendelian genes. Toyama reported that when races of silkworms which possessed characteristic but different egg markings, such as shape, color, etc., were crossed, the characteristics of the egg hybrid were like those of the maternal race only. However, when the adult F_1 were raised from these eggs, and when they in turn produced embryos, the distribution of egg characteristics was no longer solely maternal and could be explained by Mendelian mechanisms.

A similar case of "maternal inheritance" was reported in the gastropod *Limnaea peregra*. It will be recalled that the reverse forms of symmetry in mollusks had been understood by E. G. Conklin to represent a "fundamental" character of the organism due to the organization of the egg cytoplasm. However, in 1923 A. E. Boycott and C. Diver at the University of London reported results which suggested that the inheritance of dextral and sinistral coiling in the snail was an illustration of "maternal" inheritance that was nevertheless dependent upon the chromosomes. The American geneticist A. H. Sturtevant seized upon this work. Responding to the threatening belief that Mendelian genes controlled only trivial characters, he wrote in a review of the work published in *Science:*

> Further data on the case of Limnaea will be awaited with interest, for it seems likely that we shall have here a model case of the Mendelian inheritance of an extremely "fundamental" character, and a charcter that is impressed on the egg by the mother. (Sturtevant, 1923, p. 270)

Many American geneticists used the observations of Toyama, Boycott, and Divers to argue that there was an inheritance through the cytoplasm of the egg, but these cytoplasmic characters were themselves of biparental origin and therefore were actually determined by nuclear genes. Thus, critics of the predominant role of the cytoplasm in heredity maintained that hybridization experiments with egg characters, to be critical, must be carried as far as the F_3 generation in order to establish whether or not the character in question was determined by the cytoplasm. As arguments with nucleocentric geneticists continued, defensive attitudes and motivations became visibly intertwined in the formal debates. The American geneticist L. C. Dunn at Harvard, for example, argued that the claim of those who postulated the existence of cytoplasmic heredity rested solely on their refusal to accept exclusive nuclear control, or on their refusal to admit it even though they recognized it:

> The whole case of the supporters of any theory which views the cytoplasm as determinative rests on either their refusal to go back and inquire the source of this cytoplasm, or on their refusal to give due emphasis to the source, even though they recognize it. (Dunn, 1917, p. 296)

While some Mendelian geneticists claimed that the belief in cytoplasmic inheritance was an emotional one, supporters of the cytoplasm claimed that the limited techniques employed by Mendelians were in part responsible for their narrow views. As A. F. Shull (1916, p. 6) put it:

> The cytoplasm often (perhaps usually) determines the type of cleavage, the early course of development, and in large measure the larval characters, while the adult characteristics are determined by the chromosomes. With the developmental stages the student of heredity using the usual breeding methods has little to do. He may be pardoned a bias in favour of the chromosomes because he rarely studies larval characters. To the physiologist and morphologist, on the other hand, the rigid conviction of the geneticists, that the chromosomes contain all the tools of his trade has not unnaturally been viewed with skepticism.

The results of Toyama had only a limited impact on embryologists. They meant only that *some* of the differentiations in the cytoplasm in some organisms were developed anew in each generation and represented "maternal inheritance." As Conklin (1917, p. 104) described these cases:

> This is Mendelian inheritance though somewhat complicated by the fact that every ontogeny has its beginnings in the preceding generation.

Conklin pointed out that a real difference of *modus operandi* nevertheless existed between the two kinds of characters. The promorphological characters of the embryo that were foreshadowed in the egg cytoplasm before fertilization were unaffected by the sperm that subsequently entered the egg. Inasmuch as they were affected by chromosomes, it had to be those which had been derived from both grandparents. To this extent, then, he claimed the parental egg and sperm did not play identical parts in determination, the former contributing more to the heredity of the offspring than did the latter.

In a posthumous paper Boveri (1918, p. 466) also suggested the possibility that some cytoplasmic characters were established by an epigenetic (chromosomal) process at an earlier period in the history of the egg. Like Conklin, Boveri pointed out nonetheless, that the cytoplasm still played a large role in the process of heredity:

> If one designates as heredity the totality of internal conditions which achieve the unfolding of characteristics of the new individual, this gives to the cytoplasm a much more specialized significance than one often has inclined to assume; and more than ever one realizes the absurdity of the idea that it would be possible to bring a sperm to develop by means of an artificial culture medium. (translated in Baltzer, 1967, pp. 83–84)

However, as will be discussed more fully in the next chapter, "heredity" as defined by classical geneticists had a much more restricted meaning. It did not include all the elements necessary for the production and reproduction of a new individual. In short, "heredity" in the "classical" genetics paradigm meant the sexual transmission of "genes" from one generation to the next. It did not embody mechanisms of cellular differentiation and morphogenesis.

In *The Physical Basis of Heredity* Morgan (1919) continued to attack the theory that the cytoplasm was concerned with the characters of the larger phyletic groups while Mendelian genes determined only individual differences. This time Morgan ignored his previous criteria of cytological continuity and now admitted that there was no direct means of determining whether all egg characters were due to the influence of the nucleus (Morgan, 1919, pp. 226–227). Nonetheless, he claimed that a dichotomy between "fundamental" and other characters could not be discerned. In effect, Morgan claimed, in virtual conflict with the views of Shull, that Mendelian procedures were atheoretic and fully capable of detecting the existence of alternative hereditary mechanisms. His opinion, he claimed, was based on pure empiricism rather than on an attempt to exaggerate the importance of his work:

> Mendelian workers can find no distinction in heredity between characteristics that might be ordinal or specific, or fundamental, and those called "individual." This failure can scarcely be attributed to a desire to magnify the importance of Mendelian heredity, but rather to experience with hereditary characters. That there may be substances in the cytoplasm that propagate themselves there and that are outside the influence of the nucleus, must, of course, be at once conceded as possible despite the fact that, aside from certain plastids, all Mendelian evidence fails to show that there are such characters. In a word, the distinction set up between generic versus specific characters or even "specificity" seems at present to lack any support in fact. (Morgan, 1919, p. 226)

Morgan's claim relied in part on the Darwinian view that the nature of the hereditary differences between species could be elucidated by studying heredity in crosses within species. As we have seen, not only did many supporters of cytoplasmic heredity deny this doctrine, but many claimed that the Morgan school was simply not trying hard enough to obtain evidence for cytoplasmic heredity. They were simply doing what was easiest to do. As the Swiss geneticist Emile Guyénot stated the issue in his text *L'Hérédité:*

> The results of embryonic mechanics show . . . that it is necessary to enlarge the framework of heredity, and not to limit efforts to the evidently more easy study of chromosomal factors. Heredity is not a nuclear phenomenon, nor a cytoplasmic phenomenon; it remains, in its entirety, a cellular phenomenon. (Guyénot, 1924, p. 289, my translation)

The fact that the overwhelming majority of inherited differences detected from crosses between similar organisms could be attributed to chromosomal genes and that some cases of "cytoplasmic" inheritance depended on nuclear genes greatly encouraged Morgan and his followers to take on a more gene-centered and less organism-centered view of heredity and evolution. By the 1920s geneticists had shown that genes intervened in phenotypical realization at different stages of the life history of an organism. Moreover, as Morgan (1919, pp. 241–246) stressed, when discussing "the organism as a whole" in terms of "the collective action of genes," geneticists had demonstrated that a single gene may be concerned with many characters both in time and in space, and that a single character may depend on many genes.

However, Morgan's discussion of "the organism as a whole" was little more than a restatement of the particulate theory of heredity. Geneticists could not explain *why* the same genes were associated with different characters at different stages and at different loci. The postulate of genetics was that the nuclear gene system was largely immune from environmental influences. Genes themselve remained constant, in whatever place and at whatever time within the life history of the individual. As Morgan (1919, p. 241) stated, "each cell *inherits* the whole germ plasm." As long as this remained a necessary part of the gene theory, genes lay beyond the essential problem of development. Differentiation in relation to space and time—the order and control of hereditary potentialities—had to be an environmental relationship mediated through the cytoplasm.

Mendelian geneticists could say nothing about cellular differentiation or about the difficult problem of how small building blocks could be transformed into the complicated compounds and structures of the organism. Nonetheless, by 1926 Morgan was prepared to make no compromises with the idea of a universal cytoplasmic inheritance in animals. Although he admitted a limited genetic role for plastids in plants, he rejected the idea of a genetic role for other "bodies" or "units" of the cytoplasm. He put the idea of self-reproducing "materials" in the cytoplasm into the category of "maternal inheritance" or "predetermination," whereas the notion of a preorganized cell structure, he simply ignored. Morgan used the phenomena of "maternal inheritance" to argue that some criticisms of the theory that genes were the exclusive factors in heredity, and the claim that the cytoplasm could not be ignored in any complete theory of heredity, were based on a "confusion," and resulted from a failure to keep apart the phenomena of heredity and those of embryonic development (see Morgan 1926a).

In *The Theory of the Gene* Morgan addressed the issue raised by Johannsen and others who claimed the Mendelian gene mutations investigated in the laboratory were not the sort of changes that led to the origin of species in nature. In fact, Morgan (1926b, p. 67) expressed his own concern over the inability of geneticists to create new species through gene mutation in the following passage:

> Since only the differences that are due to genes are inherited, it seems to follow that evolution must have taken place through changes in the genes. It does not follow, however, that these evolutionary changes are identical with those that we see arising as mutations. It is possible that the genes of wild types have had a different origin. In fact, this view is often implied and sometimes vigorously asserted. It is important, therefore, to find out whether there is any evidence in support of such a view. De Vries' earlier formulation of his famous mutation theory might at first sight seem to suggest the creation of new genes.

Morgan came back to this issue in his presidential address to the Sixth International Congress of Genetics in 1932. This time he attempted to exclude the criticisms of the mutation-selection theory as being irrational and based on mysticism:

> Without elaborating, I wish to point out briefly that there is today abundant evidence showing that the differences, distinguishing the characteristics of one wild-type or variety from others, follow the same laws of heredity as do the so-called aberrant types studied by geneticists.
>
> Even this evidence may not satisfy the members of the old school because, they may still say, all these characters that follow Mendel's laws, even those found in wild species, are still not the kind that have contributed to evolution. They may claim that these characters are in a class by themselves, and not amenable to Mendelian laws. If they take this attitude, we can only reply that here we part company, since *ex cathedra* statements are not arguments, and an appeal to mysticism is outside of science. (Morgan, 1932, p. 288)

CHAPTER 2

Constructing Heredity

> Except for the rare cases of plastid inheritance all known characters can be sufficiently accounted for by the presence of genes in the chromosomes. In a word the cytoplasm may be ignored genetically. (T. H. Morgan, 1926a, p. 491)

In the last chapter it was shown that the cytoplasm was held to play a predominant role in the process of ontogenetic development and in organic evolution by many embryological investigators. On the other hand, American geneticists, basing their views on the sexual transmission of differences between individuals of a species, upheld the predominant, if not exclusive, role of the nucleus in heredity. The principal goal of this chapter will be to investigate how American geneticists came to ignore the possible importance of cytoplasmic inheritance, and yet rose to an authoritative position in the field of heredity by the early 1930s.

To this end, I attempt to analyze the emergence of Mendelian genetics with respect to its power relations with other disciplines in the field of heredity. It will be argued that the formation of a discipline of genetics with its own norms, methods, theories, and doctrines, based on confining the notion of heredity to the transmission of genes from one generation to the next, resulted from a strategy which investigators who supported the predominant role of Mendelian genes employed in the struggle for scientific authority.

I suggest that the genotype-phenotype distinction raised and maintained by geneticists played a polemical role in the construction of the genetic conception of heredity and in excluding contending approaches from its study. In order to understand this role, I set the distinction in the social and intellectual milieu which conditioned its construction and maintenance. First I reconstruct the broad field of heredity as it existed during the first decades of the century and briefly describe some of the prominent features of its principal disciplines and their sociointellectual interrelationships. It is shown that during the first decades of the century, several distinct notions of heredity were produced from within specific disciplines. Finally, I examine the sources of power and authority and trace the course which led Mendelian genetics to a dominant position in the field.

Disciplines in Conflict

During the first decade of the century, at the time when the Mendelian laws of heredity were rediscovered, heredity was a central concern of biology. Nonetheless, heredity was an extremely vague notion for biology as a whole. Certainly, inasmuch as heredity was thought to be responsible for the similarities and dissimilarities exhibited in successive generations, it was understood to be important for the problem of evolution, a focus of commitment for all biologists. More specifically, biological problems which were claimed to depend on heredity were varied and were concerned with such questions as the nature of heredity variations, which in turn were proposed as the basis of the mechanism of evolution and animal and plant breeding; how an organism grows, develops, and is maintained; why certain parts of the organism are capable of producing the whole (totipotency); how characters are transmitted from one generation to the next; the physical basis of inheritance; and the course of evolution.

These problems began to be the concern of several more or less sanctioned areas of practice, such as cytology, embryology, physiology, practical breeding, natural history, and biometry. Heredity was the "natural" product of different contexts of production. Indeed, how an individual understood the term resulted largely from the techniques, theories, explanatory standards, and overall objectives of his or her particular discipline. Each discipline was characterized by a range of possibilities within which the production of knowledge of heredity took place. The range of possibilities was defined not only by the current theories or beliefs about heredity, but also by the nature of the objects accessible to investigation, the equipment available for examining them, and the way of observing and discussing them.

To begin with, Mendelism was based essentially on an experimental and statistical examination of the reappearance of visible differences between individuals of a species. Led until World War I by William Bateson, Mendelism supported the notion of discontinuous evolution. Biometry, on the other hand, led by W. F. R. Weldon and K. Pearson, who supported continuous variations, was based on the Galtonian theory of ancestral heredity and on statistical examinations of visible characters within populations. Heredity, for biometricians, was a statistical law. "Heredity," wrote Karl Pearson (1900, p. 474) "is the law which accounts for the change of type between parents and offspring, *i.e.*, the progression from racial towards the parent type."

In addition to investigations grounded in biometrical theory, those of paleontology, entomology, systematics, and morphology, here referred to collectively as natural history, were based on a study of the visible characteristics of the organism. However, while biometricians constructed a statistical conception of heredity, for natural historians at the turn of the century heredity was investigated indirectly by ordering "nature" and by describing the past. Naturalists investigated heredity and evolution by compiling data illustrating relationships between new forms of living and extinct organisms brought to light by expeditions sent out from universities and museums. Unlike Mendelian investigators, who were concerned with an analysis of differences between individuals, naturalists investigated heredity as the link that was responsible for the similarities and differences that accounted for the relationships among phyologenetic groups. Of course, for many naturalists, such as A. Hyatt and E. D. Cope, who believed in the inheritance of acquired characteristics, the extraorganismic environment played a direct role in heredity (Allen, 1979).

In contrast to members of these disciplines, who were not directly concerned with the material link between generations, cytologists were highly concerned with what they considered to be "the physical basis of heredity." Cytology was based on microscopic examination of the cell, and since the 1870s cytological investigations were primarily concerned with the morphological study of eggs and oogenesis, spermatozoa and spermatogenesis, and fertilization. Following this work on the early history of the germ cells and the fertilization of the ovum, the principal goal of cytology was to reconcile cell theory with evolutionary theory. It became clear that, in the words of E. B. Wilson (1900, p. 6), "the general problems of embryology, heredity, and evolution are indissolubly bound up with those of cell structure, and can only be fully apprehended in the light of cytological research."

As discussed in the preceding chapter, by the last decades of the nineteenth century, with the help of improved staining techniques, cytologists had accumulated evidence for the physical continuity of the cell nucleus and claimed that it played a direct role in heredity. Nonetheless, heredity remained an obscure notion from a cytological point of view. Wilson (1914, p. 352) wrote:

> Our conceptions of cell organization, like those of development and heredity, are still in the making. The time has not yet come when we can safely attempt to give them very definite outlines.

The objectives of experimental embryology, beginning in the 1880s, whether it be *Entwicklungsmechanik, embryologie causale,* or the physiology of development, were to investigate the causes (physical, chemical, physiological, mechanical, etc.) of development, which embryologists claimed to be the same as those of heredity. In contrast, to the practice of Mendelian workers, that of embryologists was not concerned with the transmission of traits. Embryologists viewed heredity as a process of *production* and *reproduction*. Both "intrinsic factors" (germinal protoplasm) and "extrinsic factors" (practically all other conditions) played causal roles in the process of heredity.

To embryologists, heredity was concerned with all the morphological and physiological characters which the descendant shared with its parents. In this sense, the fact that the egg of a rat always gave rise to a rat, and that of a frog to a frog, represented a first aspect of the phenomena of heredity. Heredity involved concepts of integration, organization, regulation, temporal sequence, space differentiation, etc. On the other hand, Mendelian investigators were confined by the nature of their practice, based on experimental breeding, in viewing heredity not in terms of production and reproduction, but rather in terms of *distribution* and *exchange*. Mendelism obviously lacked the dimensions of embryology; it concerned particles or factors which were dealt with simply by counting. Conklin

(1908, p. 90) descibed "heredity" from the embryological point of view as follows:

> Indeed, heredity is not a peculiar or unique principle for it is only similarity of growth and differentiation in successive generations. The fertilized egg cell undergoes a certain form of cleavage and gives rise to cells of a particular size and structure, and step by step these are converted into a certain type of blastula, gastrula, larva and adult. In fact, the whole process of development is one of growth and differentiation, and similarity of these in parents and offspring constitutes hereditary likeness. The causes of heredity are thus reduced to the causes of successive differentiations of development, and the mechanism of heredity is merely the mechanism of differentiation.

Unlike many cytologists, and later geneticists, who centered their investigations on the chromosomes in the nucleus of the cell, embryologists centered theirs on the cytoplasm of the egg. As discussed in Chapter 1, with regard to the causes of heredity in general, many embryologists and physiologists rejected the idea that heredity could be reduced to the properties of morphological units *per se*. Instead, they supported a holistic and integrative concept of heredity, claiming that the germ cells had essential properties that were not present in any of their constitutive elements and that the essential properties of the cell system were derived from the interactions between the constitutive parts. Many postulated the existence of factors of heredity located in the cytoplasm and rigidly adhered to their doctrine that the physicochemical processes and structural organization of the cytoplasm played the preeminent role in heredity, development, and evolution.

An additional group of practitioners, who played a leading role in heredity investigations during the first decades of the century were breeders situated outside of the universities. At the time of the rise of experimental breeding based on Mendelian theory in academic institutions, plant and animal breeders came together to view heredity as an important "economic force." The "heredity values" of specially bred strains of plants and animals were claimed to be as real as the seemingly more concrete values of land or goods. The value of the "unseen carriers of heredity" was considered by some in 1910 to be "far above that of gold." Perhaps the notion of heredity for breeders was best represented by the following metaphorical passage:

> Hereditary is a force more subtle and more marvellous than electricity. Once generated it needs no additional force to sustain it. Once new breeding values are created they continue as permanent economic forces. ("Heredity: Creative Energy," 1910, p. 79)

Finally, there were eugenicists who concerned themselves with the "inborn qualities" of race. Eugenicists ostensibly applied their heredity theory to improve the "fitness" of human populations, but in reality they used it as a weapon in the class struggle. Generally, in Britain and the United States during the first third of the century eugenics functioned as a middle-class ideology to legitimate existing social order as being "natural." In the United States it was also employed as a racist program designed to decrease the number of migrants from southern and eastern Europe, before being used by the Nazis in the 1930s (see Farrall, 1979; Kevles, 1985).

So heredity theory was pursued by various groups with different interests, aims, and methods. At first glance, the differentiation of practitioners into sanctioned areas of competence may appear to be only a convenient division of labor. However, one of the central points of this book is that this is not the case. To the degree that members of each practice defined and explored heredity with their own methods and theories, each claimed the value of his or her approach to be greater than that of the others. The contending groups represented divergent views concerning what questions were important, what answers were acceptable, what techniques were appropriate, and what phenomena were interesting. The importance of these diverse biological problems and various views of heredity was not hierarchically ordered within biological research by an intrinsic logical necessity of scientific thought. Rather their importance in biology depended directly on both the technical capacity and the institutional power of the discipline from which they were produced. In this light we will turn to an analysis of the construction of the genotype-phenotype concepts.

Genotype/Phenotype: A Discursive Tactic

> The science of genetics is in a transition period, becoming an exact science just as the chemistry in the times of Lavoisier. (Wilhelm Johannsen, 1911, p. 131)

By searching for members of different disciplines claiming to study heredity, but basing their investigations on different theories of heredity, it is possible to reveal conflicting practices engaged in a struggle for authority. For example, in direct conflict with naturalists and Mendelian investigators, the biometricians Pearson and Weldon maintained that "the problem of animal evolution is essentially a statistical problem" and that a "statistical" knowledge of the changes going on in a number of species was "the only legitimate basis" for speculations as to the course of evolution (Provine, 1971, p. 31). In a particularly explicit passage representing the desire of biometricians to dominate the field, Pearson (1898, p. 397) wrote about Galton's law:

> If Darwin's evolution be natural selection combined with *Heredity,* then the single statement which embraces the whole field of heredity must prove almost as epoch making to the biologist as the law of gravitation to the astronomer.

In direct conflict with the viewpoint of biometry and that of several other disciplines concerned with the study of heredity, the embryologist E. G. Conklin (1908, pp. 89–90) argued that the most effective way of studying heredity was by embryological means:

> Heredity is today the central problem of biology. This problem may be approached from many sides—that of the breeder, the experimenter, the statistician, the physiologist, the embryologist, the cytologist—but the mechanism of heredity can be studied best by the investigation of the germcells and their development.

Not only did Conklin claim that heredity could be studied best by embryological procedures, but similary he reasoned that since the factors causing evolution and those causing development were essentially the same, "the embryologist is especially well fitted to deal with the factors of phylogeny" (Conklin, 1919, pp. 481–506). Indeed, Conklin (1912, p. 128) suggested that a knowledge of the causes of development was a prerequisite to a knowledge of the causes of evolution:

> If we are as yet unable to determine the precise manner in which the structure of the germ evolves into the structure of the adult . . . it is a small wonder that we have been unable to determine in detail the way in which one race is transmuted into another.

The distinction between the genotype and the phenotype was constructed and maintained within this sociointellectual milieu. The historical importance of the genotype-phenotype distinction, as proposed by Wilhelm Johannsen, has been the object of detailed investigations by Frederick Churchill (1974) and Garland Allen (1978, 1979). According to Churchill, there are two different views of the genotype-phenotype distinction. The first, which he argues to be characteristic of Johannsen's view, Churchill calls a "vertical analysis." Johannsen was concerned with a statistical examination of the range of phenotypic variation within a population brought about by several successive generations of inbreeding or by rigorous selection. He examined changes within a population over a period of time—in terms of what Churchill calls "vertical descent." On the other hand, Churchill has noticed that almost immediately after the terms "genotype" and "phenotype" were proposed, other biologists began to apply them to individuals as well. This interpretation Churchill has termed the "horizontal cleavage" of the individual into its genetic (hereditary) and its epigenetic (hereditary + environmental) components.

Allen (1975) has suggested a number of ways in which the phenotype-genotype distinction closed the gap between the "naturalist" and the "experimental" "traditions." Placing Darwinism and Mendelism at the center of his study, he has sought the conceptual conditions for their "synthesis." The genotype-phenotype distinction has been considered to be significant for two fundamental reasons. First, Allen (1978, pp. 55–56) argues, it cleared up a "basic confusion" in the minds of many biologists, particularly embryologists, who he claims failed to grasp the distinction between the hereditary particle itself and the recognizable adult character to which it presumably gave rise.

Second, the distinction has been regarded as being instrumental "in resolving some of the apparent contradictions between the naturalist and the experimental mode of thought" (Allen, 1979, p. 199). Allen points out that most Darwinians, such as H. F. Osborn, C. S. Minot, and V. L. Kellog, representing the "naturalist tradition," maintained that selection acted on continuous variations. For these investigators the Mendelian principles threw little light on evolutionary processes. On the other hand, Mendelians supporting discontinuous variation, such as W. Bateson, T. H. Morgan, and Jacques Loeb, who Allen argues represented the "experimentalist tradition," initially held that Darwinian selection was only of minor importance for the origin of species. Since in Mendelian terms, the genotype-phenotype distinction showed that genes do not vary the same way characters do, Allen (1979, p. 204) claims that Johannsen's work laid the basis for showing that the question of whether selection acts on continuous or discontinuous variations was "misguided."

Assuredly the above accounts help us to understand the role the genotype-phenotype distinction *might* play in making Mendelian theory seem more intellectually compatible with Darwinian theory. However, in the above presentations, there is little recognition of disciplines which produce and maintain, by controlling the production of knowledge, particular theories, methods, and dogmas, as the basis of their practice. It is necessary to consider, for instance, the social and cognitive effects of the rise of genetics on other disciplines, as well as the reasons for the construction and maintenance of the genotype-phenotype dichotomy in genetic discourse. In order to understand the significance of the distinction within the context of the struggle for scientific authority and to reveal its place in the rise of genetics, a rereading of Johannsen's celebrated paper, which appeared in the American literature in 1911, will prove to be helpful.

As a Mendelian, non-Darwinian, and supporter of discontinuous evolution, Johannsen first took issue with the word "heredity" itself. The terms "heredity" and "inheritance," he pointed out (1911, p. 129), were borrowed "from everyday language, in which the meaning of these words is the '*transmission*' of money or things, rights or duties—or even ideas and knowledge—from one person to another or to some others: the 'heirs' or 'inheritors.'" Since the second half of the nineteenth century, the transmission of property from parent to child had been a prevailing notion of heredity in some biological discussions as in jurisprudence. The observations of natural history, the neo-Lamarckian theory of the inheritance of acquired characteristics, and the contemporary biometrical definition of heredity as "the degree of correlation between the abmodality of parent and offspring." Johannsen (1911, p. 130) argued, were all based on the judicial notion of heredity as the transmission of property from parents to offspring. They were based on the idea that the personal qualities of individual organisms were the true heritable elements or traits. These cases Johannsen categorically distinguished as relying on what he called the "transmission conception of heredity."

Johannsen proposed his "genotype conception of heredity" in an attempt to break away from the dominant notion of heredity, based on the personal character as the elementary agent which is transmitted from one generation to the next, to uphold the Mendelian notion of the character as a consequence of something else which is transmitted hereditarily. The word "heredity" itself, became a stake in this "transition," and Johannsen attempted to expose the approach of biometricians and naturalists as being based on folk belief.

Between 1890 and 1910, many biologists trained in developmental physiology also pointed out the metaphorical use of the word "heredity" as the transmission of characters. They clearly recognized that parental characters were never transmitted to the offspring. Conklin, (1908, p. 90) also saw the problem as one involving the intrusion into scientific thought of ideas based on social relations: "The comparison of heredity to the transmission of property from parent to children has produced confusion in the scientific as well as the popular mind." Similarly, Morgan (1910, p. 449) wrote; "When we speak of the transmission of characters from parent to offspring, we are speaking metaphorically; for we now realize that it is not characters that are transmitted to the child from the body of the parent, but that the parent carries over the material common to both parents and offspring."

The distinction between what Johannsen called the "transmission conception of heredity" and the "genotype conception of heredity" was reinforced by the results of his pure line experiments and would provide the rhetoric for excluding natural history and biometry from the study of the fundamental basis of heredity, or what Johannsen called the "genotype." The transmission conception of heredity, represented by biometrical and naturalist approaches, was based on historical explanation. As William Coleman (1971) has stressed, the historical conception of nature was seen as identical to causal explanation and belonged to the same order of thought as the nineteenth-century notion of human history as one continuous, genetic, causal process.

Johannsen stressed that the "genotype conception of heredity" was an *ahistorical* view of heredity. To stress the ahistorical nature of the genotype conception, the ex-chemist paralleled the genotype (that is, "the sum total of all the 'genes' in a gamete or in a zygote") with a complicated physicochemical structure which reacts only in consequence of its realized state, but not in consequence of the history of its creation:

> The genotype-conception is thus an "ahistoric" view of the reactions of living beings—of course only as far as true heredity is concerned. This view is an analog to the chemical view, . . . chemical compounds have no compromising ante-act, H_2O is always H_2O, and acts always in the same manner, whatsoever may be the "history" of its formation or the earlier states of its elements. I suggest that it is useful to emphasize this "radical" ahistoric genotype-conception of heredity in its strict antagonism to the transmission—or phenotype—view. (Johannsen, 1911, p. 139)

The fundamental basis of heredity and variation would now be hidden deep within the gametes of the organism. Under the "genotype conception of heredity," only through the analytic methods of experimentation could the fundamental basis of heredity be explored. Morphologists, zoologists, botanists, and biometricians, whose studies of heredity were based on the study of visible characteristics of organisms, could offer, according to Johannsen, "no profound insight into the biological problem of heredity" (Johannsen, 1911, p. 130).

Johannsen proposed the term "phenotype" as a polemical word against the morphological, descriptive character of natural history. The notion of evolution by continuous transitions, supported by huge collections in natural history museums, would, under the grid of the phenotype-genotype distinction, be seen to be the result of the varying external conditions of life. "All 'types' of organisms," Johannsen (1911, p. 134) argued, "distinguished by direct inspection or only by finer methods of measuring or description, may be characterized as 'phenotypes.'"

Johannsen (1911, p. 138) dismissed the Galtonian law of regression elaborated by Pearson, who pretended to have established the "laws" of "ancestral influences" in statistical terms, as being based on mysticism or superstition: "Ancestral influence! As to heredity, it is a mystical expression for a fiction. The ancestral influences are the 'ghosts' in genetics, but generally the belief in ghosts is still powerful." Genetics, he argued, should be pursued "*with* mathematics not *as* mathematics." Statisticians, according to Johannsen (1911, p. 130) could not rely for clients on biologists concerned with the problems of heredity:

> Certainly, medical and biological statisticians have in modern times been able to make elaborate statements of great interest for insurance purposes, for the "eugenics-movement" and so on. But no profound insight into the biological problem of heredity can be gained on this basis.

Not only did Johannsen attempt to exclude naturalists and biometricians from the field of heredity, he also denied the scientific legitimacy of breeders. In contrast to academic scientists, who were excluded on intellectual grounds, Johannsen (1911, pp. 142–143) denounced breeders on grounds of their lack of professional training:

> The practical breeders are somewhat difficult people to discuss with. Their methods of selection combined with special training and "nurture" in the widest sense of the word are mostly unable to throw any light upon questions of genetics, and yet, they only too frequently make hypotheses as to the nature of heredity and variability. Darwin has somewhat exaggerated the scientific value of breeders' testimonies, as if a breeder *eo ipso* must be an expert in heredity.

The terms "genotype" and "phenotype" would play a dual role in Johannsen's paper. This distinction would be raised as an argument against the biometrical and morphological descriptive approaches to the study of heredity and evolution, and it would be used to defend Mendelian theory against the charge of being merely another speculative theory of heredity. As discussed in the previous chapter, many embryologists and physiologists voiced opposition to the "particulate," "mechanistic," "deterministic," "nondynamic" theories which, they argued, could not account for the orderly differentiation and behavior of the "organism as a whole." Adult characters seemed to be the product of an elaborate series of processes, and the germinal differentiations in the nucleus were too remotely connected with the end-product to think of them in terms of special particulate "determinants" except in a purely symbolic fashion. Johannsen agreed with this view and opposed the idea that the gene was identifiable with a morphological structure with an independent life of its own. However, to many biologists it seemed as though the morphological speculative theories based on hypothetical hereditary entities were extended into the twentieth century under the guise of Mendelism—which had little or nothing to say about the integration of the factors or gene into an organism during development.

Johannsen attempted to defend Mendelism against these views. As historians of biology recognized—as well as many contemporary biologists, including Johannsen himself—the distinction between the "transmission conception" of heredity and the genotype conception had been proposed by earlier biologists, such as Weismann and Galton. In fact, Allen (1979, p. 205) has argued for the importance of the continuity of Johannsen's genotype-phenotype distinction with

Weismann's distinction between the germ plasm and the somatoplasm, in placing "one more nail in the coffin of neo-Lamarckism." Yet this continuity is only apparent. Weismann's distinction between the germ plasm and the somatoplasm was raised from speculation and was summoned to ward off another speculative theory: that of the inheritance of acquired characteristics. The distinction between the genotype and the phenotype, on the other hand, was based on experimentation, and served as a polemic against descriptive, speculative, and morphological approaches to the study of heredity—categories which included Weismann's theory itself.

Johannsen explicitly stressed the issue that the two speculative Weismannian notions that elements in the zygote correspond to special organs, and that discrete particles of the chromosomes are "bearers" of special parts of the whole inheritance in question, were neither "corollaries" of nor "premises" for the "genotype conception." He reasoned that "they have no support in experience, the first of them is evidently erroneous, the second a purely speculative morphological view of heredity without any suggestive value" (1911, p. 132). On the other hand, Johannsen (1911, p. 132) maintained:

> The genotype-conception of the present day, initiated by Galton and Weismann, but now revised as an expression of insight won by pure line breeding and Mendelism, is in the least possible degree a speculative conception.

Moreover, like experimental embryologists, Johannsen (1923, p. 141) claimed that an absolute independence of germ plasm did not exist in reality. He argued that the germ plasm-somatoplasm distinction was "incommensurable" with the genotype-phenotype distinction. "The non-inheritance of acquired characteristics," he argued, "is not a consequence of this assumed independence or difference, but only a striking expression of the fact that the external conditions may easily mould phenotypes in a more or less adaptive manner, but can hardly or rarely induce changes in the genotype" (Johannsen, 1923, p. 141).

The important point about Johannsen's work, then, was that it attempted to distinguish and separate Mendelian theory from all reliance on the reductionist, morphological speculations of Weismann and other nineteenth-century theoreticians. Mendelism no longer had to rely on the discredited views of Weismann. As Johannsen (1911, p. 132) contended, "Of all the Weismannian army of notions and categories it may use nothing." Thus, Johannsen explicitly attempted to denounce and distinguish the above views from what he claimed to be the "science of genetics" and offered genetics what soon would be held as the dogma of the "genotype conception of heredity." A new common nomenclature of "the science of genetics" was introduced and the terms "gene," "genotype," "phenotype," and "biotype" were proposed. The word "gene" was presented as a harmless term to be a replacement for various other theory-ladened morphological expressions used by Mendelian researchers. (Johannsen, 1911, p. 132).

Johannsen viewed the position of genetics to be in a transition period, and drew on the history of his previous discipline, chemistry, to argue for the state of contemporary genetics. He reasoned that the "transmission-conception of heredity" represented exactly the reverse of the real facts, just as the famous Stahlian theory of 'phlogiston' was an expression diametrically opposite to chemical reality (Johannsen, 1909, p. 130).

With his "genotype conception of heredity," Johannsen had provided the discursive strategy for the exclusion of all the nonexperimental approaches from the study of the newly proposed "fundamental basis of heredity." In fact, he would go further and suggest a new genetic definition of heredity and state: "Heredity may then be defined as *the presence of identical genes in ancestors and descendants*" (Johannsen, 1911, p. 159).

If one accepts scientific authority to be a result of technical capacity and social power, the question arises: what was the technical capacity through which genetics, based on the study of the transmission of "genes" from one generation to the next, might attain social power? Part of the answer to this question is provided by William Bateson, in an address he made as the president of the British Association for the Advancement of Science in 1914:

> If a population consists of members which are not alike but differentiated, how will their characteristics be distributed among their offspring? This is the problem which the modern student of heredity sets out to investigate. (Bateson, 1914, p. 289).

Several years after his celebrated dispute with biometricians, Bateson argued for the supremacy of Mendelian analysis over the approaches of all other disciplines in the field of heredity, and claimed that a knowledge of the system of hereditary transmission stood as a primary necessity in the construction of any theory of evolution. "Formerly," the ex-embryologist argued, "it was hoped that by simple inspection of embryological processes, the modes of heredity might be ascertained, the actual mechanism by which the offspring is formed from the body of the parent" (Bateson, 1914, p. 288). These expectations were not realized by embryologists. "With the existing methods of embryology," Bateson scoffed, "nothing could be analyzed further than the physiological events themselves." Alternatively, he rationalized:

> We at least can watch the system by which the differences between various kinds of fowls or the various kinds of sweet peas are distributed among their offspring. By thus breaking the main problem up into its parts we give ourselves fresh chances. (Bateson, 1914, p. 289)

In addition to denouncing the efficacy of embryological methods for the study of heredity, Bateson maintained that cytology had failed equally. Since cytologists could not show consistent distinctions between cytological characters of somatic tissues in the same individual, Bateson (1914, p. 289) argued, they could not be expected to perceive such distinctions between chromosomes of the various types. Led by Bateson, genetics was not concerned with a cytological basis of heredity; it could say nothing about the causes of growth, development, or maintenance of an organism, and nothing directly about the course of evolution. Instead, within biology, genetics was limited to investigations of the origin and distribution of variations, which were generally recognized to be at the basis of evolution.

With the leverage of *experimentation,* genetics would attempt to work its way to a secure institutional place in the biology curriculum via evolutionary studies—a place which was predominantly established and occupied by naturalists. Experimentation was Bateson's primary polemical tool. Genetics was based on the experimental study of heredity and variations and was the means by which Bateson sought to introduce new standards of criticism into evolutionary studies. Experimentation provided the evidence for the geneticists' argument against the belief that changes in the conditions of life were direct causes of heredity modifications which altered the definitiveness of species.

Not only did the results of experimentation support the conviction that acquired characteristics were not transmitted to offspring, but they were also held up in support of the claim of Bateson and others that natural selection could not have been the chief factor in delimiting the species of plants and animals. Bateson claimed that experimentation pointed to the conclusion, in the first decade of the century, that the greater differences which characterized distinct species were due to independent "genetic factors." Generally, then, experimentation was the basis of authority upon which some geneticists argued that variations "in the old sense" were not genuine occurrences at all.

Based on the above arguments, Bateson claimed "knowledge of heredity" and the studies of genetics to be vital. Only through genetic analysis could the origin of new forms be investigated by the study of the positive separable factors. Until the "facts of heredity" are explained, Bateson (1914, p. 294) argued, one should turn aside from phylogenetic problems. He mocked the integrity of natural history, saying:

> Naturalists may still be found expounding teleological systems which would have delighted Dr. Pangloss himself, but at the present time few are misled. The student of genetics knows that the time for the development of theory is not yet. He would rather stick to the seed pan and the incubator. (Bateson, 1914, p. 293)

Bateson's plea for institutional support for genetics and his attempts to supplant natural history were not ignored by naturalists. In 1917, for example, in an article entitled "Genetics versus Paleontology," W. K. Gregory, a paleontologist at the American Museum of Natural History, attempted to vindicate his discipline.

Allen (1979, p. 181) has isolated Gregory's paper and interpreted it as part of a continuous conflict arising in the early nineteenth century between "experimentalist and naturalist traditions." The "two traditions," he argues, "seemed at the time to be in conflict over a fundamental question: which method gives the most meaningful answer to scientific, that is, biological problems?" However, Bateson is asserting much more than this. What is at stake here is a definition of science itself. What constitute legitimate scientific problems, questions, and answers? Moreover, "experimentalism" *per se* did not represent a homogeneous movement. Mendelian geneticists were also in conflict with other experimentalists, especially embryologists. Hence, the conflict may be seen more clearly as an interdisciplinary struggle with geneticists who were attempting to assert themselves at the top of the hierarchy of biological sciences—and this struggle continued throughout the century.

Gregory realized well enough the nature of the social and intellectual stakes of the struggle between natural history and genetics. In response to the passage from Bateson's text quoted above, Gregory (1917, p. 622) wrote:

> Taken in connection with other passages, this seems to imply the belief that the present is no time to investigate phylogenetic problems or to formulate any generalities concerning the origin of systematic groups of organisms. Until the facts of heredity are explained we should turn aside from most of the major problems that engaged the attention of the great comparative anatomists and paleontologists of the 19th century.

Gregory attempted to legitimate the activity of his discipline by pointing out the depth of its historical roots, at the same time claiming that Bateson was overlapping disciplinary boundaries and was "confusing two fairly distinct lines of investigation, genetics and phylogeny" (Gregory, 1917, p. 623). Although Gregory conceded that "the student of evolution may well reserve judgment as to the theories of evolution," he was quick to point out that paleontologists had their own well-established methods for their own problems, their own clientele, and their own intellectual and institutional space:

> As long as museums and universities send out expeditions to bring to light new forms of living and extinct animals and new data illustrating the interrelations of organisms and their environments, as long as anatomists desire a broad comparative basis for human anatomy, as long as even a few students feel a strong curiosity to learn about the course of evolution and the relationships of animals, the old problems of taxonomy, phylogeny and evolution will gradually reassert themselves even in competition with brilliant and highly fruitful laboratory studies in cytology, genetics and physiological chemistry. (Gregory, 1917, p. 623)

Bateson himself was successful in establishing a genetics school in a British university. During the first decade of the century he supported himself and his plants and animals on a modest income from studentships and a fellowship at his college, St. John's, Cambridge. He was appointed to a professorship of biology in 1908, but the post had been created only for a five-year period (Kevles, 1980). The hostility towards geneticists in many quarters of the life sciences would continue well into the 1950s (see Chapters 5 and 6). Although genetics may not have found much favor with many established academic biologists, it did not rely solely on sources of power and authority internal to an academic community. Bateson left Cambridge in 1910 to become director of the John Innes Horticultural Institution at Merton. He established a group of collaborators and attempted to bring geneticists and practical breeders into close cooperation. Bateson (1911, p. 10) summarized the strategy for geneticists in all countries:

> If we are to progress fast there must be no separation made between pure and applied science. The practical man with his wide knowledge of specific natural facts, and the scientific student ever seeking to find the hard general truths which the diversity of nature hides—truths out of which any lasting structure of progress must be built—have everything to gain from free interchange of experience and ideas. To ensure this community of purpose, those who are engaged in scientific work should continually strive to make their aims and methods known at large, neither exaggerating their confidence nor concealing their misgivings.

The Rise of American Genetics and the Determinist Gene

> If another branch of zoology that was actively cultivated at the end of the last century had realized its ambitions, it might have been possible to-day to bridge the gap between gene and character, but despite its high-sounding name of *Entwicklungs-mechanik* nothing that was really quantitative or mechanistic was forthcoming. Instead, philosophical platitudes were invoked rather than experimentally determined factors. Then, too, experimental embryology ran for a while after false gods that landed it finally in a maze of metaphysical subtleties. (T. H. Morgan, 1932, p. 285)

During the World War I, leadership in Mendelian studies changed hands from Britain to the United States. T. H. Morgan was much more successful in the academic marketplace than Bateson, and he created a major school of genetics at Columbia University. Morgan and his *Drosophila* group added cytological investigations to genetics, giving Mendelian genes a physical location in the chromosomes of the cell. In so doing he won the loyal institutional support of the leading cytologist E. B. Wilson. As A. H. Sturtevant (1959, p. 295), one of the early members of the school, recalled:

> No small part of the success of the undertaking was due to Wilson's unfailing support and appreciation of the work—a matter of importance partly because he was head of the department.

Morgan (1919, p. 247) redefined the aim of genetics to embrace the chromosomes:

> Our study of the germ-plasm is largely confined . . . for the present to the study of the transmission of the genes, to the kinds of effects they produce on the organism, and to the special relations of the genes in the chromosomes where they are located.

Geneticists in the United States maintained tight control over their practice and worked the chromosome theory into a secure institutional position. The academic refereed journal *Genetics,* established in 1916, played no small part in the process of maintaining orthodoxy. Following the work of geneticists using *Drosophila* and those basing their work on corn, led by R. A. Emerson and E. M. East, genetics in the United States began to take the form of a sanctioned normative practice, with its own well-defined methods and explanatory standards. Indeed, biologists carrying out studies based on breeding experiments with stocks of unknown genetic character, and not based on exact Mendelian analysis, were enthusiastically excluded from the discipline of genetics.

Among the excluded may be found, for example, the well-known evolutionary biologist F. B. Sumner (1874–1945), who according to his own testimony, called himself "a geneticist when he was sure no geneticists were within hearing" (Provine, 1979, p. 211). According to Provine (1979, pp. 233–235), during the late 1920s Sumner had developed a comprehensive view of evolution and genetics that corresponded well with the view that emerged later from population genetics in the "evolutionary synthesis" of the 1930s and 1940s. However, at the time that Sumner was beginning to recognize the merit of Mendelism for a theory of evolution and to do some novel breeding experiments with wild mammals, the genetics establishment dealt him a blow. The editorial board of *Genetics* refused to publish his results because they were not based on exact Mendelian analysis a laboratory geneticist might accomplish using stocks of known genetic character. Sumner resubmitted his paper to the *American Naturalist* in which it soon appeared (Sumner, 1928). This episode reinforced his view that Mendelian geneticists cared little about larger evolutionary considerations. Feeling the chill of rejection, Sumner wrote in a letter one of his more sympathetic colleagues:

> I think that someone would do a considerable service to science, if he would offer a satisfactory definition of the word, "genetic." Some people seem disposed to give it an extremely narrow scope. (quoted in Provine, 1979, p. 235)

Genetics may have been a discipline limited in its intellectual scope between 1910 and 1930, but it had a highly competitive technical capacity to produce results and students. Geneticists such as T. H. Morgan at Columbia, E. M. East and W. E. Castle at Harvard's Bussey Institution, and R. A. Emerson at Cornell had no problem producing geneticists at the rate of about one Ph.D. a year between 1910 and 1930 (Kevles, 1980, p. 451).

The rapid rise of genetics to a prominent institutional position in American biology was the result of several factors. Certainly, the simplicity of Mendelian analysis and the use of *Drosophila* to produce quick results can never be underrated. Rapid production plays an important role in the recruitment and institutionalization of researchers. With a prolific technique to be exploited, students would be attracted into an area where their chances of success were good. On the other hand, a theory with a large explanatory range but lacking a technique capable of producing results rapidly would obviously take longer recruiting researchers. Quick recruitment would also mean being able to take advantage of institutional opportunities as they became available.

The relationship between Mendelism and practical breeders played a significant role in the development of American genetics (Rosenberg, 1976). Mendelian analysis showed breeders that every organism showing a specific trait was not necessarily a purebred for that trait; that is, with Mendelian theory, breeders could detect whether a certain line was a purebred or a hybrid. Such detection was highly valuable in crossing plants where such traits could be followed easily, so as to improve the quantity and/or quality of the yield. Beginning in the first decade of Mendelism, major breeding work aimed at creating improved forms of pedigreed plants and animals was carried out by the United States Department of Agriculture, in state experimental stations, and under other public and private auspices. This situation had stimulated the organization of clubs devoted to the study of heredity in centers of learning across the United States, and philanthropists were encouraged to dedicate generous sums of money to foster the development of genetics in universities.

One can distinguish two phases in the United States between 1900 and 1932 characterized by different relationships between breeders and Mendelians. In a first phase, ending around 1915, there was little distinction between what were known as "genetics" and "practical breeding." In fact, the relationship among genetics, practical breeding, and, after 1905, eugenics was so intimate that the word "genetics" could have carried the broader meaning and would not have been out of harmony with the contemporary use of the term in the United States (see Kimmelman, 1983). Mendelism was a movement represented by such figures as G. H. Shull and C. B. Davenport at the newly founded Carnegie Institution of Washington Station for Experimental Evolution, and E. M. East, first at the Connecticut Experimental Station and after 1909 at Harvard's Bussey Institution with W. E. Castle. This was a period when Mendelian studies were directed toward the solution of specific socioeconomic problems as much as toward understanding problems of evolution. It was a time when Mendelian investigators recognized no distinction between "pure" and "applied" science and attempted to establish the widest possible support for their work.

This phase was characterized by the establishment of several institutional structures common to investigators in government experimental stations, private research institutions, and universities. In 1903 the American Breeder's Association was founded. Its function was to "bring the practical breeder in close contact with scientists," and its aim was "to achieve scientific and economic results of the highest order." By 1914 the American Breeder's Association was retitled the American Genetics Association. The *American Breeder's Magazine* was founded as the official organ of the association in 1910. The next year it carried the subtitle *A Journal of Genetics and Eugenics*. This was a period when the figures of Charles Darwin and Gregor Mendel were joined with that of the famous breeder of shorthorn cattle, Amos Cruikshank, to form an institutional triumvirate.

During a second phase, after 1915, genetics in the United States became clearly distinguishable from practical breeding and eugenics. This was when genetics, led by the Morgan school, took the form of a discipline and attempted to establish a secure position for itself in American universities. This period was characterized by the rapid growth of genetics research in universities, the establishment of an academic journal, *Genetics,* distinct from the periodicals for practical breeding work, and the emergence of a purely academic genetics society, quite separate from the American Genetics Association. The Genetics Society of America was founded in 1932, as an offshoot of the American Society of Naturalists.

Although during this period genetics distanced itself from practical breeding, their relationship certainly did not come to an end. Practical breeding programs based on Mendelian principles were constant sources of power for genetics as a university discipline. Practical breeders employed genetic methods and theories in the solution of specific economic problems, and the rise of agricultural programs provided geneticists with an ample source of students to help fill their courses at the universities. Indeed, after 1915 the relationship between breeders and geneticists was no longer based on equal authority, as it had been previously, but became a relationship between professional or expert and client, or producer and consumer of knowledge of heredity.

Against the protests of nongeneticists, by the middle of the 1920s geneticists came to occupy a dominant position in the field of heredity in the United States. Based on the Mendelian-chromosome theory, genetics had established itself as a progressive scientific activity. Mendelian theory could be used to account for almost every case of inheritance within interbreeding groups of plants and animals, and furthermore had a corresponding physical basis, the chromosomes, whose fate in the cell at fertilization could be readily followed by cytological means.

Although American geneticists came to view genes as discrete material units located at definite points on chromosomes, their physicochemical nature remained indefinite and undefined. In *The Theory of the Gene* (1926b), Morgan (p. 25) attributed five principles to the gene derived from purely numerical data: segregation, independent assortment, crossing over, linear order, and linkage groups. He also strongly defended the constancy and integrity of the gene during development (1926b, p. 27) and upheld Weismann's doctrine of an independent germ line to deny the inheritance of acquired characteristics. "Between the characters that furnish the data for the theory, and the postulated genes, to which the characters are referred," Morgan (1926b, p. 26) wrote, "lies the whole field of embryonic development."

Although geneticists could say nothing about the way in which genes were connected with the end-product or character, American geneticists often claimed full authority over the field irrespective of their ability to deal with it effectively. They simply redefined the concept of heredity to suit their practice. As L. C. Dunn (1917, p. 299) claimed:

> The working of the effective method is known for heredity, if heredity be properly only concerned with the way in which hereditary factors are distributed in the germ cells. For development, the mechanism is but grossly known, but we have learned enough . . . to foster a suspicion that one day the governance of the chromosomes over development will be explained in physico-chemical forms.

As genetics rose to a prominent position in the field, so did its restricted view of heredity. Conklin (1919, p. 487) saw social forces playing a role in the distinction between development and heredity clearly when he wrote:

> Development is indeed a vastly greater and more complicated problem than heredity, if by the latter is meant merely the transmission of germinal units from one generation to the next.

Other embryologists would not concede the genetic distinction between heredity and development. Lillie (1927, p. 362) struggled to maintain what he called "the physiological conception of heredity as repetition of life histories." In keeping with Lillie's view of heredity, Albert Brachet (1935, p. 3) wrote:

> For the embryologist the word heredity takes on a very broad meaning; heredity is the totality of the developmental potentialities in the fertilized egg; it is the ensemble of the causes which make the egg produce, when in adequate environmental conditions, following a succession of well-defined processes, a new organism having all the characters of the species to which it belongs. Thus understood, heredity is the real object of embryology—to know it at the same time in its origin, in its manifestations and in the mechanisms which it puts in place in order to realize its final goal. (my translation)

These statements did not simply mean that geneticists were restricting themselves to the problem of hereditary transmission. They meant more—that the corpuscular gene theory in itself did not provide an effective working hypothesis to approach the problem of development. They also meant that genetics was dealing mostly with superficial and trivial matters which could have little to do with the grand process of evolution. As discussed in the last chapter, the idea that heredity could be reduced to individual self-perpetuating determinants had long been rejected by embryologists and physiologists. If the germ contained thousands of packets of chemicals massed in a haphazard way but arranged in a definite manner, as American geneticists claimed, then, many physiologists and embryologists claimed, some sort of principle was required to hold such a "swarm" together within the bounds of an individual and to direct their work. Moreover, it seemed self-contradictory to explain embryonic differentiation and regulation by the behavior of genes which were the same in every cell. As long as the constancy and integrity of the nuclear genes remained a necessary part of the gene theory, other heredity properties and materials in the cytoplasm would be required to account for cellular differentiation.

Many European geneticists, including Johannsen, Bateson, and Correns, who had attempted to avoid a determinant conception of the gene also protested against the Weismannian mechanism and reductionism concealed within the corpuscular theory of heredity. In their view, the genotype could be dissected into distinct particles only for analytic purposes. Even then, some maintained the existence of an unknown property of the genotype that would resist such dissection. Johannsen (1923, p. 139) summarized the major criticism clearly in one sentence when he attacked Morganist genetics: "The Mendelian units as such, taken per se are powerless."

Ironically, Johannsen's genotype/phenotype distinction offered geneticists the conceptual space or route by which they could bypass the organization of the cell, regulation by the internal and external environment of the organism, and the temporal and orderly sequences during development. Although the genotype/phenotype distinction represented an implicit theoretical acknowledgement of the beginning and end of a production, in practice, Mendelian geneticists ignored developmental processes and the possible influence of extragenic conditions in the production of characters. The presence of genes was inferred by experimental and statistical manipulation of phenotypes. As a result, adult characters were often presented as being the direct result of genes. The claim of a direct causal relation between gene and character is well expressed in the term "determiner." An explicit example of this view is offered by L. C. Dunn (1917, p. 286) who wrote about the word "determined:"

> It does not mean that the character itself is present in the germ in any form, but rather that it is represented by substances or forces which not only *stand for* the character but in some way bring about its expression.

Yet, the notion that the genotype or genes controlled the phenotype was pure tautology in the typical breeding experiments. The problem of identifying genes with traits—viewing genes as determiners rather than instruments of production—was raised by the embryologist Lillie in a letter to Julian Huxley (March 19, 1928):

> If you will excuse a paradox gene theory is essentially a theory of phenotypes, i.e. something always static for as soon as it changes it is already another phenotype.

It was around this time, after some fifteen years of maintaining a restricted program, that Morgan began to call for the extension of the principles of genetics to problems of development. In 1926 he brushed aside the need for postulating cytoplasmic hereditary properties, advocated by embryologists. He dogmatically asserted that the cytoplasm could be ignored genetically and maintained that the application of genetics was a most promising method of attack on the problems of development (Morgan, 1926, pp. 491–496). The following year Lillie, who would lead the embryologists' protest against the application of genetics to problems of development in the following decades (see Chapter 3), wrote, with some regret, "Genetics has become quite a unitary science and the physiology of development is at most a field of work" (Lillie, 1927, p. 362). Confronting the dominant discipline of genetics, Lillie attempted to defend the integrity of his practice. In direct conflict with the call of Morgan, he wrote:

> The progress of genetics and of development can only result in a sharper definition of the two fields, and any expectation of their reunion (in a Weismannian sense) is in my opinion doomed to disappointment. . . .
>
> Instead of distorting our workable conceptions to include that which they can in no wise compass, may it not be profitable, for a while, to admit that more lies without than within our confines of mechanism and statistics? (Lillie, 1927, pp. 367–368)

Ten years later Ross Harrison (1937, p. 372) would continue to express his hostility to the gene theory as applied to development:

> The prestige of success enjoyed by the gene theory might easily become a hindrance to the understanding of development by directing our attention solely to the genome, whereas cell movements, differentiation and in fact all developmental processes are actually effected by the cytoplasm. Already we have theories that refer the process of development to genic action and regard the whole performance as no more than the realization of the potencies of the genes. Such theories are altogether too one-sided.

Many nongeneticists between the two World Wars believed, or at least hoped, that genetics was a fad and that it would pass as soon as the limited confines of the statistics and pure mechanism that were used to construct it were fully recognized. They frequently called upon historical arguments based on patterns of changes in biological explanation and interests to legitimate their claims. The following passage by Joseph Needham (1919, p. 457) is illustrative:

> We are all out of balance. Some of our laboratories resemble up-to-date shops for quantity production of fabricated genetic hypothesis. Some of our publications make a prodigious effort to translate everything biological into terms of physiology and mechanism—an effort as labored as it is unnecessary and unprofitable. Why not let the facts speak for themselves? They go from one extreme to another. In my high school days we did nothing but dissecting; later came morphology and embryology, then experimental zoology, then genetics, and the devotees of each new subject have looked back upon the old with something like that disdain with which a debutante

> regards a last year's gown. Natural history and classification are perhaps long enough out of date, so that interest in them may again be revived.

Natural history, which had occupied a prominent position in biology, was perhaps most dramatically affected by the rise of genetics. By the 1920s, naturalists viewed their discipline to be threatened with extinction in American universities. It was this situation which led the American naturalist William Morton Wheeler to write his diatribe of 1923 entitled "The Dry-Rot of Our Academic Biology." In despair, Wheeler (p. 62) confessed:

> My mental condition is, no doubt, partly due to the disappointing spectacle of our accomplishments as more or less decayed campus biologists in increasing the number, enthusiasm and enterprise of our young naturalists.

Wheeler (1923, p. 62) also turned to the history of science to argue for the central position and the need for the breadth of natural history in the biological sciences:

> History shows that throughout the centuries . . . natural history constitutes the perennial rootstock or stolon of biologic science and that it retains this character because it satisfies some of our most fundamental and vital interests in organisms as living individuals more or less like ourselves. From time to time the stolon has produced special disciplines which have grown into great, flourishing complexes. . . . More recently another dear little bud, genetics, has come off, so promising, so selfconscious, but alas, so constricted at the base.

Wheeler expressed a common viewpoint of nongeneticists and many European geneticists, who by the early 1920s saw little future in Mendelian genetic methodology, explanation, and theory. However, American geneticists, led by Morgan and H. J. Muller, saw the situation differently. They believed that Mendelian genetics would become the stolon that would bind the whole field of biology into a unified discipline which would some day rival the physical sciences. Morgan responded to his critics with some historical interpretations of his own to justify mechanistic interpretations and American geneticists' operational approach:

> Realizing how often ingenious speculation in the complex biological world has led nowhere and how often the real advances in biology as well as in chemistry, physics and astronomy have kept within the bounds of mechanistic interpretation, we geneticists should rejoice, even with our noses on the grindstone . . . that we have at command an additional means of testing whatever original ideas pop into our heads. (Morgan, 1932, p. 264)

While Wheeler had an interest in the diversity of living individuals, H. J. Muller, who would later develop his own loyal following, came to view the gene itself as living, forming the stolon of all living individuals, including Wheeler. In 1926 in a paper entitled "The Gene as the Basis of Life," Muller formulated a theory of the gene which became highly acclaimed by Mendelian geneticists during the 1940s and 1950s. To Muller, who would in 1946 be awarded a Nobel Prize for his studies of artificial mutation of X-radiation, genes possessed the unique fundamental property of identical reduplication. Muller claimed that the ability of genes to vary (mutate) and to reproduce themselves in their new form conferred on these cell elements the properties of the building blocks, required by the process of evolution.

Taken together with the countless proofs that Mendelian genes controlled an amazing variety of functions, this concept reinforced the claim that the nucleus was the "governing body" of the cell. The simultaneous absence of an equally demonstrated, universal cytoplasmic inheritance influenced him and other Mendelian geneticists to look at the rest of the cell as a by-product of genic activity. In Muller's words:

> Genes (simple in structure) would, according to this line of reasoning, have formed the foundation of the first living matter. By virtue of their property (found only in "living" things) of mutating without losing their growth power they have evolved even into more complicated forms, with such by-products—protoplasm, soma, etc.—as furthered their continuance. Thus they would form the basis of life. (Muller, 1929, p. 921)

The artificial production of mutations by irradiation gave Mendelian genetics a new burst of life in the late 1920s and 1930s, providing it with one of its most important analytical devices and one of its most important sources of material for investigation. Geneticists no longer had to wait for mutations to arise spontaneously but were able to produce unprecedented numbers of new mutations. The study of mutations by mutagenic radiation was quickly extended from X-rays to include gamma rays, beta rays, cathode rays, and ultraviolet light. A focal point of radiation genetics was the evidence it provided on the nature of mutation itself. At an early stage in the work Muller had argued that the mutations produced were of two kinds: chromosomal rearrangements and gene mutations or "point mutations" which he claimed were due to "reconstructions of the gene." The kinds of mutations produced were neither specific nor directed by the mutagenic agents. As will be discussed more fully in later chapters, the randomness of the kinds of gene changes seemed to afford little possibilities for the directed hereditary changes thought to be required to account for the orderly nature of cellular differentiation during ontogeny. Nonetheless, Muller denounced the claims of critics who argued that the artificially produced mutations had little bearing on the constructive evolutionary processes and on their physical basis. He and his followers claimed that the mutations which radiation geneticists produced in the laboratory were of the same kind as those that arose spontaneously in nature. The artificially produced mutations included among them "artificial building-blocks of evolution as good as the natural stones" (Muller, 1930, p. 224).

By the late 1920s, genetics in the United States had risen to such a dominant institutional position that the idea of modifying the American Society of Naturalists to a "genetics society" was seriously considered. The American Society of Naturalists was an "umbrella society," made up of members of the various biological disciplines, and had as its scope of interest "organic evolution" (Conklin, 1934, pp. 385–401). In a letter sent to members of the executive committee of the society, the secretary of the American Society of Naturalists referred to the idea of forming a new "professional genetics society" as follows:

> There are many, however, who deprecate the multiplication of societies any more than is absolutely necessary, having in mind the complications and conflicts of interest which are already characteristic of the convocation week meetings. In this

> connection it has been suggested that if the American Society of Naturalists were to be modified so as to become a genetics society, it might meet the need that seems to be felt for a distinctive genetics organization; on the other hand, certain rather radical changes particularly of policy, would be necessitated. (Cole, October 25, 1929)

Genetics had a monopolistic strategy in the United States; it did not result from a simple convenient functional division of labor in the field. The particular form this discipline took (its norms, methods, theories, dogmas, journals, and societies) was shaped by a specific social and intellectual milieu. It can be understood as a strategy employed by those supporting Mendelian theory who struggled for scientific authority in the field. Mendelian geneticists constructed a discipline and a concept of heredity which they believed gave them the competitive advantage in this struggle. As we have seen, both the discipline and the genetic concept of heredity were restricted largely to the sexual transmission of hereditary factors from one generation to the next. Problems of development which could not be dealt with profitably through Mendelian procedures were ignored.

As we shall see in the next chapter, in Germany during this period, genetic investigations took a different form. Many biologists protested against the sharp distinction between heredity and development constructed by Mendelian geneticists. Many German geneticists challenged the determinist nature of genes, their exclusive role in heredity, and the intellectual hegemony of Mendelians. They opposed the disciplinary nature and restricted scope of Mendelism. Instead, they continued the nineteenth-century attempts to construct a unified theory of heredity that would at once embrace development and evolution. Many upheld the cytoplasm—between gene and character—as playing the predominant role in heredity.

CHAPTER 3

Challenging the Nuclear Monopoly of the Cell in Germany

> It may seem at present that the entire problem of heredity is wholly governed by the action of these chromosomes and the genes which they contain, for most geneticists have been inclined to confine their attention to the role of the genes, while the plasm has been for the most part neglected. A sound construction of genetics, however, requires a study of the cell as a whole, in which the process of inheritance is located. Geneticists forget that the genes as such can show their effects in the phenotype and only through the instrumentality of the plasm. It is fitting, therefore, that attention be drawn to a number of facts which point to a more or less independent action of the plasm as a determiner of or a contributor to inheritance. (M. J. Sirks, 1938, p. 113)

The role of the cytoplasm in the transmission of hereditary characters remained a much-debated question between the two World Wars. The view of the celebrated biologist Jacques Loeb and others (see Chapter 1) that the cytoplasm was responsible for the transmission of the more important characters of the organism while the nucleus imprinted varietal or species differences represented one extreme. That of Morgan—who, as we have seen, maintained that the cytoplasm could be ignored genetically—represented the other. By the mid-1930s there were few reports of cytoplasmic inheritance in the genetic literature. Almost all of them were reported in higher plants and most concerned the maternal inheritance of chlorophyll variegations as first reported by Correns and Baur in 1909 (see Chapter 1). The majority of geneticists who discussed these cases tended to accept the hypothesis of Baur that they were due to inheritance through the plastids or chloroplasts. However, this interpretation was not considered definitive.

Some of the issues to be considered are well illustrated by discussions at the Department of Plant Breeding of Cornell University in Ithaca, New York. The corn genetics group at Cornell under R. A. Emerson and the cytologist L. W. Sharp was rivaled in the United States only by Morgan's "fly room." During the late 1920s and early 1930s, the group included a number of students who would later become distinguished geneticists, including Barbara McClintock, George

Beadle, and Marcus Rhoades. During the 1920s there were two reports of maternal inheritance of chlorophyll in maize by E. G. Anderson (1923) and Milislav Demerec (1927). Although both authors claimed that inheritance through the plastids was likely to account for such cases, they claimed that other interpretations could not be ruled out. Demerec suggested that parasitic viruses could account for some cases, and with advice from the fly room offered by A. H. Sturtevant, he suggested that nuclear gene interpretations could be made applicable. In view of these possibilities, Demerec (1927, p. 149) stated that "it might be well to defer judgement as to which of the explanations is most probable until more experimental evidence has been accumulated."

However, the nature of the possible causal elements in such cases was not limited to cytologically visible cellular bodies; the fluid cytoplasm itself was also considered. One of the earliest genetic interpretations along this line in the United States concerned a case of male sterility stumbled upon by Emerson in a plant arising from an ear of maize he and the hybrid corn seed breeder G. D. Richey collected in Peru. It was investigated as the Ph.D. thesis of one of Emerson's students, Marcus Rhoades. Rhoades (1933) concluded that an unknown nonparticulate quality of the egg cytoplasm played the chief role in the transmission of this character and suggested that the same interpretation could be made also for some cases of chlorophyll characters, as suggested by Correns (1909).

Although Rhoades challenged the orthodox genetic view in America that chromosomal genes played the exclusive role in the transmission of characters, he did not attempt to establish methods for systematically investigating the genetic role of the cytoplasm. In 1935 he joined the United States Department of Agriculture as a research geneticist in the Division of Cereal Crops and Diseases at Ames, Iowa, where he participated in setting up the Iowa Corn Field Test. At the same time he did basic cytogenetic work on maize chromosomes. His work was later recognized to be second only to that of McClintock (Fox Keller, 1983). The mechanism underlying cytoplasmic male sterility in corn remains a mystery to this day. However, during the 1950s plant breeders began to use it in hybrid seed production as a means to overcome the labor-intensive process of hand-detasseling.

In the meantime, the theoretical significance of cytoplasmic inheritance remained a subject of controversy. Rhoades himself only returned to the problem of cytoplasmic inheritance during the early 1940s at the Department of Botany at Columbia University where he reported a case of a gene that affects the formation of plastids in maize. In summarizing the case, he wrote (1943, pp. 228–229), "Although induced by a nuclear factor, . . . the mutated plastid, like a Frankenstein monster, is no longer under the control of its master." Rhoades's analogy reflects well the attitude of many American geneticists towards the cytoplasm. But extranuclear inheritance was less likely to be perceived as monster-like the further one got from the Morgan school and American genetics in general, during the 1920s and 1930s.

In fact, many European geneticists actively sought out evidence in support of cytoplasmic inheritance. Their technique involved reciprocal crosses between different varieties, species, and genera of higher plants. Usually hybrids from reciprocal crosses were equal. That is, the characteristics exhibited by the offspring were the same whether the female of strain "A" was fertilized by the male of strain "B" or vice versa. Occasionally, differences in reciprocal hybrids were detected, especially in crosses between more distantly related types, such as species and genera. In the zygotes, the largest amount of cytoplasm is contributed by the female germ cell. In some cases the male parent contributes no cytoplasm. Differences in reciprocal crosses then could be attributed to the cytoplasm if the character in question persisted through several generations of cross-breeding.

At the John Innes Horticultural Institution, Merton, England, R. J. Chittenden (1927) reinterpreted a case of male sterility in flax, first reported by Bateson and Alice Gairdner (1921) to be due "to the cytoplasm or some cytoplasmic constituent other than the plastids" (Chittenden, 1927, p. 342). After discussing the conflicting views of Loeb and Morgan, Chittenden concluded that species and races could differ in cytoplasm as well as nuclear constituents, and, like the chromosomes, "the cytoplasm was capable of variation and may have a definite effect on the ultimate result of the reaction of nucleus and cytoplasm" (Chittenden, 1927, pp. 342–343). In a series of papers published in 1931, M. J. Sirks at the *Genetisch Instituut,* Groningen, Holland, described several cases of the influence of the cytoplasm on the expression of various characters based on reciprocal differences in crosses between two subspecies of *Vicia faba* (see Sirks, 1938).

However, nowhere was the role of the cytoplasm in heredity discussed more widely than in Germany (Saha, 1984; Harwood 1984, 1985). Indeed, between the two World Wars there was a decisive split between the genetics community in the United States and that in Germany. Many genetic investigators led by Carl Correns, Fritz von Wettstein, Otto Renner, Peter Michaelis, Freidrich Oehlkers, and others who worked on higher plants attempted to challenge Morganist views, or what they called "the nuclear monopoly." In order to better understand the focus on cytoplasmic inheritance research in Germany during this period, it is necessary first to situate the genetic work within the social and intellectual milieu specific to Germany.

The Conditions for the Possibility

> The interpretation was that the nucleus alone is important. The Nobel-prize-winner Morgan (1926[a]) said: "The cytoplasm may be ignored genetically." The nucleus was thought to produce the cytoplasm it needs. Therefore the first task was to prove the existence of a cytoplasmic inheritance unequivocally in a suitable object. (Peter Michaelis, 1965, p. 83)

The Mendelian-chromosome theory was introduced to Germany and was popularized primarily by Hans Nachtsheim, who wrote several reviews on the chromosome theory and who in 1921 published a German translation of Morgan's 1919 text *The Physical Basis of Heredity*. It was also strongly supported by Erwin Baur, who carried out extensive classical genetic experiments on plants during the 1920s and 1930s and was one of the most consistent Mendelians and Darwinians among the continental geneticists. Baur's textbook *Einfuhrung in die ex-*

perimentelle Vererbungslehre was considered one of the best of the early genetics textbooks and was repeatedly revised and republished up until 1930.

Baur was particularly instrumental in the institutionalization of genetic studies in Germany. As early as 1909 he founded the journal *Zeitschrift für induktive Abstammungs- und Vererbungslehre* (ZIAV), which provided a swift and satisfactory means of publication for biologists who carried out cross-breeding analysis. Later he founded *Bibliotheca Genetica* and, in the domain of applied genetics, *Der Züchter*. In 1911 Baur was called from his post as professor of botany at the University of Berlin to the *Landwirtschaftliche Hochschule* in Berlin. In 1913 a new chair was established for Baur as professor of genetics in that institute and a new institute for genetics was decided upon and provisionally established at Potsdam. By 1921 the erection of a permanent *Institut für Vererbungsforschung* had begun at Berlin-Dahlem. In 1928, largely through Baur's influence, the *Kaiser Wilhelm Gesellschaft zur Förderungder Wissenschaften* erected its own institute for the study of breeding problems, especially in relation to their practical applications. The new *Institut für Zuchtungsforschung* at Munchenberg was planned and developed by Baur, and by the 1930s it was one of the largest of its kind in the world, having over 85 acres of land, accommodating about 30 investigators, and employing a large number of technical assistants (Scheibe, 1961).

However, with the exception of Baur and his associates, Curt Stern, and a few others who carried out Mendelian investigations, classical genetic studies of chromosomal gene transmission were not highly represented in Germany, particularly in the universities. This is not to suggest that experimental analysis of hereditary mechanisms was not highly represented in Germany. It is to say rather that heredity itself continued to include development, which had typified studies of heredity during the 1880s and 1890s. Many German biologists who investigated the role of the cytoplasm in heredity claimed that Morgan's theory of the gene was inadequate not simply because geneticists did not explain how genes directed the formation of the characters said to be under their control. They believed that genes and chromosomes could not fully account for development and evolution.

As discussed in the previous chapters, many nongeneticists predicted the collapse of the exclusive chromosome theory. Geneticists had failed to show that gene mutation could create new species. Moreover, Morgan's own theory of the integrity of the nuclear genes and his doctrine of nuclear equivalence in the face of cellular differentiation seemed to many to preclude genes from playing a primary role in ontogenetic develpment. Nuclear genes were responsible for segregation patterns between individuals, whereas the differentiating cytoplasm was held to be responsible for segregation patterns within individuals. Both embryologists and neo-Lamarckian naturalists embraced the pliable cytoplasm as the principal seat of heredity.

As Mayr (1980) and Rensch (1980) have stressed, the idea that the cytoplasm controlled the fundamental characteristics of higher taxonomic groups while nuclear genes controlled relatively trivial characteristics was maintained by many German neo-Lamarckians during the 1920s, including the zoologist Ludwig Plate, the paleontologist Franz Weidenreich, and the celebrated systematist Richard von

Wettstein. It was equally maintained by many German embryologists (Hamburger, 1980).

The prevalence of investigations of the nucleus in the United States, and of the cell as a whole and the role of the cytoplasm in Germany, in part reflected different institutional and ideological constraints of academic practice. The major structural differences in university politics of the two nations are well known: the German institute versus the American department, the power of the German professoriate, and the state control and/or regulation of university appointments and research budgets in Germany versus laissez-faire politics in American academic science. The effects of German academic politics and institutional structure on the development of various disciplines have been discussed by several writers, most prominent among them Ben-David (1968–69), McClelland (1980), Ringer (1969), Ash (1980), Weindling (1981), and Saha (1984). The nature of the constraints can be best discerned when compared to the organization of genetics in the United States.

As discussed in the last chapter, the growth and maintenance of a nucleocentric view of heredity relied on two primary strategies. It required the formation of a separate discipline with its own objectives, theories, societies, and refereed journals. Through this process American geneticists tended to dissociate themselves from the larger realms of biology, creating and solving their own problems, constructing their own definition of heredity, and analyzing chromosomes, while remaining largely immune from the attacks of their critics. It also required social, intellectual, and economic support from agriculturalists in the public and private sectors.

Both of these sources of power for Mendelism were lacking in German universities. First, as Ben-David and others have stressed, the institutionalization of a new discipline was much more difficult in Germany than in America. The gross differences between the structure of the German university institute and the American department account for part of the reason for the different degrees of difficulty. American university departments were generally characterized by a group of professors with more or less equal institutional security and possessing a relative degree of individual academic autonomy to direct their own research and to compete for students and funding. In German institutes these privileges were embodied in only one full professor who possessed the bureaucratic power to allocate facilities and funds to junior staff, thus dictating the direction and nature of research in his or her own institute. Professors also taught the large introductory courses, for which they were rewarded financially by obtaining part of their income from registration fees.

Junior staff often received their income solely from student fees and tended to be left in charge of smaller, more specialized classes. Their only recourse to a better income and the security to carry out innovations of any kind was to acquire their own chairs and institutes and train their own following of students. However, as Ash (1980, p. 259) has stressed, this process took a long time—an average of sixteen years by 1909—and only half of those who completed their doctorates between 1860 and 1909 ever advanced so far. The case of Correns is exemplary. He worked for 16 years as *Privatdozent,* paid only from teaching fees, possessing no official status, assistants, or students before entering the ranks of the German "mandarins" as professor of botany at the University of Münster in 1909 (Saha, 1984, pp. 187–188).

The conservative effects of this system were strengthened by the Ministries of Culture, which legislated over the establishment of new chairs in the universities in their states. Educational and financial officials of the individual states generally accepted the recommendations of university faculties when they made appointments to existing chairs. However, when the establishment of a new chair was at stake, financial and cultural considerations often took precedence. The reluctance to establish new chairs was also shared by members of university faculties, who tended to prefer accommodating new domains within existing institutes by offering their representatives temporary teaching contracts or nonbudgetary associate professorships (Ben-David, 1971, pp. 131–132). This system of control was in sharp contrast to that which operated in the United States, where geneticists enjoyed a variety of patrons who found social and economic value in Mendelism.

The system of control in Germany, which tended to discourage specialization, fitted well with the role of the universities during the Wilhelmian period, which was to train Germany's educated elite, who reciprocated as articulate and cultured spokespersons for German political programs (McClelland, 1980, pp. 325–326). Ash (1980, p. 257) has argued that the classification of disciplines, the order of knowledge, and the degree of state support awarded them in Wilhelmian Germany relied on their promise to enhance the quality of education for Germany's cultural elite. There was an enormous expansion of the German universities and funding for science between 1870 and 1914 (Pfetsch, 1974, p. 52). Many new disciplines broadened their institutional basis and various new disciplines were established. Genetics, however, was not among them. In fact, there was only one chair of genetics established before 1945.

Investigations of heredity were carried out primarily in botany and zoology institutes, where they remained constrained by the general aims and objects of the older disciplines. A professor of botany typically taught plant physiology, development, anatomy, systematics, and phylogeny, and a definition of heredity would have to be equally embracing. The institutional and ideological constraints of German scientific practice were recognized and generally endorsed by many German professors. As Saha (1984) has argued, they were also supported by Correns and other German biologists who investigated the role of the cytoplasm in heredity, who were unable to accept disciplinary boundaries and who protested against what they saw as the limited theoretical scope of Mendelism.

American geneticists, characterized by the Morgan school, took what might now be seen as an operational approach to their work, defining heredity and the gene in terms of the experimental operations by which they might be demonstrated. German investigators took a conceptual point of view and attempted to define the genetic system in the broadest way, trying at once to conceive of its function in the organism, its place in evolution, its phylogenetic significance, and as many other aspects as possible. The tendency of German investigators toward unifying theories has also been traced to a Kantian influence which Saha (1984) and others claim was deeply embedded in the German *Wissenschaft* tradition.

These collective social and intellectual forces helped to keep the possibilities for cytoplasmic inheritance alive and its investigations in public view in Germany. One can also understand the concentrated work on the cytoplasm and the attempts to challenge Morganist views by attempting to provide definitive "proof" of cytoplasmic inheritance as a strategy to compete for recognition and prestige in the international field. Indeed, the expression "nuclear monopoly" (*Kernmonopol*) itself, commonly used by German investigators, embodies the idea of prestige and recognition acquired by Morgan and his followers. However, although many leading German biologists recognized the difficulties which the Mendelian-chromosome theory posed for macroevolution, the belief in a primary role for cytoplasmic inheritance did not reflect a unified movement. In Germany, orthodoxy was maintained by powerful professors, not be refereed journals, and there were almost as many different published views on the relative importance and nature of the cytoplasm and nucleus and their roles in heredity and evolution as there were individual professors.

Investigators at the *Kaiser-Wilhelm-Institut für Biologie* at Berlin-Dahlem provided some of the seminal ideas on nucleo-cytoplasmic relations which were discussed throughout the 1920s, 1930s, and 1940s in Germany. Founded in 1913, the institute provided conditions for systematic research on heredity free from the overwhelming teaching responsibilities in the universities. In 1914 Correns was called from his chair of botany at Münster to be its first director. The institute was structured not around disciplines or specialities *per se,* but around four men, each with his own division (*Abteilung*): Correns (botany and genetics). Richard Goldschmidt (zoology and physiological genetics), Hans Spemann (embryology), and Max Hartmann (protozoology and "general biology").

The Gradual Disappearance of Victor Jollos

> Following Fisher, . . . the majority of geneticists of our day . . . are inclined to minimize the possible effect of "directed mutations" and to attribute the course of evolution, and especially the so-called "orthogenetic" evolution, only or chiefly to the influence of natural selection. From a purely mathematical point of view this may be justified, but in my opinion the biologist cannot neglect clear experimental facts. (Victor Jollos, 1934c, p. 487)

Hartmann's division was particularly productive during the 1920s and 1930s and emerged as one of the leading centers for genetic work on unicellular organisms. Many protozoologists in Hartmann's division worked out the taxonomy and life cycles of various forms of microorganisms, investigating cell structure and the difficult problem of sex determination. Some of the first systematic genetic studies of microorganisms also emerged in Hartmann's division. During the late 1920s, Franz Moewus began to domesticate various forms of microorganisms for genetic use. He mapped genes in the unicellular alga *Chlamydomonas* and studied their physiological effects. By the mid-1950s Moewus would come to represent one of the most ambitious cases of fraud in the history of science (see Chapter 7).

Some of the most important and controversial early genetic work indicating a role for the cytoplasm in heredity also came to the center of investigations in Hartmann's division. These were based on studies of the inheritance of environmental effects led by Victor Jollos. Beginning in 1913, Jollos and later several other protozoologists who extended his work at Berlin-Dahlem, demonstrated that environmental factors such as higher temperatures or chemicals could induce hereditary changes in *Paramecium* which were transmitted for hundreds of generations in vegetative reproduction after the removal of the inducing agent (see Hämmerling, 1929). Some of the induced hereditary changes, such as resistance to heat and arsenic, represented specific adaptive responses. In other cases, the induced hereditary modifications were less specific.

When attempting to interpret the induced environmental changes, and particularly when trying to establish their seat in the cell, Jollos and his followers attached much significance to peculiarities which some or all of these changes showed. First, after hundreds of cell generations passed under conditions that were free from the agent that produced them, the acquired modifications gradually faded away. Second, the inherited environmental modifications very commonly were found to disappear when the microorganisms were allowed to reproduce by conjugation. Based on these considerations, Jollos, who led the theorizing on these matters, assigned the modifications to the cytoplasm instead of the nucleus. He reasoned that a change in the gene or mutation was a permanent change; it would not disappear after many generations in an altered environment. But a change in the cytoplasm would, in the course of time, be overcome and dominated by the unchanged nucleus, bringing about a return to the original characteristics. Jollos called the lasting environmental modifications *Dauermodifikationen*.

Jollos's work had a profound impact on perceptions of the role of the cytoplasm in heredity and its investigations throughout the century. As Rensch (1980) and Harwood (1985) have pointed out, during the 1920s it was widely embraced by neo-Lamarckian naturalists in Germany. As will be discussed in the next chapter, some leading geneticists in the United States also viewed it as providing evidence for the inheritance of acquired characteristics. A few hoped that with repeated analogous treatment the adaptive changes would sooner or later appear as stable cytoplasmic transformations or nuclear mutations that would be inherited sexually as well as asexually.

Although theoretically *Dauermodifikationen* could result in lasting changes bringing about new specificity, Jollos himself downplayed their evolutionary significance *per se*. This is not to say, however, as Harwood (1985, p. 293) has claimed, that "unlike many of his admirers Jollos was a selectionist." In fact, Jollos opposed the all-exclusive role of selection on random mutations in directing the course of evolution. Jollos was looking for agents in the natural environment that would have a directing influence on the mutations produced by them, such that the mutations would occur in a step-by-step orthogenetic series, as had been claimed by some paleontologists from a study of fossils. Paleontologists had made many claims of such orthogenetic evolution, such as the reduction of the limbs in snakelike lizards. The possibility that changes in temperature were the natural environmental factors directing such change was an old idea. Temperature changes are ubiquitous, and paleontologists repeatedly claimed that periods of rapid evolutionary change coincided with, or followed, periods of marked climatic fluctuations.

Until the late 1920s the only reliable technique for artificially inducing mutations was to expose germ cells to X-rays or radium, as shown by H. J. Muller and others. However, in 1929 Richard Goldschmidt reported a technique for producing large numbers of mutations in the offspring of *Drosophila* by heat treatment of larvae. During the following years Jollos applied Goldschmidt's technique to examine "the problem of whether environmental factors which produced mutations might also have an immediate *directing* influence on the mutations produced by them" (see Jollos, 1934c). Between 1930 and 1933 he confirmed and extended this work in an elaborate series of experiments in which he claimed to have provided experimental evidence for environmentally directed mutations in an orthogenetic series.

Jollos (1934c) concluded that repeated exposure to sublethal temperature induced simultaneously the same phenotypic changes in the following order of frequency: (1) certain particular somatic or cytoplasmic modifications in the heat-treated generation, (2) *Dauermodifikationen,* (3) mutations of the same type, and (4) finally, more and more extreme alleles of the same genes, such as slightly spread to completely spread wings. When explaining the parallelism of the somatic variations in the heat-treated generation and the mutations, Jollos assumed that the genetic element altered by heat treatment in the case of somatic variations in the heat-treated generation was a "gene product," a specific cytoplasmic substance produced by corresponding genes in the nucleus. He further reasoned:

> Since alteration of this specific "gene-product" and that of the corresponding gene itself, caused by the same environmental factors, has the same effect on the development of the individual fly, we may conclude that the corresponding genes and "gene-products" have the same or very similar structure. The much greater frequency of alterations of the "gene-product" (leading to "modifications") than of the gene itself (leading to corresponding mutations) may be attributed to the better "insulation" of the genes due to protection by the chromosome cover (Jollos, 1934c, p. 491)

Jollos (1934c, p. 492) was reluctant to interpret *Dauermodifikationen* in terms of gene products, claiming that parallels between *Dauermodifikationen* and nuclear mutations were yet comparatively few, and called for further experiments.

Jollos did not have the opportunity to continue his experiments in a systematic way. He was forced to leave Germany with the rise of Hitler's regime and was invited to the United States as one of many refugee scholars of Jewish extraction, and he arrived at the University of Wisconsin in 1934. He soon found himself in a milieu which was hostile to both him and his work. By 1940 he was in a desperate situation, having no laboratory facilities and no means for further existence. He died the next year, leaving his family in poverty.

The issues underlying Jollos's controversial and tragic plight in the United States are complex and some of them highlight the cultural and intellectual differences between the American and pre-Nazi German biological communities. First, Jollos's work did not enjoy the recognition and acclaim it obtained in Germany.

Almost immediately he found himself engaged in a controversy with American geneticists who tested the heat-treatment technique on *Drosophila* but who defended the exclusive role of natural selection in directing the course of evolution.

Harold Plough and Philip Ives (1935) at Amherst College conducted the most extensive investigations of the effects of the heat treatment on *Drosophila* in the United States. They confirmed the induction of modifications in heat-treated generations and reported changes inherited through the nucleus and cytoplasm. However, they observed no cases of *Dauermodifikationen* and no significant correspondence between cytoplasmic and nuclear mutations. They discredited Jollos's claims that high temperature would bring about orthogenetic evolution and they endorsed the neo-Darwinian view supported by the statistically based population genetic theories of R. A. Fisher, Sewall Wright, and J. B. S. Haldane. They quoted Fisher (1930, p. 48):

> It is scarcely possible . . . to ascribe to mutations any importance in determining the direction of evolutionary change; their importance in evolution lies in playing the very different role of maintaining the stock of genetic variance at a certain level, which level in its turn is a factor in determining the speed, though not the direction of evolutionary progress.

Jollos, however, continued to insist on the validity of his results and interpretations, and like many of the first generation of geneticists who also opposed the all-exclusive role of natural selection in directing evolution, he denied the power of statistical or mathematical reasoning over experimentation.

Protozoan geneticists in America such as H. S. Jennings and his former student T. M. Sonneborn (see Chapter 4) were very appreciative of Jollos's work and attempted to help him obtain a position. As Sonneborn (February 12, 1940) wrote to Jollos:

> You must believe me when I tell you that I am prompted to take these efforts for you because I am a great admirer of your work. For years I confess I was skeptical about much of your work, but in recent years my own work has led me more and more to appreciate the greatness of yours.

However, Jennings had just retired and young Sonneborn was not yet an authoritative figure in American genetics. They were not successful in their attempts to secure a position for Jollos.

Some biologists at Wisconsin also recognized and applauded Jollos's broad grasp of biology. Jollos was particularly well educated. He had all the medical training except for the intern work that went with the M.D. degree. He had worked on a variety of basic biological problems, including cellular structure, physiology, protozoology, and parasitology, in addition to genetics. He had spent a couple of years at the Egyptian University of Cairo organizing and heading a large department of zoology and had teaching experience in various fields of zoology besides protozoology and genetics.

Despite Jollos's scientific accomplishments and distinctions, all attempts to find employment for him in an American university failed. This failure was owing in no small way to some trouble that developed at Wisconsin, where Jollos had managed to antagonize almost everyone with whom he had come in contact. At

Jollos's request, Lowell Noland, who knew the situation well at Wisconsin, summarized the major issues (Noland to Jollos, February 23, 1940). In teaching and seminar work several students in his classes complained that he was "too brutally and unsympathetically critical of other points of view" and criticized the "manner by which he rode rough-shod over any opposing points of view." Jollos lost sympathy from faculty who complained that he expected the same facilities and supplies that regular professors in the university demanded. From the beginning they claimed that his status at Wisconsin was a kind of relief measure to take care of him temporarily, and they found it presumptuous that he considered that the university "owed" it to him to continue its support for his research. Outside the university, Jewish groups in the city complained that although Jollos accepted financial assistance from them, he was careful to avoid having any connection with them. But Noland fully recognized that the fault was not on Jollos's side. There was also a "hypersensitivity" and anti-Semitic feeling at the University of Wisconsin and elsewhere which played a significant part in Jollos's plight.

Jollos himself responded to the charges raised against him. He fully recognized that he had made some mistakes but said that it was difficult to avoid them in academic and social life when he was suddenly transferred into a new environment with different traditions and customs. In his opinion his criticizing of "wrong scientific ideas" and the correcting of evident mistakes in seminars and other discussions were probably the most serious ones. He had no intention of hurting the feelings of students but claimed that he "simply followed in this respect the common tradition in German Universities." As he wrote Noland:

> It worked in Germany very well. It saved time, and the students learn much in this way. They don't mind to be corrected in the presence of their fellow students, and they use their right to criticize the opinions of others, including those of the professor. I realized only after a long time that it is not done here, and that in general you have to be, here, much more cautious in disagreeing with others. (Jollos to Noland, February 25, 1940)

It was also difficult for Jollos to step down from a position of considerable social and financial importance in Germany to a lesser one in the United States. He pointed out that his "Americanization took and takes much longer time than in many other cases." However, Jollos was under the impression that the university was interested in his work and expected him to continue his research. He claimed that he was in no actual need at the time of the invitation and that he had a good chance for a suitable position in Edinburgh. He did not apply for help from America and the invitation came unexpectedly. This explained his "presumptuous" demands for laboratory equipment and supplies. Even then, his requests for facilities were modest. Moreover, his research on the influence of cosmic radiation on heredity was financed by a grant from the National Research Council under the stipulation that the university would provide laboratory facilities. When all this was considered, Jollos claimed the bad feelings among the staff and "loss of sympathies" could not be justified.

Jollos also attempted to rectify the criticism by Jewish groups, which was in his opinion based on a complete misinterpretation of his attitudes and behavior.

First, he claimed that he had never in his life tried to avoid connections with Jewish people or to conceal his Jewish descent. It would also be particularly ridiculous to do it in Wisconsin, where everyone knew that he was ousted by the Nazis just for his Jewish extraction. However, Jollos pointed out that he was not a "Jew" in the sense this term was usually used. He never belonged to a Jewish community but was raised in the Protestant faith, and consequently did not know Jewish traditions. He claimed that he tried to emphasize these facts to avoid the impression that he was accepting Jewish help on false pretenses. He also did not seek out help from Jewish groups. This was done for him by others in the university who thought he would want to join their community. Jollos wrote to Noland:

> The conditions in pre-Nazi Germany were very different in this respect. There was no sharp social separation of Jews, "Non-Aryans" and Gentiles in academic and other educated circles. They were not interested in racial or church connections of others, and considered it bad taste to inquire about them. (Jollos to Noland, February 25, 1940)

Despite Jollos's efforts to correct the "misunderstandings" that developed at Wisconsin, no American university would hire him. Noland listed the remaining real issues which made obtaining a job more than difficult for Jollos: younger men would be preferred to older men since they were more "adaptable" and did not have "their lines of interests and research already planned out." Jollos was fifty-three, and he challenged genetic and evolutionary orthodoxy in America. One also had to consider Jollos's health. He had diabetes and "difficulties" with his eyes and heart. There was the "problem" of his "foreign extraction" and last, but not least important, his race: "There is some prejudice against the Jewish race" (Noland to Jollos, February 1, 1940).

As we shall see in the next chapters, many of the geneticists who challenged orthodoxy after World War II were of Jewish extraction, but most, like Jollos, did not identify themselves with Jewish communities or religious groups. Some, like T. M. Sonneborn, also had difficulties in obtaining a position in America during World War II because of alleged "strong Jewish traits." Jollos's open criticisms of the all-exclusive role of natural selection were silenced, but his work on *Dauermodifikationen* was continued during the 1940s and 1950s by Sonneborn and his co-workers, who tried to be more cautious than Jollos when interpreting their results and challenging genetic orthodoxy in America (see Chapter 4). In the meantime, *Dauermodifikationen* and the role of the cytoplasm in development and evolution remained at the center of theoretical discussions in Germany.

"Subsumed Under a Single Formula": The Case of Richard Goldschmidt

> If we remember . . . reaction velocities, the wonderful consistency of all the facts and their connection through a rather simple idea becomes . . . apparent. (Richard Goldschmidt, 1932, p. 355)

> Ontogeny is a moving equilibrium, which involves all fundamental physiological process at each stage, and it can no more be envisaged under a single formula than can the conception of life itself. Genetics on the other hand is subsumed under a single formula. (F. R. Lillie, 1927, p. 365)

Richard Goldschmidt was a prominent figure in German zoology before he was forced to leave his country with the rise of the Third Reich. Like many German biologists between the two World Wars, Goldschmidt came to oppose the all-exclusive role of natural selection in the origin of species and higher taxonomic groups. By 1933 he began emphasizing "the importance of rare but extremely consequential mutations affecting rates of decisive embryonic processes which might give rise to what one might term hopeful monsters" (Goldschmidt, 1933, p. 546). Goldschmidt maintained his belief in macromutations at the University of California, Berkeley, and persistently claimed that macromutations overcame "the great difficulties which the actual facts raise for the neo-Darwinian conception as applied to macro-evolution" (Goldschmidt, 1940, p. 249). Unlike Jollos, Goldschmidt managed to survive in America as one of the few outspoken geneticists who opposed classical neo-Darwinism views.

However, unlike neo-Lamarckians who claimed the cytoplasm as the basis for macroevolution, Goldschmidt attempted to accommodate his evolutionary views within the confines of nucleocentric genetics. Following his studies of sex determination in 1911, Goldschmidt formulated one of the first general theories of heredity attempting to reconcile genetics and embryology. His primary goal, which he called "physiological genetics," was to replace the "static" and "symbolistic" Mendelian language with a dynamic viewpoint of the physiology of development and with "more definitive physico-chemical conceptions." He proposed his theory of genic action and genic expression at the Marine Biological Laboratory at Woods Hole in 1932.

Goldschmidt's primary theoretical tactic was to bring the dimension of time into genic action in an attempt to account for embryological regulation and differentiation. He claimed that this could be done by two assumptions. First, he suggested that genes which had enzyme properties operated by controlling the speed of developmental processes. Second, he proposed a "lock-and-key" theory of the cytoplasm as substratum which possessed spatial properties. As Goldschmidt (1932, p. 355) put it: "There is in addition, the differentiation of the substratum in three dimensions of space without which the reaction system which produces the right thing at the right time could not be imagined to produce it also in the right place." The cytoplasm allowed the products of genic reactions to act or not to act or to act differently in different regions, allowing gene-controlled reactions to have different consequences in the different areas of the developing organism.

Although Goldschmidt acknowledged that the process by which cytoplasmic spatial differentiation occurred was unknown, he claimed that one simply had to be content to know that cytoplasmic activities and differentiations were ultimately under the dictates of the nucleus. As he put it:

> The causation of this change belongs to the domain of physiological genetics, and is adequately understood by the system of timed velocities. But in what this change consists and what are its consequences in regard to determination—this is the proper domain of experimental embryology. (Goldschmidt, 1932, p. 355)

Genetics (genes) was concerned with causes, embryology (cytoplasm) was concerned with effects. In fact, Goldschmidt opposed the idea that the cytoplasm determined any hereditary traits by its own action. In his view, (Goldschmidt, 1938, p. 279) maternal influences were due to "a differential action of different plasmata as substratum upon the action of the genes in controlling the differentiation of definite characters." As a substratum, the cytoplasm possessed specificity, but the specificity was not of a type similar to genes but simply a physical or chemical property—a matter of permeability, or a chemical causing a change in pH. In the physiological genetics of Goldschmidt (1934, p. 15) the most probable action of the cytoplasm was in the inhibition or augmentation of velocities of gene-controlled developmental reactions.

Even when confronted with his own apparent evidence for cytoplasmic inheritance, Goldschmidt remained faithful to his view of the cytoplasm as a substratum for genic action. By the mid-1930s, he had brought forward extensive genetic evidence from his work which indicated that the factors controlling female sexuality in moths were based on cytoplasmic properties uninfluenced by genes. "The proof that this primary property is inherited within the cytoplasm," Goldschmidt (1938, p. 279) argued,

> forces one either to assume cytoplasmic genes, which is improbable, or to attribute to the specific condition of the cytoplasm a specific influence upon the action of the sex-determining genes.

Goldschmidt (1938, p. 279) followed a suggestion of T. Dobzhansky and attempted to bring cytoplasmic sex determination into line with nuclear control. He assumed that the cytoplasmic female factor was something that controlled the level of "the threshold for the action of male determining genes" and thus shifted the "balance toward femaleness."

Goldschmidt maintained his claim that the cytoplasm functioned only as a substratum for genic actions well into the 1950s. However, while he, like many American geneticists, claimed that embryology was concerned with the phenotype and genetics with the causes of development, many embryologists, including E. E. Just (see Chapter 5), Viktor Hamburger, and Hans Spemann, continued to take the opposite view. They claimed, in effect, that genetics was concerned largely with phenotypic analysis and that embryology was concerned with the causes of development. F. R. Lillie, who opposed the possibility of a single theory of development, summarized some of the major criticisms of nucleocentric views when he attacked Goldschmidt's "unifying" theory at Woods Hole in 1927.

First, Lillie stressed that the belief that each cell received the entire complex of genes was an almost universally accepted genetic doctrine. As long as this was a necessary part of the chromosome theory, the phenomenon of embryonic differentiation lay beyond the scope of Mendelian genes. If each gene reacted with a specific substratum, as Goldschmidt proposed, the substratum would necessarily have to be always present at the appropriate time and place during development. It seemed self-contradictory to attempt to explain the genetic restriction of somatic cells and cellular differentiation by the behavior of genes which were believed to be the same in every cell. As Lillie (1927, p. 366) argued:

> This seems to me to postulate the process which it is invoked to explain, *viz.*, ontogenetic segregation and the time relationship of specific events.

Second, Lillie criticized the type of explanation Goldschmidt offered. By postulating specific genes for all differentiating characters of each stage of development, and latency for all genes except those postulated for the specific event, Goldschmidt's views were similar to the deterministic theory of Weismann. In heredity, in general, Lillie stressed, what was inherited was not the character but only a specific form of reaction to environment. Goldschmidt's theory was physiological only in name. Lillie (1927, p. 366) charged, "In its essence the theory is deterministic and not consonant either with sound physiology or sound genetics." Finally, whether one adopted a nucleo-cytoplasmic-relations theory or any other theory, Lillie claimed, there was still nothing in the known principles of genetics nor of physiology that gave one a clue as to the nature of the definite ordered sequence of developmental events.

Hans Spemann: The Magnetic Order of Cells

> Underlying the scientific discussions of these problems there was . . . an inclination of the Germans for a metaphysical underpinning and the often subconscious need to combine their scientific thinking with a metaphysical *Weltanschauung*. Their metaphysics came, for the ones I knew, from *Naturphilosophie* of the early nineteenth century—Goethe—and from Kant. (Viktor Hamburger, 1980b, p. 303)

During the 1930s embryologists continued to work largely in isolation of genetic research and theories. Many insisted that their aims, concepts, and techniques were fundamentally incompatible with those of genetics. Embryologists concerned themselves primarily with the higher levels of interaction at the supracellular level and with epigenetic mechanisms such as induction, gradient fields, and morphogenetic movements. The school of Hans Spemann played a leading role in fostering the development of embryology away from genetics.

Spemann was a former student of Boveri (see Chapter 1) and remained a director at the *Kaiser-Wilhelm-Institut für Biologie* during World War I, before he succeeded Weismann as professor of zoology at Freiburg in Breisgau where he remained until his retirement in 1938. He was awarded a Nobel Prize for his embryological research in 1935. Spemann's most celebrated work was on embryonic induction through tissue interaction in the embryo. His work on the "organizer," which could somehow create organizational fields in the developing embryo, had a great impact on embryologists (see Haraway, 1976, pp. 115–116).

As Viktor Hamburger (1980a) has pointed out, although embryologists of the 1930s and 1940s seldom published their evolutionary views, many continued to have misgivings about pure selection theory to account for all adaptations. Boveri (see Chapter 1) and the Belgian embryologist Albert Dalcq (1949, 1952), like Goldschmidt and others, upheld the idea that macromutations occurring in the early stages of development would lead immediately to new organization and were the basis of the origin of higher taxa. Boveri summarized a common embryological sentiment when he wrote:

> It is not the evolutionary changes of organisms per se that stimulate our curiosities so powerfully, but that the changes are teleologically measured by human standards, or more concretely: not the small modifications are important for us whereby a new species can be distinguished, but those big steps call for an explanation according to which the water animals become land animals, land animals again water animals, crawling animals become flyers, nonvisual ones become visual, and instinctive drive becomes reasoned action. (Boveri, 1906, p. 16, translated in Hamburger, 1980b, p. 305)

German embryologists generally did not confine their evolutionary thinking to unitary explanations that would operate in all cases of adaptation. When, for example, Boveri asked what caused the large changes in organization he found himself in the domain of metaphysics arguing for psychic explanations. The holistic and organismic outlook of Spemann led him to combine Lamarckian principles with psychic forces when he felt they were necessary. Spemann drew on ideas from comparative anatomy, from the influential nineteenth-century school of Carl Gegenbaur, and from the idealist morphology of Goethe who, according Hamburger (1980b, p. 306), was still widely read in Germany. Both Spemann and Boveri were also influenced by the earlier writings and vitalist views of August Pauly (1905). Pauly, however, could not accept pluralistic principles operating in evolution. He was the strongest defender of an all-exclusive psychic principle—a conscious psychic striving towards a certain goal on the part of the organism and all its elements.

At the same time, embryologists such as Spemann took a keen interest in morphogenetic fields as originally proposed by Paul Weiss, claiming that they represented an additional vital principle operating in organisms. As discussed in Chapter 1, Child and his followers had proposed that gradient fields were the result of metabolic activity alone. Spemann, on the other hand, was one of several embryologists who sought a structural foundation for fields. As he wrote in his well-known text *Embryonic Develoment and Induction,* "I cannot escape the impression that P. Weiss imagined the field as being, not only a concept, but also a reality rather independent of its substrate" (Spemann, 1938, p. 301). He further inquired: "What . . . is the source of this field? This question includes the problematic character of the conception of the 'gradient fields.' For if the analogy with physical fields—for instance, a magnetic field—is to be maintained, the field is bound to a source" (Spemann, 1938, p. 322). In Spemann's opinion the typical character of morphogenetic movements in the egg was "inconceivable without the assumption of a directive structure which . . . is stamped upon the germ" (Spemann, 1938, p. 326). He referred to the work of Conklin (see Chapter 1) on a mosaic of different cytoplasmic substances in ascidian eggs, and he stressed that its disturbance by centrifugation also disturbed the development of the different regions of the germ. Thus, he remarked, "Even before the beginning of cleavage, a distribution or a segregation of the different egg substances has taken place which afterward is fixed and made definite" (Spemann, 1938, p. 209).

Spemann well realized that cellular differentiation in the face of an unchanged nuclear material represented a paradox for geneticists and embryologists. Weismann's idea that a sorting out of nuclear substances occurred had been clearly refuted. However, this did not mean that alternative forms of nuclear differentiation were ruled out. As Spemann (1938, p. 210) argued, "By refuting this fundamental hypothesis one cannot conclude that each cell of the body will contain the whole undiminished idioplasm; for genes may be lost or become ineffective in other ways besides that of elimination out of the cell." Spemann's famous "constriction" experiment, which suggested to many that nuclear differentation did not occur early in development, has already been noted in Chapter 1. Nonetheless, it remained possible that nuclear differentiation did occur later in development. The techniques required to test this hypothesis were not developed until the late 1950s (see Chapter 7). In the meantime a primarily cytoplasmic basis for cellular differentiation remained highly plausible.

Both Hamburger and Spemann maintained that cellular differentiation was predominantly cytoplasmic in operation well into the 1930s and 1940s. In a letter to Spemann, Lillie (December 28, 1931) wrote:

> As you know, I have quite a similar outlook on development as your own in spite of terminologies. "Embryonic segregation." I think of as a "mechanism of control" set in the midst of the autogenetic process, which differs from the Mendelian mechanism, and is presumably cytoplasmic in operation.

Lillie (May 2, 1944) wrote to Hamburger, who requested offprints of his polemical article of 1927:

> "The progress of genetics and of physiology of development can only result in a sharper definition of the two fields, and any expectation of their reunion (in a Weismannian sense) is in my opinion doomed to disappointment. . . ." I am happy that you appear to be in general agreement with this statement.

The role of the cytoplasm in heredity continued to be investigated by embryologists who found it impossible to reconcile the atomistic configuration of nuclear genes with the regulative and orderly aspects of ontogenetic development. Whether or not the cytoplasm possessed its own hereditary materials and properties remained an open question, and the belief that the cytoplasm controlled the fundamental characteristics of the organism persisted. However, as long as heredity investigations were concerned largely with combinations of visible characteristics, rearrangement of those fundamental characteristics common to all members of a species could only be expected from cross-breeding individuals with widely different characteristics.

Merogony experiments, in which an enucleated egg of one species is fertilized by the sperm of another, initiated with the classical experiments of Boveri, continued to be carried out by many embryologists, including Fritz Baltzer, a student of Boveri, Paula Hertwig, and Gunther Hertwig (see review by E. M. East, 1934). However, the embryological approach to the nucleus-cytoplasm problem posed difficulties in terms of the required proof of cytoplasmic inheritance. The exchange of nuclei, with assurance that the chromatin was that of only one parent, was a difficult task. It was not always certain that the nuclear material ws entirely removed from the egg. In other cases, the development of undoubted merogonous embryos was impeded, and comparison with the "normal" of the maternal and paternal species could not be made with any exactitude. There were, however, a few important exceptions where, with novel techniques, merogony experiments furnished some clearer evidence indicating that certain characters were attributable to the cytoplasm.

Richard Harder (1927) at the *Botanisches Institut der Technischen Hochschule,* Stuttgart, reported cytoplasmic effects in a cross between two genera of higher fungi. Another widely discussed case of the influence of the cytoplasm in heredity was reported ten years later by Ernst Hadorn (1937), student of Fritz Baltzer and grandstudent of Boveri. Hadorn fertilized two species of *Triton,* exhibiting clear differences in the character of the epidermis, and succeeded in removing the female nucleus before nuclear fusion. The haploid embryo developed only to the blastula stage. But when a portion of the presumptive epidermis was grafted to a normal larva of another species, it maintained its identity and developed to the adult state. The epidermis resembled the parent which contributed the cytoplasm, indicating the decisive effect of the cytoplasm.

In the United States, during the 1930s and the 1940s, Ethel Brown Harvey (1932, 1935, 1942) presented another series of merogony studies which caused a great deal of excitement among American embryologists. By centrifuging echinoderm eggs in a medium of the same density as the eggs, Harvey demonstrated that it was possible to obtain large quantities of nonnucleated egg fragments of constant size without contamination with nuclear fragments. After fertilizing these with the sperm of another species having distinct morphological differences, Harvey concluded that the early stages of development were cytoplasmic rather than nuclear. The rate of cleavage, the size of the blastula, and later of the gastrula and early pluteus appeared to be due to the cytoplasm. For Harvey (1942, pp. 224–225) the conclusion was clear enough: early inheritance was cytoplasmic rather than nuclear:

> The chemical materials used in forming the early developing morphological structures such as the skeleton of the pluteus are furnished in large quantity by the cytoplasm of the egg, whereas the nuclear material is small in amount, and might be considered more in the nature of the enzyme.

The evidence provided by merogony experiments was controversial, however. In others cases of merogony the cytoplasm seemed to have no effect at all upon the character of the offspring. There was also the argument of "predetermination," as illustrated by the maternal inheritance of the direction of shell coiling in snails.

Assembling the *Plasmon*

> The hypothesis has been advanced that the nuclear genes transmit only the traits of the species, whereas the cytoplasm passes on the traits of the genus or family. . . . We therefore investigated the Epilobium hybrids that showed the greatest amount of reciprocal difference after crossing of the most distantly related species. (Peter Michaelis, 1954, p. 296)

The limitations of the gene-selection theory for heredity and evolution and the theories postulated by naturalists and embryologists were highly considered by many leading botanists in Germany who investigated the role of the cytoplasm in heredity. Correns (Saha, 1984) and the botanist Hans Winkler (1924), who had carried out merogony experiments early in his career, endorsed the embryological idea that the cytoplasm was responsible for the "fundamental" characteristics of the organism while nuclear genes were concerned with trivial traits. At the same time, whether the cytoplasm was composed of self-reproducing genic entities or acted as a physiological or organized whole remained a matter of controversy among German botanists throughout the 1930s and 1940s.

Winkler (1924) was one of the first to adopt the former view. He imagined that the cytoplasm was made up of a myriad of self-reproducing submicroscopic entities, which he called "plasmagenes" (*plasmatischen Gene*). Unlike nuclear genes in chromosomes, which were distributed equally during somatic cell divisions, Winkler supposed that plasmagenes would be differentially distributed to daughter cells. Since a precise distribution mechanism comparable to that for chromosomes was lacking in the cytoplasm. Winkler argued that hundreds or thousands of copies of each plasmagene had to exist. Only in this way could one account for their reliable transfer.

Correns, on the other hand, countered this claim with the statement that one should speak of plasmagenes only if many different ones could be found to be sorted out in the course of cell division. Like several other early supporters of Mendelian theory, Correns was well aware of the difficulties to be encountered with deterministic theories of heredity, having investigated plant physiology and development early in his career. At the extensive gardens, greenhouses, and laboratories of the *Kaiser-Wilhelm-Institut für Biologie,* Correns labored in domains far off the main track of classical genetics until his death in 1933.

After his first published report of non-Mendelian inheritance of chlorophyll characteristics in 1909, Correns studied variegation in a number of plant genera, distinguishing between Mendelian and non-Mendelian forms. Whereas Baur had proposed that plastids themselves were responsible for the cases of non-Mendelian inheritance of variegation, Correns denied the self-determination of plastid development in his cases. Instead he postulated that non-Mendelian inheritance of the variegation was due to the labile state of the cytoplasm, which could cause the plastids to develop into normal, green bodies or abnormal, white ones (Correns, 1937, p. 46). The failure to detect cells with white and green plastids mixed and to see a clear boundary between mutant and normal tissues provided him with objections to Baur's particulate theory.

Correns brought together and summarized the results of his work on "non-Mendelian inheritance" at the Fifth Berlin Congress of *Vererbungswissenschaft* in 1928, of which he was vice-president. Recognizing the certainty that the chromosomes in the nucleus were also bearers of heredity, he prudently and systematically explored the possible mechanisms which might account for non-Mendelian segregation patterns. It was well known that inactive chemicals (Sudan red), active chemicals (antigens), parasites, and symbionts could be transferred from cell to cell, incorporated into the egg cytoplasm, and then transferred from one generation to the next, thus giving an illusion of hereditary transmission ("pseudo-inheritance"). There were also the cases of "predetermination" or "maternal inheritance" and, of course, plastid inheritance, as well as the "plasmagene" hypothesis of Winkler.

Correns was convinced that the developmental processes characteristic of an organism, and their normal integration in time and space, were dependent upon the cytoplasm. In his view the cytoplasm did not act merely as a substratum for autonomous genic action, but both cytoplasm and nuclear genes interacted as a whole, and both parts of the cell played direct roles in developmental processes. The genes would interact with the cytoplasm quantitatively at certain specified times and places. The developmental activation of genic effects, he claimed, was due to qualitative changes in the cytoplasm (Correns, 1928, pp. 161–168).

Correns's views on cytoplasmic inheritance were most forcefully promoted by one of his most influential students, Fritz von Wettstein. Although Wettstein never attained the international recognition of Correns or many others of his German counterparts, he was certainly one of the most prominent figures in German botany between 1925 and the end of World War II (Stubbe, 1950–51; Melchers, 1961a). Wettstein's family tree was well ornamented with some leading naturalists, such as his grandfather, Anton Kerner, and his better-known father, Richard von Wettstein, a neo-Lamarckian and an inventor of phylogenetic plant systems (Schmidt, 1931).

After completing his *Gymnasium* in Vienna and after four years as an officer in the German army during World War I, Wettstein quickly climbed the ranks of German biology. His scientific career began when he was called to be an assistant to Correns, the outstanding German botanist, and a good friend of his father. Shortly after completing his doctoral dissertation under Correns in 1925, at the young age of 29, Wettstein was appointed full professor (*Ordinarius*) of botany at Göttingen. In 1931 he was called to Germany's largest botanical institute and garden at Munich. Three years later, he succeeded Correns as a director of the *Kaiser-Wilhelm-Institut für Biologie,* where he would remain throughout World War II.

Brought up in the systematic world of his father, Wettstein inherited a strong interest in phylogenetic questions and the possibility of the inheritance of acquired characteristics (Wettstein, 1935). The possibility of the inheritance of acquired characteristics was much more evident in plants than in animals. If embryologists had difficulties in distinguishing between the somatic cells and germ cells in animals, a sharp dichotomy was almost impossible to make in plants, where the germ cells were clearly not separate from somatic tissues. New germ cells were clearly not separate from somatic tissues. New germ cells were differentiated every year from embryonic or meristematic cells. There was, therefore, little theoretical reason for denying the inheritance of acquired changes. Fritz von Wettstein's son Diter, a well-known geneticist at the Department of Physiology, Carlsberg Laboratories in Copenhagen, gives the following description of the evolutionary views of his father, grandfather, and Correns:

> As to my father and grandfather's opinion on evolution and the question of Darwinism and neo-Lamarckism . . . no question that hereditary adaptations like the vicaric species gave my grandfather, father and Correns a lot to think and experiment about. My grandfather being the inventor of the phylogenetic plant system and Correns being the outstanding Mendelian geneticist with the clearest mind were very good friends and held each other in greatest respect. I don't think that they really differed in their views on evolution, possibly they gave different emphasis to various mechanisms operative in evolution. Naturally my father was under the influence of my grandfather's ideas, as they did a lot of botanizing and discussing together also after my father had left Vienna in 1919, but this caused no interference with his working relationship with Correns. (letter to the author, February 3, 1981)

As professor of botany at Göttingen, Wettstein was brought into direct contact with the well-known zoologist Alfred Kuhn. In 1927 Kuhn had found clearly transmissible differences between reciprocal crosses in strains of the wasp *Habrobracon,* selected for differences in pigmentation, which he attributed to cytoplasmic inheritance (see Caspari, 1948). Following the views of Correns, Wettstein (1926, p. 253) claimed that the gross effects of cytoplasmic inheritance could be examined most effectively only by crossing organisms with very distant genetic constitutions. Such crosses were not viable in most organisms. However, they were possible in certain mosses, and the results were striking. No reciprocal differences could be detected in hybrids resulting from crosses between different varieties. Reciprocal crosses between different species, however, showed differences in several morphological characteristics, some of which, such as leaf shape and length of midrib, were always identical to the female parent, which transmitted the cytoplasm. Reciprocal crosses between different genera revealed even more strikingly different hybrids which, again, showed predominantly maternal characteristics (Wettstein, 1928). These results clearly seemed to support the theory that the fundamental differences between species, genera, and higher taxonomic groups were based on cytoplasmic differences, and thus only the differences between varieties and strains were due to genes.

In order to guard against the possibility that the results in question were due to a "delayed nuclear effect" or "predetermination," Wettstein conducted a series of backcrosses by which the nuclear genes of one species were implanted in the cytoplasm of another species. The reasoning behind the experiments was similar to that for merogony experiments. It may be briefly outlined as follows: When the resulting hybrid of a cross between species A as female parent, and species B as male parent, is continually backcrossed to the male of species B, the nucleus becomes more and more similar to that of the male parent with each backcross generation. The A cytoplasm of the former mother, however, would remain unchanged, if the pollen does not transfer cytoplasm. After numerous generations, a homozygote nucleus of the male B parent would lie in the A cytoplasm.

Wettstein's experiments with species of mosses, backcrossed over a number of generations (resulting in organisms with mostly paternal genes, but with a cytoplasm derived from the original female plant), convinced him that the cytoplasm possessed inheritable elements through which it played a direct role in the development of characters. It was clear to him that the cytoplasm was able to react in its own peculiar way to the activities of nuclear genes, and could produce certain characteristics entirely by its own properties. On this basis, Wettstein (1926, p. 263) applied the term *Plasmon* to the "genetic element of the plasm," as contrasted to the term *Genom,* by which Winkler in 1924 had denoted the whole collection of genes contained in the chromosomes. Between the mid-1920s and 1933, several of his students continued to extend the kinds of *Plasmon*-controlled characteristics in mosses. This work was quickly associated with that of various other botanists outside of his institute whose work also indicated a hereditary role for the cytoplasm, and an interaction between cytoplasm and genome.

Otto Renner, who worked on *Oenothera* at Jena, Ernst Lehmann at Tübingen, and Peter Michaelis, who worked on the willow-herb *Epilobium,* led some of the most extensive investigations of cytoplasmic inheritance outside of Wettstein's institute during the 1920s and 1930s. Their experiments were based on crossing different varieties and the most distantly related species and studying the hybrids that showed the greatest amount of reciprocal differences. Their results were consistent. Reciprocal crosses showed dissimilarities in a variety of characteristics: in growth, size of petals, development of anthers, degree of fertility, as well as physical differences such as permeability and viscosity (Caspari, 1948; Michaelis, 1954). However, their interpretations were inconsistent.

Renner and his collaborators and Michaelis and his students, supported by Wettstein and Correns, interpreted their observations in terms of a *Plasmon*. They argued for a determinative role of the cytoplasm independent of the nucleus, which could influence the action of specific chromosomal genes. Lehmann and his students, on the other hand, defended the integrity of the gene and its monopoly over the physiological processes of the cell, and gave a Mendelian interpretation of the results. In Lehmann's view, each nuclear combination influenced the production of a specific cytoplasm. He claimed that the lack of reciprocity could be explained by the existence of a delayed nuclear effect resulting from the transmission of gene-produced hormones, or of self-reproducing cytoplasmic products of nuclear genes.

Between 1924 and 1935, Renner investigated chlorophyll characters in *Oenothera* and attempted to distinguish between the genetic effects of the *Plasmon* and the chloroplasts or what he called the *Plastidom* (see Renner, 1936) The atypical genetic behavior of *Oenothera,* which led de Vries to propose his mutation theory, had attracted a great deal of interest during the first decades of the century. Renner was already celebrated for his contributions toward reconciling the puzzle. *Oenothera* was a complex heterozygote, and this made possible Renner's experiments on plastids, since it easily permitted the combining of entire genomes of one type with different cytoplasms. Moreover, since in *Oenothera* plastids were often transferred by both pollen and seeds, Renner was able to construct organisms with mixed plastids from different species. The mixed plastids were found to segregate and produce variegation, thus showing that they might be physiologically different. From an impressive body of data Renner concluded that plastids were genetically different as autonomous self-duplicating bodies, which might show changes comparable to gene mutations.

Although Renner's studies were later considered classical, his hypothesis of a *Plastidom* was certainly not accepted by all botanists and geneticists during the 1930s. K. L. Noack (1931), for example, continued to insist that plastids as such were not the primary cause of this type of variegation, but that some metabolic disorder was produced by a disharmony between the nature of the cytoplasm and the genes contained in the nucleus. On the other hand, Renner's claim for genetic properties of plastids was supported by work led by Julius Schwemmle and his school (see Caspari, 1948), who extended Renner's studies of species crosses in *Oenothera* and studied plastid behavior and nonparticulate *Plasmon*-inherited characters simultaneously. It was also supported by Wettstein and Michaelis, who claimed that cellular differences could occur by plastid segregation and a labile cytoplasm functioning side by side and interacting with each other and nuclear genes.

In 1937 Wettstein brought together the diverse botanical evidence to discuss both the genetical and the developmental significance of the cytoplasm. Breaking up the *Idioplasm*—that is, the totality of the genetic material of the cell—into its constituent parts, the *Plastidom,* represented the idiotype of the plastids, and was distinguished from the plasmon—"the structure of the cytoplasm"—which in turn was distinct from the genome of the nucleus. By 1937 Wettstein and most other *Plasmon* investigators came to oppose the strict distinction theorized by Loeb, Winkler, and many others that the cytoplasm controlled the characteristics of higher taxonomic groups, while the nuclear genes were concerned only with varietal or species differences. The results from investigations in *Oenothera* and *Epilobium* seemed to demonstrate the fallacy of this idea. Reciprocal differences were sometimes reported to be as large between varieties of one species as they were between different species. Wettstein claimed that the greater differences among higher taxa were due to the interactions of both an increasing number of different genes and increasingly different plasmons. The resulting phenotypes were due to the interaction of plasmatic elements with nuclear genes. Wettstein stated his view as follows:

> Chromosomes and cytoplasmic permeability, growth and gastrulation, chlorophyll formation and pigmentation, hairiness and habitus, all of these traits are the product of the cooperation between the *genom* and the *plasmon*.
>
> One should therefore finally dispense with the entirely wrong opinion, that race—and species—characteristics are determined by nuclear genes and more profound characteristics of organization (=traits of higher taxonomic groups) by the plasm. This is basically wrong and shouldn't be discussed again and again.
>
> The cooperation (between the *Plasmon* and the *Genom*) is the essential point. (Wettstein, 1937, pp. 245–246, my translation)

Like many embryologists, Wettstein saw the primary significance of the cytoplasm to be in its organizing properties, as contrasted with the atomistic and undirected nature of the genes. He regarded the *Plasmon* as the structure which determined the series of developmental changes in the organism. Through the cytoplasm a definite developmental scope was given. The phenotypical circumstances of the cytoplasm would change under the steering of the genes and external conditions until development was completed.

Investigations of cytoplasmic inheritance were not the only studies carried out under Wettstein's direction. In fact, of the twenty-eight doctoral dissertations written by his students, only five were directly concerned with cytoplasmic inheritance. According to one of Wettstein's students, Georg Melchers (interview, December 12, 1981), during the rise of the Nazi regime and World War II, Wettstein, who was not a Nazi, struggled to bring back genetically trained soldiers from the war, claiming that they were needed to investigate what he said were economically important problems related to virus research in plants. Many of his students would, in fact, later pursue careers as plant breeders and in plant virus research.

With the sudden death of Wettstein in 1945, the leadership of *Plasmon* investigations in Germany changed hands. The most extensive investigations of cytoplasmic inheritance in Germany were carried out by Michaelis at the *Kaiser-Wilhelm-Institut für Zuchtungsforschung,* in 1951 renamed *Max-Planck—Institut für Zuchtungsforschung* (*Erwin-Baur Institut*) in Köln-Vogelsang. Melchers recalls (interview, December 12, 1981) that during the rise of the Nazi regime, Michaelis and his students carried out their investigations in isolation, both socially and intellectually, from those headed by Wettstein at the *Kaiser-Wilhelm-Institut für Biologie. The Kaiser-Wilhelm-Institut für Zuchtungsforschung* was headed by a powerful Nazi, W. Rudorf. Although in Hitler's Germany outstanding scientists were forced to leave their native country, Michaelis enjoyed support from his institutional affiliation with Rudorf. Although the principal objective of the institute was the study of breeding problems in relation to their practical applications, Michaelis's work was directly concerned with the *Plasmon* and the problems of development and macroevolution.

Between 1923 and 1960, Michaelis and his collaborators published more than fifty long papers on the role of the cytoplasm in heredity. In order to demonstrate the genetic specificity of the cytoplasm, Michaelis carried out an elaborate series of backcrosses. In more than 100 different nucleus-cytoplasm combinations, the cytoplasm remained constant—in some cases for more than twenty-five backcross generations (see Michaelis, 1954). In Michaelis's view, since all the socially accepted possibilities of error had been excluded, his results represented "proof" that the cytoplasm retained its specificity.

Michaelis constructed a highly theoretical genetic system. His "life-system of the cell" was composed of the nucleus and its genes and of the "submicroscopic structure of the cytoplasm" and its components. Each had special functions; each showed a special behavior during reproduction, transmission, and mutation, but all were coordinated with each other and all worked together in a balanced reaction. All properties manifested as phenotypes were the products of these interactions.

During the 1940s Michaelis included in the concept of the *Plasmon* semiautonomous self-perpetuating cytoplasmic genetic particles. This theoretical move was due in part to a series of cytoplasmic investigations in the United States and France (see Chapters 4 and 5). With the incorporation of microorganisms for genetic study during World War II, the "plasmagene theory" based on self-perpetuating cytoplasmic bodies such as plastids, mitochondria, and other submicroscopic genetic entities came into prominence. In an extensive review summarizing his experimental and theoretical work and translation into English, Michaelis (1954, p. 290) put forward his conception of the *Plasmon* as follows:

> I propose . . . to include all extranuclear hereditary elements of the cell in the term *plasmon* and to subdivide this into (1) the cytoplasmon, that is the element of the cytoplasm, and (2) the plastom, that is, the hereditary elements of the plastids, etc.

In experimental investigations, however, Michaelis avoided using the concept of plasmagene "in order to discourage any comparison with the nuclear gene" (1954, p. 289). Moreover, he was opposed to conclusions which, based on a study of phenotypes, attributed the cause of cytoplasmic inheritance solely to cytoplasmic units such as plastids and perhaps mitochondria. "All such conclusions," he repeatedly argued, right up the mid-1960s,

> are incorrect, as in all properties of a genetical system, the chlorophyll deviations of the plastids, for example, may not only be produced by nuclear genes, by a real plastid inheritance, but also by cytoplasmic inheritance and perhaps by chondriosome inheritance. Conclusions from the phenotype are very risky and incorrect. . . . Non-Mendelian inheritance shows only an extra-chromosomal inheritance. Maternal inheritance proves only differences between the sum of the egg cytoplasm and the spermatozoid, not more. (Michaelis, 1965a, p. 87)

Indeed, in most experiments on cytoplasmic inheritance, especially those in higher plants such as *Epilobium,* one could not distinguish between manifestations of "single plasmagene" units. Without the appearance of cytoplasmic segregation of discrete particles, it was only possible to detect the combined effect of all the hereditary components of the cell that did not exhibit Mendelian segregation. In other words, it was only possible to analyze the genetic effects of the total protoplasm of the egg cell. Through reciprocal crosses one could not predict with any degree of certainty which cytoplasmic components had the capacity to be identically reproduced and which components possessed hereditary potency.

Like many other *Plasmon* theoreticians, Michaelis had an integrative view of the genetic system which caused him to object to phenotypic studies that attributed the cause of cytoplasmic inheritance to cytoplasmic units:

> We can assume that the cytoplasm is not simply the sum of cytoplasmic units but a complicated hereditary system in which the units participate—in various, still unknown ways—in the composition and structure of the cytoplasm. (Michaelis, 1954, pp. 289–290)

Michaelis (1954, p. 290) thought it possible that the genetic properties of the cytoplasm could be changed not only through *mutation* of individual "plasmagenes" or nuclear genes, but also through an *alteration* of the whole system. He expressed his holistic views as follows:

> Not only do the various components of the cells form a living system, in which the capacity to live, react, and reproduce is dependent on the interactions of all the members of the system; but this living system is identical with the genetic system. The form of life is determined not only by the specific nature of the hereditary units but also by the structure and arrangement of the system. The whole system is more than the sum of its parts, and the effect of each of the components depends on and is influenced by all previous reactions, whose sequence is in turn determined by the whole idiotype. (Michaelis, 1954, p. 320)

This hypothesis, in Michaelis's view, had the greatest significance for understanding both development and macroevolution. First, he claimed, the developmental potencies of cells often become limited through unequal distribution of cytoplasm. Situated between the inflexible nuclear genes and the intraorganismic environment, the more easily alterable *Plasmon* could respond to intersystemic changes. Only in this way, he claimed (Michaelis, 1954, p. 369), could an orderly and harmonious alteration of the whole system be possible.

Contrary to the expectations expressed by some naturalists that environmentally directed adaptive hereditary changes in the cytoplasm should be readily produced, Michaelis (1954, p. 356) failed to provide any noteworthy examples. However, he claimed that this did not mean that the role of the external environment in directing *Plasmon* alterations could be overlooked with regard to evolution (Michaelis, p. 368). In 1912 the botanist Charlotte Ternitz provided what Michaelis viewed as a useful example of directed alterations as they related to the evolutionary significance of the *Plasmon*. In the green alga *Euglena,* decreased multiplication of plastids was shown to result from cultivation under conditions of reduced light and appropriate nutrients. The retardation of plastid multiplication could ultimately cause the elimination of plastids, which resulted in the production of what Michaelis (1954, p. 389) saw to be an animal species parallel to the green flagellates. There were also the controversial cases of *Dauermodifikationen* in protozoa, as first demonstrated by Jollos. As will be discussed more fully in the next chapter, these cases became exemplars of Lamarckian principles operating through the cytoplasm for many geneticists. To Michaelis, paleontologists were the chief authorities on evolution, and only they could truly judge the evolutionary significance of these mechanisms:

> Their true significance must be elucidated by further investigations on the frequency of these *Plasmon* alterations; and in the end the paleontologists have the last decision as to whether the experimental findings can explain their observations. At present it would be just as erroneous to place too great hopes in the new possibilities as to underrate them. (Michaelis, 1954, p. 389)

The alterability of the *Plasmon* under the influence of the whole system and of the environment provided potentialities for phylogenetic development that seemed to Michaelis to be otherwise impossible. Taking a predominantly selectionist viewpoint, he (1954, pp. 388–394) suggested that the integrated genetic system of genome and *Plasmon* combined with intrasystemic selection could account for several objections of paleontologists to mutation and selection. *Plasmon* alterations could be directed by intrasystemic selection where the possibilities of alteration were limited by the structure of the system and the comparative stability of the nuclear genes. These two factors alone, he argued, served to explain orthogenesis.

Generally, Michaelis believed that the *Plasmon* might resolve some of the problems paleontologists encountered when accounting for macroevolution. First, it will be recalled that many paleontologists and taxonomists considered it impossible to account for, in terms of selection and mutation, the origin of complicated organs in which the formation and selective value of the individual parts depended on the presence of all the other numerous parts. Michaelis reasoned that this difficulty would be overcome by an integrated system of nuclear genes and *Plasmon*. Secondly, many paleontologists found that in most cases the quality and quantity of new modifications were determined not by environment and selection but autogenetically through the organism. According to Michaelis, intraindividual selection made this finding understandable, since it could bring about preadaptation.

The fact that most gene mutations scarcely changed the fundamental structure of the main branches of the phylogenetic tree led some paleontologists to assume that there was a special mechanism for macroevolution which was distinct from microevolution. As Michaelis saw it, addition of genic mutations could account for microevolution; the harmonious combination of genic mutations in an organized system of the genome and *Plasmon* could be the basis of macroevolution:

> It would not be correct, however, to attribute micro-evolution to nuclear genes and macro-evolution to the *Plasmon*. The essential factor is not the material localization of the hereditary elements but the manner in which they are combined into a system. It might be more correct to interpret micro-evolution in terms of the alteration of secondary and peripheral reaction chains, and macro-evolution in terms of the reconstruction of primary links in the reaction chain during the embryonic stage. Micro-evolution might also be explained by mutation, and macro-evolution by the complicated combination of important mutants of the different carriers of heredity. (Michaelis, 1954, p. 392)

"The Scotch Verdict, 'Not Proven'"

> Because these units [genes] ordinarily behave as if they form an isolated system unaffected by the cytoplasm, most geneticists have been led to assume that inheritance is completely under nuclear control and that the only remaining problem of heredity is the manner in which this control is exercised. . . .
>
> In taking this view of the nuclear plasma problem, biologists have followed established precedent. It has been found in science that when a sub-universe of discourse can be dissociated from a larger universe and a means of studying behavior found which is but slightly affected by uncontrollable factors, the results usually have high value in prediction. This is the reason for the extended progress of physics and chemistry. It is also the reason for the rapid progress in gene analysis. (E. M. East, 1934, p. 289–290)

The investigations of cytoplasmic inheritance and *Plasmon* theory were not well received by geneticists outside Germany for a complex of social, technical, epistemological, and ideological reasons, all of which have to be taken into consideration when understanding their responses. First, most genetic investigations were based on an analytic method of investigating single differences that arose through exchange of nuclear genes within a system that was kept constant. A large body of breeding data and cytological observations was interpretable in terms of the grouping and distribution of genes. In contrast to the reductionist procedures of Mendelian genetics, *Plasmon* investigations were based on a synthetic method in which as many different Plasmons as possible were interchanged and the differences that arose as a result were compared to one another. In sum, investigations of the *Plasmon* and those of the genes were based on two competing theoretical assumptions: one holistic or integrative, the other essentially reductionist.

Second, investigation of the role of the *Plasmon* was very difficult from a technical standpoint. It involved complex experiments and was possible only in certain materials. It is noteworthy that in one of the best investigations, that of Michaelis, twenty years passed before the first extensive theoretical report. To carry out such time-consuming and tedious investigations in the United States, for example, would be a difficult task indeed. In fact, Rhoades's study of male sterility in maize was the only well-known and extensive investigation of cytoplasmic inheritance reported by an American geneticist prior to World War II. His results were not published in an American journal, however. They appeared in the British *Journal of Genetics*.

American geneticists had established a system of norms that were compatible with reductionist doctrines. This included rapid production of results based on studies which could be carried out easily by established procedures. There were few attempts to reconcile genetics and embryology in the United States and physiological genetics was not highly represented. Cytoplasmic inheritance lacked the socioeconomic applicability of transmission genetics and nuclear genes. A second generation of American scientists had been imbued with Morganist doctrines, which included the claim that chromosomal genes and their products act on the cytoplasmic substratum of the cell in some yet unknown way and control developmental processes.

Moreover, during the 1930s, investigations of cytoplasmic inheritance continued to be perceived as a threat to the primary role of nuclear genes in heredity and to the social recognition and prestige of most American geneticists. The evidence for cytoplasmic inheritance remained allied with the possibility that nuclear genes played only a minor role in evolution and that the "fundamental" traits of the organism were outside of Mendelian genetic analysis. When American geneticists reviewed the evidence for cytoplasmic inheritance during the 1930s, they looked for possible objections and attempted to replace cytoplasmic interpretations with diverse *ad hoc* nuclear (Mendelian) explanations. Some would not even accept the evidence that chloroplasts played a genetic role as integrated constituents of plant cells. Surprisingly, there was no hereditary phenomenon that could not be explained, with due allowances, by Mendelian genes.

When the third edition of *Principles of Genetics,* by E. W. Sinnott and L. C.

Dunn was published in 1939, it included a brief new chapter on "cytoplasmic inheritance." In the view of the authors, the physical continuity of cytologically visible bodies was a necessary precondition for hereditary determinants. They were ready to accept cytoplasmic control over the development of characters that had a cytological basis in the cytoplasm, such as plastids: "Cytoplasmic inheritance of plastids is easy to understand because of the obvious mechanism involved" (Sinnott and Dunn, 1939, p. 247). They discredited all other cases of maternal inheritance or reciprocal differences in plants and animals which involved no clear cytologically visible mechanism for material continuity.

They further argued that since the majority of cases reported by geneticists conformed to Mendelian laws, the apparent exceptions to Mendelism were merely cases where the "genic control" was "masked in some way" (Sinnott and Dunn, 1939, p. 251). They asserted that cytoplasmic differences could be due to the genotype of the maternal nucleus ("predetermination") or to developmental incompatibilities between genes and cytoplasm. They upheld the case of maternal inheritance of the direction of coiling in snails (see Chapter 1) to exclude all examples of reciprocal differences lacking a clear cytological mechanism. When discussing the organization of the egg cytoplasm, which was known to be associated with the polarity, symmetry, and rate of cleavage of the developing embryo, the authors wrote:

> It seems rather unlikely that such traits as these are controlled by a specific mechanism carried in the cytoplasm when they can be explained equally as well as early induction effects of maternal genes brought about before the reduction division. Such an hypothesis may explain some of the cases of cytoplasmic inheritance which have been described above. (Sinnott and Dunn, 1939, p. 252)

In their well-known textbook *An Introduction to Genetics* (1939) A. H. Sturtevant and George Beadle also interpreted most of the evidence for cytoplasmic inheritance in terms of "predetermination" or "delayed nuclear effects" and accepted only the evidence for chloroplasts as possessing permanent genetic properties. Although they recognized that there were a number of examples described in the literature which "taken at face value" did establish the existence of permanent autonomous cytoplasmic elements, they dismissed them on technical grounds, stating:

> In none of these does the evidence seem to us conclusive. Some of them are based on too few individuals to be critical; others involve complex experiments in which there are many opportunities for unsuspected errors to creep in, and in still others the possibility that the observed effects may be due to plastids or to ordinary genes has not be excluded (Sturtevant and Beadle, 1939, pp. 331–332)

E. M. East (1934) at Harvard's Bussey Institution wrote the most extensive critique of cytoplasmic inheritance in the 1930s. Trained originally as a chemist, East came to Mendelism and genetics through his attempts to improve the protein and fat content of corn through breeding techniques at the Connecticut Agricultural Station in New Haven. In 1909, at the recommendation of William Bateson, East was invited to join the faculty of Harvard's Bussey Institution. A pioneer in the development of hybrid corn and an expert on inbreeding and cross-breeding in general, by the 1920s East emerged as one of the most highly regarded geneticists in the United States. However, East's work was not only concerned with genetic principles as they related to corn breeding. Between 1919 and 1931 he dedicated several texts to social issues, whereby he attempted to demonstrate the relevance of Mendelism and natural selection for understanding social issues and for framing social policy. Underlying all East's writings on human affairs was his hereditarian conviction that "social progress depends primarily upon the genetic constitution of the people of which a society is composed" (East, 1923, p. 195).

East was decidedly opposed to any type of inheritance outside of Mendelian segregation. His primary discursive tactic was to dismiss the question of the relative importance of the nucleus and cytoplasm in heredity on formal methodological grounds. Unlike Morgan (see Chapter 1), however, East did not use empiricism simply to deny the distinction between "fundamental" and "trivial" traits. Instead, he attempted to push it outside the realm of science as an illegitimate question. East admitted that the possibility existed that the cytoplasm controlled characteristics of higher taxonomic groups while the nucleus controlled only characteristics of the variety or species. However, in his opinion Loeb's hypothesis "was not a very satisfying scientific proposal," since it could not be proved or disproved by existing techniques:

> If true the fact could not be demonstrated, since parental *differences* only can be detected and followed. If untrue, evidence for its falsity ordinarily could be obtained only indirectly, and with difficulty, by following the non-generalized differences between the greater taxonomic groups; and such crosses are rarely viable. (East, 1934, p. 291)

He dismissed the idea of Boveri and Conklin that the cytoplasm of the egg determined the early stages of development, while only later differentiations were influenced by the sperm, as being "still less satisfactory as a working hypothesis" (East, 1934, p. 292). He argued that it was useful in discriminating between problems of inheritance and problems of development, but it was not helpful in determining whether or not the cytoplasm served as "an independent vehicle of inheritance."

East recognized that there were no embryological facts incompatible with the supposition that the cytoplasm contained hereditary potentialities which were transmitted from one generation to the next. On the other hand, he claimed that there were also no embryological facts incompatible with the assumption that the nucleus had complete control of both heredity and development. He claimed that one could interpret the situation by supposing that the nucleus was in charge of cytoplasmic differentiation, and that the development pattern was caused by the reactions thus made possible, when account was taken of cell succession and position. East failed to offer a nucleocentric interpretation of cellular differentiation. Instead, he simply dismissed all the embryological experiments which had been cited as arguments against this second view as being susceptible of "reasonable interpretations" which were in harmony with the theory of nuclear control (East, 1934, p. 293). He claimed that most merogony experiments produced indecisive results, and where results were of a "more decisive nature," the maternal effects could be due to "predeterminations" or "delayed genic action." The only cases of "true merogony" which he considered to be beyond error were cases which indicated nuclear control (East, 1934, p. 300).

East also attempted to undermine the evidence for plastid inheritance. He claimed that the physical continuity of plastids in higher plants was questionable and if they did reproduce themselves they could be considered as symbionts, as others had claimed, and be dismissed from genetics (East, 1934, p. 403). Non-Mendelian inheritance of chlorophyll characteristics could be due to a disease affecting gene-controlled plastid development—a view he himself favored. He supported his position by pointing to the relative scarcity of examples of plastid inheritance. Of some 400 cases concerning variegations in plants that had been analyzed by the 1930s, only about forty were reported to be non-Mendelian (East, 1934, p. 403).

East went further in his attacks on cytoplasmic inheritance and questioned the actual motives and competency of the investigators themselves. Suspecting that some German researchers may have been attempting to demonstrate the inheritance of acquired characteristics, he endeavored to cast a shadow of notoriety on cytoplasmic inheritance research. As previously mentioned, there were several types of phenomena reported where there was direct transfer, from cell to cell, of "alien matter" capable of producing morphological changes. For example, a number of so-called alga-like symbionts and parasites were capable of being transmitted from cell to cell, giving an illusion of hereditary transmission. "It is not to be supposed," East warned,

> that modern biologists will cite such instances, when recognized, as examples of heredity. But since an earlier generation of students used them, before their cause was discovered, to support arguments on the inheritance of acquired characteristics, it is well to be cautious in citing similar, though less obvious, cases as being illustrations of non-Mendelian heredity. (East, 1934, pp. 409–410)

East also found the evidence in favor of a *Plasmon* in plants to be unconvincing. The nucleus was involved in all of the reciprocal differences; cytoplasmic differences could therefore be recognized only indirectly and interpretations based on nuclear control could not be ruled out. "Weighing all the evidence for and against the *Plasmon,* therefore," East concluded, "one is forced to the Scotch verdit 'not proven'. But to-morrow or the next day this verdict may be incorrect" (East, 1934, p. 431).

In the 1930s, the evidence brought forward in favor of cytoplasmic inheritance did not conflict only with the doctrines of classical Mendelian genetics. It also faced the development of a "new evolutionary synthesis" based on Mendelian mutations and recombinations, together with natural selection. Evolutionary theory of geneticists in England and the United States was slowly turning away from what the naturalist Ernst Mayr (1980) has called "typological thinking," which characterized the evolutionary perspective of the first generation of geneticists. During the first third of the century, Mendelian geneticists had viewed species and populations not as highly variable aggregations consisting of genetically unique individuals, but rather as uniform types. In contrast to the "typological" or "essentialist" perspective, the new synthetic view of evolution, led by the work of R. A. Fisher, Sewall Wright, J. B. S. Haldane, T. Dobzhansky, and others, was based on the statistical analysis of the frequency of genes in natural and artificial populations. Nongenic cytoplasmic inheritance seemed to threaten the significance of the merger of Mendelian genetics and selection theory and therefore had to be denied.

In 1938 J. B. S. Haldane attempted to refute the thesis that nuclear differences accounted only for variations within a species while the more fundamental differences between species depended on the cytoplasm. Although he was willing to accept "rare" cases of cytoplasmic inheritance, such as that reported by his colleague R. J. Chittenden in England, he emphasized that they were not responsible for species differences (Haldane, 1938, pp. 89–90). Claiming that the majority of differences between crossable species were determined by the nucleus, he confined cases of cytoplasmic inheritance to plants and emphasized their rarity. The possibility also existed that some results in plants were due to virus infection, which, he claimed, could not be clearly distinguished from plasmatic inheritance. He further denied the evidence from merogony experiments on echinoderm eggs by Harvey and others, and, like other defenders of the nuclear monopoly, stated that they were understandable in terms of maternal inheritance, whereby the maternal nuclear genes determined the cytoplasmic architecture of the egg (Haldane, 1938, p. 88).

Similar views were maintained by Dobzhansky, who, in the first edition of his classic text *Genetics and the Origin of Species* (1937), reported the widespread "defeatist attitude" of some writers who maintained a distinction between continuous variation, which was thought to be environmental in origin, and discontinuous variation:

> Continuous variability was declared different in principle from the discontinuous one. It was said that only the latter is clearly genic, while the former was alleged to be non-Mendelian and to be due to some vague principle which assiduously escapes all attempts to define it more clearly (Dobzhansky, 1937, p. 57)

Arguing against the sharp distinction between continuous and discontinuous variability and the demise of blending inheritance with the recognition of the "particulate" theory of the gene, Dobzhansky (1937, p. 182) wrote:

> Only in the obscure realm of cytoplasmic inheritance a situation approaching blending may be obtained, although too little is known about it to make any conclusion secure. With the obsolescence of the blending inheritance theory, one of the greatest impediments to the progress of evolutionary thought was removed.

Sewall Wright's views on the rôle of the cytoplasm in heredity were more complex and are treated in the next chapter.

While many American geneticists continued to be hostile to the idea of extranuclear inheritance, embryologists such as Ross Harrison continued to maintain its necessity. Harrison (1937, p. 372) put his case as follows:

> Whether we accept the *Plasmon* concept or not, we are obliged . . . to assign to the cytoplasm of every egg specific characters, which are different in each species

> of organism. In the egg there are characteristic local differentiations, which are frequently of the nature of inclusions, but after these are all accounted for, the specific character of the cytoplasm still persists in the ground substance.

After World War II genetic investigations of cytoplasmic inheritance came to the fore in the United States and France, led by the work of T. M. Sonneborn, Boris Ephrussi, and others (see Chapters 4 and 5). During the 1950s the work of the German botanists wsa more favorably introduced into the Anglo-American literature. Marcus Rhoades (1955) wrote an extensive review which included the work of the German botanists. T. M. Sonneborn (1950e) also cited their work as supporting his conclusions of cytoplasmic inheritance. In England C. D. Darlington summarized the work of the German botanists in his celebrated text *The Evolution of Genetic Systems* (1958).

Other American geneticists, such as Ruth Sager, who during the 1960s became one of the most prominent geneticists in the domain of cytoplasmic inheritance (see Chapter 7), were unwilling to read the "almost unintelligible" German literature after World War II, but learned of it only indirectly through reviews by other Americans. In France and Belgium, where the anti-German sentiment was perhaps the strongest after World War II, Boris Ephrussi, Jean Brachet, and others refrained from citing the confusing German literature in their discussions of cytoplasmic inheritance.

Though some geneticists outside Germany would later cite the German theoretical writings on "nonparticulate" inheritance, the notion of the *Plasmon* was not well integrated into their research. During the 1940s and early 1950s, the German literature was often reinterpreted in terms of self-perpetuating "plasmagenes," which were used to explain non-Mendelian results in microorganisms (see Chapter 4). With the rise of molecular biology in the 1950s and 1960s and the transformation of the chromosome theory of inheritance into the nucleic acid doctrine, the genetic bases of characteristics considered by German investigators to be controlled by *Plasmon* would again be reinterpreted. They were seen primarily as the result of the actions of nuclear-based "regulatory genes" or of cytoplasmic, nucleic acid genes (see Chapter 7).

CHAPTER 4

T. M. Sonneborn: Making Plasmagenes in America

The war is over here, but as you know, we never did any suffering to speak of. The suffering is with the soul as it is everywhere. Just now, the future of our economic situation is so black just because of an inadequate general understanding of what it all means or is going to mean. Nevertheless, the lid is off and spending this Christmas season was sensational and depressing. It is just uneducated, unfeeling America but there isn't much real gaiety with it. Thus, we are not blowing bubbles even if there is no gas rationing. The upswing in science, though, is sensational. If it lasts, science in the United States will receive much support. It is having its effects on those that are making plans for the future. Big ideas are hatching and they certainly look good on paper. The genetics groups are feeling it too. You undoubtedly know that Beadle and his whole group are moving to Cal. Tech. this coming July. I had a long talk with Beadle a few weeks ago. The set-up at Cal. Tech. will be splendid. Much emphasis will be put on chemical aspects of genetics with the cooperation of the department of chemistry there—Pauling in particular. He has been anxious to get this tie-up with the Beadle group. I hope to go out there next fall for a while and will look at the consequences with considerable interest. The emphasis in genetics seems to be swinging into genic action with much attention to the cytoplasm and its part in genic action. I suspect that the cytoplasm will now begin to play its proper part in our thinking of genic action and genic control. All at once, a number of people seem to be reacting to it and finding relationships that have existed all along but have not been appreciated. It is not just the microbiology work that has brought this forward. It will appear in such ancient genetic organisms as *Drosophila* and maize. Nevertheless, microbiology is doing its part. Demerec is on that nearly altogether—when he can take time off from attempting to run, single handed, a series of three ring circuses. He has brought in Luria and a group here at Carnegie and another group at the Bi. Lab to work on bacteria. There will be a symposium at the Bi. Lab. this coming summer and it will be on microbiology. I have heard that you might be interested in this phase. Is there any chance that you would come over here for something like this?

Caspersson was here for quite a while. He was the first of the foreign visitors since the war but more and more are coming. . . .

Our department is having a renovation done on it. So many of the staff members of former times have left and no new ones have been taken in. We are now in the

> process of reconsidering many things along with personnel. It has taken some time to shake down the good from the bad but we really have made some progress. No new appointments have been made but we are hoping that they will be soon.
>
> Joint seminars have started with the group at Columbia University. When we all get together, it makes quite a mob. Nevertheless, not a great deal new is hatching from this mob. . . . Dunn is busy with this and that, as usual only much more so. Dobzhansky is much the same, still enthusiastic and his problems with evolution and the fact that they take him back to California each summer. Marcus [Rhoades] is much the same in the laboratory but he has bought a house in Hastings on the Hudson. . . .
>
> This isn't much of a letter but I did not want any more weeks and months to pass without some word from here. I know you have heard from many people, including Demerec, so much that I may have told you is old news. I expect to see Muller in two weeks and will hope for some direct and fresh news from him. . . . (P.S.) You undoubtedly heard that Morgan died first of this month. I add this just in case it might have escaped you. (Barbara McClintock to Boris Ephrussi, December 29, 1945)

This long passage from a letter from Barbara McClintock to Boris Ephrussi provides a vivid glance at the changing social and intellectual milieu which brought cytoplasmic inheritance research to the fore of genetic discourse in the United States. The main objective of Mendelian genetic research up until the mid-1930s was to study the transmission of visible differences exhibited between organisms capable of being cross-bred. These investigations entailed a study of the sexual transmission of genes between individuals, a study of their correlations with adult characters, and an analysis of the relations of the genes in the chromosomes where they were located.

Following World War II, the face of genetics quickly changed. Many young geneticists were no longer content to analyze hereditary transmission of morphological secondary characters, such as wing shape or tail length, far removed from the intracellular level where the genes were located and where they exerted their primary effects. Instead they turned to investigate genic control over the physiological properties and chemistry of the organism. This shift in interest brought with it the domestication of new organisms for genetic use. The higher organisms, especially *Drosophila* and maize, had proven to be very useful for establishing the mechanism of Mendelian inheritance. However, when geneticists attempted to bridge the gap between gene and character, the higher organisms of classical genetics were quickly outcompeted by rapidly reproducing microorganisms such as bacteria, fungi, algae, and protozoa. These organisms allowed investigators to avoid the complexities of tissue differentiation and integration of multicellular organisms.

The domestication of microorganisms for genetic use brought with it a change in the genetic concepts themselves, which included an extension of the genetic meaning of "heredity" itself. As discussed in Chapter 2, prior to World War II, when Mendelian geneticists confined their interests to the sexual transmission of differences between individuals, they constructed a definition of "heredity" to suit their practices. They restricted their meaning of heredity to a correlation with sexual reproduction. When genetic investigations centered on cross-breeding analysis and hybridization, microbes, which seemed to reproduce solely vegetatively, were excluded from the domain of genetics.

However, by the late 1930s and 1940s, microorganisms became accessible to hybridization techniques. Their genetic study, together with the detection of hereditary differences among somatic cells of higher organisms by experiments in grafting and tissue culture, led geneticists to assume that heredity in microorganisms was fundamentally comparable to "heredity at the cellular level" in higher organisms. Furthermore, in bacteria there were means other than sex for transferring genetic material form one generation to the next. Viruses could act as vehicles of inheritance and geneticists began to construct a theory of "infectious heredity." Thus, the genetic notion of heredity was extended. It came to mean the ability to perpetuate its like through successive generations possessed by every cell. It was within these two concepts of "cell heredity" and "infectious heredity" that the cytoplasm found its chief genetic significance during the 1940s and 1950s.

When microorganisms were domesticated for genetic use, many strange phenomena were reported, including relations between the environment, genes, and cytoplasm which could not be easily accommodated within the classical conception of the gene. To many biologists the rise of microbial and biochemical genetics seemed to mark the beginning of a major revolution in genetic thought with the cytoplasm at the fore. Investigations of the role of the cytoplasm in heredity were carried out by many biologists in various countries in the decades following World War II. But the most prominent were led by Tracy Sonneborn in the United States and by Boris Ephrussi (see Chapter 5) in France, who played the leading roles in directing cytoplasmic inheritance to a prominent position in genetic discourse.

Tracy Sonneborn (1905–1981) is well recognized today as the founder of "ciliate genetics," a burgeoning speciality of modern experimental biology (Nanney, 1982a,b; Beale 1982). It was Sonneborn who first introduced both unicellular organisms and systematic investigations of cytoplasmic inheritance into the domain of American genetics. In 1937 he showed that unicellular organisms could be cross-bred and used effectively for genetic analysis. Using the ciliate *Paramecium aurelia* as a subject, he subsequently demonstrated that like higher organisms, unicellular organisms possessed genes which behave in the classical Mendelian way. Sonneborn quickly rose to prominence after World War II as one of the leading authorities on the genetics of microorganisms. His subsequent investigations of the principles of action and interaction of genes, cytoplasm, and environment led him up the institutional hierarchy of American genetics. Among several other significant positions, Sonneborn came to occupy the historically important seat of president of the Genetics Society of America at the time of the Lysenko affair (see Chapter 6).

On the other hand, throughout his research career Sonneborn remained aloof from the mainstream of Mendelian genetics and consequently has managed to elude historians of twentieth-century biology. Following the World War II Sonneborn became the most vigorous and skilled apologist for the biological significance of cytoplasmic inheritance and resisted what he considered to be the "obsession" with the gene. Beginning in the early 1940s he and his co-workers at Indiana University investigated heredity in *Paramecium* and demonstrated several genetic mechanisms involving the interaction of genes, cytoplasm, and environment in organismic and cellular heredity, cellular differentiation, and evolution. They reported various novel modes of hereditary transmission, some involving visible particles and others not due to any visible cytoplasmic particle. Later, during the 1960s, they systematically analyzed a mechanism of hereditary transmission involving the inheritance of supramolecular patterns concerning the organization of semiautonomous organelles in the cell cortex (see Chapter 8). The investigations of Sonneborn and his collaborators on cytoplasmic inheritance had a great impact on genetic thought during the 1940s and 1950s.

In this chapter, research on cytoplasmic inheritance will be investigated with three primary aims in view. First, in order to understand how Sonneborn came to investigate cytoplasmic inheritance, I will follow the development of his research activity up to his first encounter with the theories, phenomena, and organisms associated with cytoplasmic inheritance. Second, the institutional strategies which led to the development and maintenance of the research program on cytoplasmic inheritance will be analyzed. Finally, I will investigate the development of research and theories of cytoplasmic inheritance, their rapport with concepts of onotogenetic development and organic evolution, and their reception by classical and physiological geneticists.

Learning to Dissent

> A cell of a complex organism contains a thousand different substances, arranged in a complex system. This great organized system was not discovered by chemical or physical methods; they are inadequate to its refinement and delicacy and complexity. (H. S. Jennings, 1931, p. 25)

> I became a Jennings protégé with all that implies as to his influence on me. (T. M. Sonneborn, 1978, unpublished autobiography, p. 12a)

Sonneborn was introduced to biology in 1922 at Johns Hopkins University in Baltimore, the city in which he was born, a little more than a decade after the Morgan school at Columbia, using *Drosophila,* had begun to establish the Mendelian-chromosome theory of heredity. Genetics was emerging as a vital part of the American biology curriculum, and part of Sonneborn's introduction to biology consisted of a study of that fly, its structure, behavior, and genetics. But as a graduate student of Herbert Spencer Jennings (1868–1947), Sonneborn's firsthand experimental study of heredity was unlike that of many of his contemporary geneticists.

Jennings graduated from Harvard in 1896 after studying under the direction of the naturalist E. L. Mark at the Zoological Laboratory of the Museum of Comparative Zoology (Sonneborn, 1948b, 1975). His introduction to experimental methods came primarily through his contact with the embryologist C. B. Davenport, then a Lamarckian and an anti-Weismannian, who later became a classical geneticist after the rise of Mendelism. After graduating from Harvard, Jennings began to study the response of the ciliate *Paramecium* to stimuli. He had become acquainted with this organism during his previous work for the U.S. Fish Commission's Biological Survey of the Great Lakes (1890–1893). During the first decade of Mendelism, Jennings pursued his classical study of protozoan behavior, directed a survey of the Great Lakes, and coauthored the well-known text *Anatomy of the Cat*. Like most biologists of his generation, Jennings also gave much attention to the fundamental problems of evolution, heredity, and development. With the rise of Mendelism, he turned to the statistical study of heredity and selection in *Paramecium*. His "pureline" studies were widely hailed as the American corroboration of Johannsen's results (Provine, 1971). By 1910, Jennings had risen to an authoritative position and succeeded the highly influential biologist W. K. Brooks (1848–1908) as professor of zoology and director of the Zoological Laboratory at Johns Hopkins, a post he held until his retirement in 1938.

Although Jennings's prominent position in American biology made him a valuable member of the original editorial board of the journal *Genetics*, he remained outside and independent of the restricted goals and research programs followed by Maize and *Drosophila* geneticists. Jennings's approach to heredity investigations was somewhat less pragmatic and more critical of the orthodoxy that was quickly solidifying in genetic research. First, he attempted to confront mechanisms of evolution directly and investigated selection, mutation, and the inheritance of acquired characteristics. His early genetic work contributed to the demise of the unit factor and representative particle interpretation of Mendelism. It focused on the interaction of genes and environment in the determination of the phenotype. From the end of World War I to the early 1930s, Jennings published only one laboratory investigation. Putting experimentation aside, he spent his time developing and publishing his views on evolutionary processes and directing the research of his many students.

Jennings's experimental and speculative investigations were not in character with those of an American geneticist, and many of his fundamental biological beliefs were incompatible with the tenets of contemporary genetics. Most striking and most important for the development of the investigations of his students was his contention that "life" could not be reduced solely to the physicochemical level. Jennings's work on the behavior of *Paramecium* led him to believe that the responses of the organism were a function of its gross structure, important aspects of which were the cell's asymmetry and correlated spiral movements (Sonneborn, 1975, p. 173). He defended the existence of a level of complexity above the physicochemical, based on a higher level of organization which possessed different properties and modes of functioning. Such a holistic, organicist view was gathering support in the 1920s from leading physiologists and embryologists, including J. S. Haldane and Joseph Needham (Haraway, 1976). As a distinguished American biologist supporting a belief in a level of organization above the physicochemical, Jennings was led into public confrontation with one of the most vigorous apologists of mechanistic reductionism, Jacques Loeb. Loeb and his student Garrey aggressively criticized Jennings, simplistically interpreting his holistic position as representing vitalism (Sonneborn, 1975, pp. 172–174).

Jennings opposed the geneticists' belief that spontaneous gene mutations together with Darwinian selection could sufficiently account for evolution. During the 1920s he came to support a growing alternative theory, that of emergent evolution (Jennings, 1927). Evolution, according to this view, represented a series of ascendant, integrated, and complex organisms, each having emerged from the less complex and less integrated. Integration would bestow upon the organism, by the nature of the interactions between different elements, a new character that could not be predicted from the study of lower levels.

Sonneborn became indoctrinated into the experimental study of heredity in the 1920s when, as discussed in Chapters 1 and 2, many biologists were doubtful of the future of Mendelism. During these early years he also became skeptical of the belief that the gene was the source of all heredity. However, his critical position did not rest solely on the theoretical influence of Jennings or on the speculations and experimentation of other biologists of the 1920s. As early as 1928, he brought forward surprising experimental evidence from his doctoral work to substantiate his claims (Sonneborn, 1930b, c). His doctoral work was part of a research program led by Jennings, based on investigations of the Lamarckian notion of hereditary environmental effects (Sonneborn, 1930b, p. 57).

Sonneborn's thesis work concerned hereditary variations of the microscopic flatworm *Stenostomum,* brought about through exposure to the well-known poison, lead acetate. "Double monsters" were created which reproduced "true to type" during asexual reproduction. Sonneborn reasoned that the nuclear genes in the cells of the "daughter" animals were identical with those of the parent, and that the structure of the parent cell somehow determined the formation of identical structures in its descendants. To Sonneborn the results were startling in two ways. First, they seemed to support the existence of self-perpetuating somatic parts and the notion of a supramolecular level of hereditary structure. There seemed to be more to heredity than the simple transmission of genes and chromosomes from parents to offspring. Second, the effect of lead acetate in producing the inherited variations seemed to him to indicate that certain environmental conditions could produce genetic changes by direct action on hereditary material. These early encounters with hereditary modifications had a lasting impression on Sonneborn's biological beliefs and future experimental work. According to his own testimony:

> It has also—and did even in my graduate student years—made me have a closer than distant relation with Lamarckism, with the inheritance of acquired characteristics. At my final examination for the Ph.D. degree, Jennings seized the opportunity to alert me to at least one of the dangers of such an association. He asked me who in the 20th century had claimed to obtain positive results on this, to describe and criticize their work, and to tell what became of them. The latter was the payoff: all had come to a bad end,—Tower had gone crazy, Kammerer committed suicide, etc. etc. At the end of my account, Jennings said "Let that be a lesson to you" or something to that effect. That was a sobering thought; but it had the effect only of putting me on guard, not diverting my attention from the possibility. (Sonneborn, 1978, unpublished autobiography, p. 19)

Sonneborn judged these influences to be two of the *Leitmotifs* of his investigative and speculative career, and led him to be what he called "a life long critic of what has seemed to be a blind and erroneous faith in the gene as the source of all heredity" (Sonneborn, 1978, unpublished autobiography, p. 19).

This is not to say that Sonneborn was uncritical of the experimental work in favor of the inheritance of acquired characteristics, however. In fact, as early as 1931, he published a critique of the Lamarckian experiments of the well-known psychologist William MacDougall, who claimed to have induced a progressive series of adaptive hereditary changes concerning behavior modification in rats (Sonneborn, 1931). Nonetheless, Sonneborn's openness to investigations of the inheritance of acquired characteristics remained secure, as the following passage from a letter he wrote to MacDougall reveals:

> I was, of course, much interested in your letter and particularly in your fine understanding of the spirit in which my criticisms of your work were made. My own experimental work has been devoted largely to an examination of the effects of certain environmental conditions on the hereditary constitution of Protozoa and lower Metazoa, so that I am deeply interested in all work on the genetic effects of any environmental condition. (Sonneborn to MacDougall, December 17, 1931)

Jennings (1937) also reported a case of nonparticulate, nongenic heredity based on what was considered to be a supramolecular level of structure in the shelled rhizopod *Difflugia*. The variation concerned the inheritance of the so-called "mouth" and "teeth" of the shell. *Difflugia* constructs its shell by cementing sand grains together with a cellular secretion. The mouth is merely a circular aperture in the shell, and the teeth are a circlet of small projections from the rim of the mouth. During vegetative reproduction, half of the cell mass extends through the mouth, and a new shell with mouth and teeth forms on it before the cell divides in two. Remarkably, the mouth and teeth of the new shell are formed in juxtaposition to those of the existing shell, each tooth of the new shell forming in the space between two of the teeth of the old shell. Jennings' detailed experiments included the demonstration that when some or all of the teeth and some of the adjacent shells were removed there were correlated effects on the shell of the daughter shell, which were inherited for a few generations. When interpreting "the changes in *Difflugia* teeth that persist for some generations," Jennings (1931, p. 336) concluded that "it is a formed product of the cytoplasm that is modified—namely the skeletal structure constituting the shell mouth and teeth."

Sonneborn's and Jennings's experimental evidence for nongenic heredity went unnoticed by geneticists. By the late 1920s the atmosphere and momentum of genetic research changed dramatically. A burst of activity in genic studies sparked primarily by two developments gave nuclear genes a new lease on life: first, the revolutionary success of H. J. Muller and collaborators in producing mutations by X-rays and ultraviolet radiation and the following intensive study of genic mutations; and second, the discovery of polytene chromosomes of the insect salivary gland in *Drosophila,* and the following-up of cytological maps and the localizing of each gene at a particular band. This period is also characterized by the development of mathematical genetics by Fisher, Wright, and Haldane, which was, however, too esoteric for most biologists.

The sudden break and extension of genic studies was not the only reason for the lack of attention given to the reports of Sonneborn and Jennings. Their work did not meet the required genetic standards for basing a claim of non-Mendelian heredity. Classical genetics was based on the study of sexually reproducing hybrids. The mechanism of Mendelian heredity had been well established by a combination of cross-breeding analysis supplemented by cytological observation. Within the accepted standards of contemporary genetics, a demonstration of nongenic inheritance first required proof by classical breeding experiments that the trait under study was not due to genes. Moreover, as Sonneborn recognized, "This meant that one also had to show of course that the organism inherited other marker traits in the standard genic way" (Sonneborn, undated autobiographical essay, p. 8).

Cross-breeding analysis had never been effectively achieved in microbes. In most cases, they seemed to reproduce solely vegetatively and showed little sign of sexuality. Their lack of cellular differentiation made it impossible to distinguish between somatic and germinal elements, between character and factor, between genotype and phenotype, development and heredity. In fact, protozoologists and bacteriologists had come to some consensus that the heredity of microbes had nothing in common with that of higher plants and animals.

After completing his Ph.D. in 1928, Sonneborn received the most coveted postdoctoral fellowship available in the United States, a fellowship of the National Research Council, which provided the necessary conditions for him to stay and work at Johns Hopkins for another two years. The absence of sexual reproduction in *Stenostomum* led him to abandon that organism and began a search for a genetically suitable unicellular organism. First, Sonneborn had turned to the ciliate *Colpidium,* an organism which he had used as food for *Stenostomum.* Again he was able to produce double monsters and demonstrate their inheritance (Sonneborn, 1932). But once again, genetic analysis was limited by failure to obtain sexual reproduction. Finally, in 1930, he was led to the ciliate *Paramecium.* Jennings received a sizeable grant from the Rockefeller Foundation for work on the genetics of *Paramecium* and offered Sonneborn a research assistantship in the new research program. The offer of a full-time research assistantship was welcome as it came during the Great Depression when university posts were scarce (Sonneborn, 1978, unpublished autobiography, p. 30).

This was not the first time Jennings had turned to investigate the hereditary properties of *Paramecium.* Between 1910 and 1916, he had tried to cross-breed members of different clones of *Paramecium* in order to test Mendelism. Unfortunately, technical limitations prevented him from establishing experimentally whether Mendelian laws or some different laws held for unicellular organisms. There was no lack of *Paramecium* differing in visible hereditary characters, but the different types refused to mate with one another. Each mated only with its own kind, even when both kinds were mated at the same time in the same culture vessel (Sonneborn, 1978, unpublished autobiography, p. 38).

Several practical and theoretical issues led Jennings to return to *Paramecium* as a genetic object. First, unicells reproduced faster and could be raised in greater quantities and hence offered a potentially lucrative tool for rapid progress in genetic analysis. Second, *Paramecium* was a well-known biological organism which owed its fame to a series of protozoological investigations carried out since the middle of the nineteenth century, and Jennings himself had long experience in studying its behavior. Moreover, as suggested by Sonneborn, perhaps one of the most important reasons for choosing *Paramecium* was that it conjugated. Conjugation was closely comparable to sexual reproduction in higher organisms. Sexual reproduction in turn represented a *sine qua non* for any genetic organism.

Jennings's return to the genetics of *Paramecium* was also motivated by the possibility of definitively demonstrating the inheritance of acquired characteristics. Since the nineteenth century pathologists and bacteriologists had been familiar with the acquirement of inherited immunity in parasitic protozoa and in pathogenic bacteria because of its practical importance. According to Sonneborn, the bacterial work with its common Lamarckian interpretation had a strong influence on Jennings, who was familiar with bacteriology, having taught a course on it in the late 1890s. However, the absence of sexuality in bacteria during the first four decades of the twentieth century was enough to discourage their use for genetic purposes.

Nonetheless, Jennings appreciated the evolutionary possibilities and probed the situation in protozoa, directing his students to do the same. Cases of the inheritance of acquired environmental effects in protozoa included degenerative changes induced by unfavorable conditions and the acquirement of immunity to certain injurious agents, such as high or low temperatures, or to injurious concentrations of chemicals. Jennings (1930, p. 86) wrote about the inheritance of acquired characteristics in protozoa:

> How are to be explained the features of these mutations, if that is what they are? How are to be explained the acquirement under the influence of particular agents, of new methods of metabolism, the acquirement of immunity, of increased resistance, that precisely fit the agent that induced them?

The flurry of investigations led by Victor Jollos encouraged Jennings's return to the genetics of *Paramecium* (Sonneborn, 1978, unpublished autobiography, pp. 34–35). As discussed in the last chapter, in a series of papers in the 1920s and 1930s, Jollos and his collaborators in Dahlem reported results whereby protozoa were modified in many ways, by the direct action of environmental conditions, and inherited the acquired changes for hundreds of generations in vegetative reproduction. But the changes frequently disappeared if the protozoa were allowed to reproduce by conjugation. Jollos assigned the inherited modifications to the cytoplasm instead of to the nuclear constituents, and he called the lasting modifications *Dauermodifikationen*. For Jennings (1940, p. 48) the results of the studies of protozoa were clear enough:

> The "doctrine of the inheritance of acquired characteristics" finds its last refuge in the genetics of Protozoa. It is a fact that Protozoa are modified in many ways by the action of environmental conditions, and it is known that the modified characteristics so induced are inherited for long periods in vegetative reproduction—for hundreds of generations.

However, as Jennings and others recognized, the inheritance of acquired characteristics could find little credence among geneticists unless it could be demonstrated by the procedures established by geneticists, procedures which had not been sufficiently worked out in microorganisms. Jollos left the study of protozoa and turned to *Drosophila* to continue his experiments on the problem of "directed mutations." Jennings, however, continued to explore *Paramecium*. His primary aim was to establish whether the environmental effects in protozoa were merely long-lasting modifications of the cytoplasm which would eventually fade away, or permanent heritable modifications which could be transmitted by the nucleus as well. Jennings (1940, p. 54) outlined his research strategy as follows:

> What is required is to induce environmental modifications—acclimatization or the like—in a certain race, then to cross this race with another which lacks the modification. In the conjugation of the two races only nuclei with their chromosomes are transferred from one race to the other. If the modifications have affected the nuclei they should be transferred by conjugation from one race to the other. But if they affect only the cytoplasm they will not be thus transferred.

According to Sonneborn (undated autobiographical essay, p. 7), after two years of work Jennings became discouraged with his investigations of heredity in *Paramecium* and abandoned them, complaining that *Paramecium* was not in condition for genetic work. Conjugation ocurred only within strains; the necessary crosses between different strains could not be made. For Jennings, nearing retirement, the construction of a genetic organism out of *Paramecium* was a risky enterprise and perhaps at best a long-term task. He returned to other problems using the rhizopod *Difflugia*.

Young Sonneborn, on the other hand, was relentless. He was not the only assistant Jennings had on the project, however. There was also Jennings's "secretary-assistant" Ruth Lynch and, at least at the beginning, a graduate student, Daniel Raffel. A brief glance at the relationship among Raffel, Sonneborn, and Jennings, as recalled by Sonneborn, provides a glimpse into the social and intellectual context within which the research took place. Sonneborn (1978, unpublished autobiography, pp. 31–33) describes Raffel as "a very bright and forceful man" who attempted to interpret all biological diversity as due to genic mutations. According to Sonneborn, Jennings and Raffel had a "knock-down conflict" over priority and interpretation of phenomena. Raffel subsequently left Johns Hopkins University and went to Russia to work with H. J. Muller on a genic problem called "position effects." Raffel never obtained a university position and lived out his life as a teacher in a private school in Baltimore and as a cattle breeder. Sonneborn, however, stayed on to study *Paramecium* and Jennings:

> As his research assistant, I of course felt obliged to do what he wished done and in the way he wished it to be done. In fact, . . . I made a point of observing closely how he went about scientific work. I kept a notebook just on that. He was the only famous scientist I knew well and I wanted to find out the basis of his fame, what qualities and approaches and methods he brought to his work. I learned a great deal, just keeping my eyes and ears open, not by questioning him. (Sonneborn, 1978, unpublished autobiography, p. 33)

After an initial two summers of work at Woods Hole, Jennings gave Sonneborn almost complete freedom to carry out his research. Progress on the control of mating was slow, but Sonneborn's persistence was repeatedly reinforced by instances of partial success and by many publishable results obtained along the way (Sonneborn, 1933, 1936). Finally, after five years, his efforts paid off in full. In 1937 Sonneborn put together the formula which allowed him to construct sex-like "mating types" at will. Different genetic strains of *Paramecium* could be mated as desired when they were brought under the proper conditions of temperature, nutrition, and light.

The revolutionary achievement resulting in the control of mating in *Paramecium* led, for the first time, to routine cross-breeding analysis in a unicellular organism in the United States. As Sonneborn (1937, p. 385) proclaimed in the conclusion of his first report of his findings:

> It may perhaps be said that with the present work, the genetics of *Paramecium* enters the quantitative and predictable stage, with tools and methods of analysis which should lead rapidly into a systematic coherent body of knowledge in close touch with the rest of genetic science.

Sonneborn's work caused a great stir in the biological community of the United States, and very quickly both he and *Paramecium* became well known in genetic circles. Sonneborn gave a demonstration of the mating type work at Woods Hole in the summer of 1937. The leaders of genetics and protozoology attended, including Morgan and G. N. Calkins. The *Baltimore Sun* ran a full-page feature on Sonneborn and his work (Sonneborn, 1978, unpublished autobiography, pp. 52–58). Sonneborn's work represented a harbinger of the great changes in genetic investigations which would come to pass during World War II when similar conditions controlling sexuality were found to be operative in other unicells.

After learning to control mating in *Paramecium,* Sonneborn carried out standard Mendelian analysis on the microorganism. *Paramecium* did contain nuclear genes that exhibited typical Mendelian behavior, but this was not the whole of the matter. Sonneborn quickly encountered various strange and novel phenomena which challenged genetic orthodoxy. The first bizarre case concerned the inheritance of sexuality itself, which defied the established "laws" governing heredity (Sonneborn, 1938, 1939). The inheritance of mating types involved some complicated intermingled relationship between nucleus, cytoplasm, and environment of an unprecedented character which could not be reconciled with the contemporary body of genetic knowledge. In fact, the inheritance of mating type would baffle researchers for many years to come (see Nanney, 1954, 1057).

With a new organism to exploit, Sonneborn was open to several avenues of investigation. By the early 1940s, he maintained three lines of investigation. One was a study of the inheritance of antigenic properties, especially in relation to "the inheritance of environmental effects" and a "test of Jollos' ideas on *Dauermodifikationen.*" The second was a study of evolutionary processes in *Paramecium aurelia*. And the third was an analysis of a "new system of determination and inheritance of characters" that he was just beginning to work out. As Sonneborn viewed it in 1943, the new system of hereditary determination was the most interesting and important, since it concerned various genetic problems. It seemed to hold the key to an understanding of "so-called cytoplasmic inheritance," "the inheritance of environmental effects such as shown in bacterial adaptation and the inheritance of antigenic properties of *Pneumococcus,*" and the problem of somatic cell differentiation during development (Sonneborn to Dunn, November 15, 1943).

The study of the abnormal case of inheritance launched both Sonneborn and the cytoplasm to the center of gentic controversy. The new system of determination concerned some strains of *Paramecium* which produced a poison called paramecin that killed *Paramecium* of certain other strains. The "killer" trait, first reported by Sonneborn in 1943, was reasoned to be due to a cytoplasmic genetic substance he called *Kappa*. Sonneborn and his collaborators Ruth Dippell and John Preer soon concluded that *Kappa* could mutate like a gene (i.e., mutant *Kappa*s controlled production of different kinds of paramecin that killed sensitive animals in different ways) and depended for its persistence on a dominant nuclear gene "K" (Preer, 1946, 1948a,b, 1950; Dippell, 1948, 1950). Stranger still was the relation of *Kappa* to the effects of environmental conditions. The growth rate of *Paramecium* could be controlled by various environmental factors such as nutrition and heat. If the cells were kept in a medium where fission was rapid, they tended to multiply more rapidly than the cytoplasmic factor *Kappa,* and finally the large majority of cells ceased to be killers. In such a way, then, they demonstrated that the concentration of these factors within the cells of a clone, and thus the character of the cell (killer, sensitive, resistant), could be controlled by environmental conditions.

Sonneborn's first announcement, in 1943, of the presence of a semiautonomous cytoplasmic substance in *Paramecium* at first seemed to indicate the beginning of a revolution in the domain of genetics, especially after World War II, and was the subject of heated discussions by many investigators in various domains for more than a decade.

Plasmagene Theory

> Sonneborn's recent work on cytoplasmic inheritance in these forms has been characterized by geneticists as the most exciting and profitable investigation of the past year in genetics. (F. B. Hansen, "Resolved," April 4, 1945)

Genetics was breaking away from the program of classical genetics, and geneticists, following World War II, were rapidly turning to the study of genic action and genic control using microorganisms. A rigorous program of biochemical genetics of microorganisms began to emerge at the Biology Division of the California Institute of Technology, which Morgan had established in 1928. During the late 1930s one of Morgan's students, Carl Lindegren, worked out the genetics of the bread mold *Neurospora* and began to map chromosomes in a typical Mendelian fashion. Shortly thereafter, George Beadle and Edward Tatum (1941) began their celebrated work on *Neurospora,* developing a technique for biochemical genetics that soon became standardized for work on other microorganisms such as yeast, bacteria, and algae (see Olby, 1974).

The pattern of research characteristic of biochemical genetics of microorganisms relied on the cooperation of geneticist and biochemist. The geneticist isolated mutants that were found to be unable to grow or that grew poorly on a well-defined medium, and the biochemist sought the reasons for this inability. It was hoped that this double study would lead to a biochemical description of genetic control. Beadle and his associates proposed that nuclear genes act by determining the specificity of a particular enzyme and thereby control in a primary way enzymatic synthesis and other chemical reactions in the organism. The program of biochemical genetics in *Neurospora* would soon lead to direct conflict with microbial genetic work on the cytoplasm.

In the meantime, during the early 1940s the well-investigated case of "killer" and its cytoplasmic factor *Kappa* in *Paramecium* was welcomed by many as a possible model for understanding nucleo-cytoplasmic relations and genetic regulation. Central to all discussions was the elusive problem of cellular differentiation—how cells which were generally assumed to possess the same nuclear genes could become phenotypically different during development. By the 1940s and 1950s, ample evidence had accumulated for the inheritance of differences among somatic cells, or what was beginning to be known as "cell heredity." Experiments with tissue cultures, it will be recalled, had shown that at least some of the differences among cells of one body persisted when the cells were taken out of the body and allowed to multiply in a test tube. Many geneticists began to recognize that the differences in cell lineages which occurred in the course of development of higher organisms were due not merely to the immediate reactions of essentially similar cellular protoplasms to different local environmental conditions. The fact that different cell types when taken from the same soma could maintain their diversity through countless generations in tissue culture indicated that "heredity" could not be restricted to a correlation with sexual reproduction. It was becoming recognized by geneticists that inheritance on the cellular level in higher organisms was comparable to inheritance in unicellular organisms. In this regard the work of Sonneborn and his collaborators on *Paramecium* had a remarkable influence on many leading geneticists, especially outside the United States.

At the same time, it was difficult to discuss the problem of cellular differentiation without discussing the relative "importance" of the nucleus and cytoplasm. As the British developmental geneticist C. H. Waddington (1940 p. 53) warned, any discussion of the relative roles of the nucleus and cytoplasm in development invited the question whether, in the final analysis, evolutionary importance could be attributed to cytoplasmic characters. The American geneticist Sewall Wright, (1941, p. 501), already celebrated for his contributions to population genetics, stressed the evolutionary stakes clearly:

> The usual and most probable view is that cellular differentiation is cytoplasmic and must therefore persist and be transmitted to daughter cells by cytoplasmic heredity. The chief objection is that it ascribes enormous importance in cell lineages to a process which is only rarely responsible for differences between germ cells, *at least within a species* (emphasis mine).

In 1939 and in later writings the British cytogeneticist and theoretician C. D.

Darlington upheld the view that self-duplicating cytoplasmic determinants or "plasmagenes" could be a basis of cellular differentiation and important for various other biological problems. While German *Plasmon* theorists who challenged the "nuclear monopoly" conceived of the cell in terms of a cooperation of its interacting parts, Darlington (1944) attempted to construct a political model of the cell in terms of a hierarchy of controlling genetic particles which also reflected his perceptions of the power relations in the field. He postulated the existence of three systems of "government" within the cell, based on differences in size, structure, and complexity of the genetic factors, each with different modes of control. The first system was that of the nuclear genes, whose "equilibrium" was "mechanical" and "predominates in the government of heredity as well as in the government of the cell."

His second system was "corpuscular," possessed a "physiological equilibrium," and could only be recognized in green plants having plastids. Based on the investigations of O. Renner in Germany and Y. Imai in Japan, Darlington claimed that the cytologically visible plastids represent genetic elements or "plastogenes" which were distributed unequally at cell division and unequally in sexual transmission—largely maternally. The final system constituted what he called the "undefined residue of heredity." It was not associated with any visible bodies in the cell and was supposed to be purely maternal in transmission. This cytoplasmic mode of inheritance was labeled the "molecular system," and he thought it required chemical rather than mechanical or physiological equilibrium for its continuance. Sonneborn's case, Darlington speculated, fitted into this last category—the molecular system. He suggested that it represented a *protein* which could be made outside the nucleus and be transmitted through the egg; it was what he called a "plasmagene."

During the 1940s Sewall Wright became more and more concerned with cell heredity and physiological genetics. His theorizing on the problem of cell heredity and physiological genetics was highly considered by cytoplasmic geneticists throughout the 1940s and 1950s. Wright (1941, p. 497) claimed that there could be no doubt that there were qualities of the cytoplasm that were autonomous. He suspected they were due to self-duplicating nucleoproteins which were autonomous with respect to basic structure. Like Darlington, he called them "plasmagenes."

In direct conflict with East, Dunn, Beadle, Sturtevant, and others who claimed that the work of *Plasmon* investigators could be accounted for without reference to cytoplasmic genetic elements, Wright (1941, p. 501) claimed that unless it is demonstrated that there is ultimate replacement of cytoplasmic substances by substances of nuclear origin, "it is superfluous to trace them all to nuclear genes." At the same time Wright recognized the scarcity of the cases of cytoplasmic heredity reported and sought a theoretical model of nucleo-cytoplasmic relations that would account for the apparent dilemma. One possibility of reconciling the results was that plasmagenes in the cytoplasm of the egg were transmitted through the germ line in an inactive form and therefore normally escaped genetic detection by cross-breeding individuals of a species. During the course of development, however, plasmagenes (cytoplasmic proteins) would be activated by prosthetic groups emanating from the nucleus. As Wright (1941, p. 503) further explained:

> Under this viewpoint, differences in local conditions may bring about differential accommodation of metabolic products arising from the interaction of cytoplasm with nuclear products and environment, and eventually lead to the elaboration of new plasmagenes in the cytoplasm in particular regions of the organism. In fact, in eggs of the less regulatory sort, irrevocable differentiation occurs in certain regions of the cytoplasm before cleavage and so cannot be a consequence of nuclear differentiation of any sort. As development proceeds, each step in the regional differentiation of cytoplasmic heredity increases the diversity of local environments and so facilitates further differentiation.

While this model could account for cytoplasmic characters which could be perpetuated across the germ line for many generations, there were also other cases such as *Dauermodifikationen* in which the cytoplasmic characters faded away. To account for these instances, Wright (1945, p. 297) proposed that nucleoproteins could be produced as wholes by "special genes" and that they retained their genic property in the cytoplasm, but were subject to decay, at least along the germ line.

Finally, Wright suggested a final model based on self-perpetuating metabolic patterns which he thought could function side by side with plasmagenes to account for cellular differentiation. The cell as a whole could be considered as a "single gene":

> Persistence may be based on interactions among constituents which make the cell in each of its states of differentiation a self-regulatory system as a whole, in a sense, a single gene, at a higher level of integration than the chromosomal genes. (Wright, 1945, p. 198)

The notion of the cell as a self-regulatory system as a whole, and that of stabilization of a particular mode of differentiation by means of self-duplicating proteins (plasmagenes) within the cell, were not supposed to be mutually exclusive. Both mechanisms could operate simultaneously in development (Wright, 1945, p. 300). In fact, Wright claimed that Sonneborn's experiments on *Kappa* particles in *Paramecium* represented a remarkable demonstration of the autonomy of cytoplasmic particles, in the sense that, while dependent on specific genes for persistence and mutiplication, they could not be produced *de novo* by Mendelian genes. "These results, wherever they may lead," Wright (1941, p. 300) claimed, "are obviously of the greatest significance for the fundamental questions of physiological genetics."

All at once a number of investigators began to react to Sonneborn's reports and find similar phenomena in other organisms. As early as 1945, Sonneborn was asked to chair a conference on extrachromosomal inheritance sponsored by the American Cancer Society. A list of the potential topics and participants at the conference gives an idea of the immediate impact of his work and provides testimony to the extent to which cytoplasmic inheritance attracted attention at this time. At Stanford University, Norman Horowitz believed he had a case of cytoplasmic inheritance in *Neurospora*. Even in the 'old' classical genetic organisms, cases of cytoplasmic inheritance came to the fore. Even one of the members of the original *Drosophila* group at Columbia got into the action. A. H. Sturtevant, who had been so critical of the evidence for cytoplasmic inheritance during the 1920s and 1930s, now thought he might have a case of cytoplasmic inheritance in *Drosophila:* Max Delbrück at Vanderbilt University was investigating the relationship between viruses and extranuclear factors in *Drosophila:* Marcus Rhoades began to pay closer attention to his own cases in maize and summarize other evidence for cytoplasmic inheritance involving plastids in other higher plants (Rhoades, 1946).

Beginning in 1945, Carl and Gertrude Lindegren, as well as Sol Spiegelman (based on independent studies), all then at Washington University in St. Louis, reported a series of results in yeast that also startled geneticists. They reported that an enzyme, melibiase, could arise and be perpetuated in the cytoplasm as long as the substrate (the sugar melibose) on it which it acted was present. The proposed genetic particle was reported to be able to maintain itself in the cytoplasm, even in the absence of the nuclear gene necessary for its initial formation. Similar phenomena were later reported to occur also in bacteria (see Monod, 1947). Although the details were variable and somewhat conflicting, one major radical generalization emerged. As Goldschmidt (1958, p. 204) phrased it:

> A microorganism, unable to metabolize an unusual substrate, learns to do so by producing the necessary adaptive enzyme, if kept for a more or less long time in the new substrate.

Many geneticists during the 1940s and 1950s believed that these studies, like those of Sonneborn, would shed light on the problem of *Dauermodifikationen*. In fact, Carl Lindegren's continued work on "adaptive enzymes" led him to dissent from the doctrines of the Morgan school in which he was trained throughout the late 1940s, 1950s, and 1960s. Lindegren emerged as a staunch defender of the importance of cytoplasmic inheritance and the inheritance of acquired characteristics. By the 1960s, however, he was dismissed to the periphery of American genetics where he criticized what he called the anti-intellectual, atheoretic, and doctrinaire climate of American biology (see Lindegren, 1966).

Lindegren began to theorize on new conceptions of heredity very early in his research on yeast. In 1946 he ambitiously formulated what he called a "new gene theory." To him, the fact that chromosomes were called, metaphorically, "bearers" of heredity and that the term "gene" was considered to be synonymous with "locus" was actually prophetic of the "real" nature of the hereditary mechanism (Lindegren, 1946, p. 68). Lindegren claimed that his experiments showed that the locus on the chromosomes, which he called the "chromogene," was simply a locus for the active element, the "cytogene," which he believed was capable of self-duplication in the cytoplasm independently of the "chromogene." This duality of the gene, he surmised, was corroborated by the work of Sonneborn on the *Kappa* substance in *Paramecium*. "Plasmagenes" or viruses, in Lindegren's view, were simply modified cytogenes which could be transmitted without recourse to a chromosomal locus. Although he later rejected the idea, Lindegren's first view was that the chromosomes depend upon the cytoplasm for the cytogenes, which would "run back" to the chromogenes from the cytoplasm after cytoplasmic synthesis, in order to "get a ride" through cell division (Lindegren and Lindegren, 1946, p. 125).

Spiegelman's continued work on what he called "the gene-enzyme relationship problem," involving the phenomenon of "adaptive fermentation" in yeast, led him to propose in 1946, and in later writings, "the plasmagene theory of gene action." In his view, genes continually produced at different rates partial replicas of themselves which enter the cytoplasm. The replicas were believed to be nucleoprotein in nature. They were supposed to possess, to varying degrees, the capacity for self-duplication and to control the type and amounts of proteins and enzymes synthesized in the cytoplasm. The results of competition between these cytoplasmic self-duplicating units would change under various conditions, and thus determine the enzymatic makeup of the cytoplasm.

The unique feature of Spiegelman's theory was the combined mechanism for nuclear genic action and cellular differentiation: "While supplying a link between gene and enzyme it at the same time predicts that cells with identical genomes need not possess identical enzymatic constitutions" (Spiegelman and Kamen, 1946, p. 583). Spiegelman saw a difference between his conception of "plasmagene" and that of others. Wright and Darlington had proposed the term to account for what they considered to be "a whole host of apparently unconnected phenomena which could not be encompassed within the usual classical Mendelian concept of what a gene is, and what it does, and how it does it" (Spiegelman to Sonneborn, September 22, 1947). Spiegelman (September 22, 1947) described his position to Sonneborn as follows:

> I believe that I have proposed, not a plasmagene theory but a theory of gene action which involves plasmagenes. . . . The critical difference between the plasmagene concept which I have used from others that I have seen, is that in the theory of gene action which I proposed at Cold Spring Harbor, the plasmagene is not a special, or unique, or isolated, cytoplasmic component, in the sense that it is outside the normal physiological processes. On the contrary, it is assumed to be an integral part of the enzyme-synthesizing system and is presumed to be the *normal* link by means of which genes can effect control over protein formation in the cytoplasm.

In 1947 Peter Medawar, who would later (1960) share a Nobel Prize for his work in immunology, also addressed the theoretical importance of the cytoplasm in cell heredity. The next year, he and R. E. Billingham reported a case of pigmentation in spotted guinea pigs which they claimed depended on a semiautonomous cytoplasmic genetic particle and represented a manifestation of cytoplasmic inheritance at the cellular level. Nowhere, however, was the impact of Sonneborn's work felt more than in France.

In November 1943, the leading physiological geneticist in France, Boris Ephrussi, wrote to Sonneborn stating that his work on *Kappa* might very well represent "the most important development in genetics made after the establishment of the fundamental principles." Shortly after the war, under Ephrussi's institutional leadership, French investigators systematically explored and brought forward diverse evidence for cytoplasmic genetic entities. Philippe L'Héritier continued his study of cytoplasmic inheritance of sensitivity to CO_2 in *Drosophila*. André Lwoff related his observations of the protozoan kinetosome (the granule that lies at the base of each cilium and flagellum) to Sonneborn's work. Ephrussi and his collaborators began to work systematically on a case of respiratory deficiency in yeast which he claimed to have a cytoplasmic basis, perhaps in mitochondria (see Chapter 5).

Assuredly, Sonneborn's work on *Paramecium* did bring the notion of cytoplasmic inheritance to a prominent position in genetic discourse. The embryologists' idea that the cytoplasm of the egg was indeed heterogeneous and that different cell lines take their origin in portions of cytoplasm of different compositions was clearly entering post-World War II genetic research. However, by 1946, what at first appeared to be a new quiet and orderly revolution in genetic thought in America, with cytoplasmic inheritance at the fore, began to fade away. Results that at first seemed to be examples of cytoplasmic inheritance turned out in the end to be explained in Mendelian terms. Mendelian genes were judged to play hitherto unexpected tricks that seemed to imitate cytoplasmic heredity. Sturtevant's case involving "maternal inheritance" of sex determination in *Drosophila* turned out to be "just plain pre-meiotic effects of an ordinary gene" (Sturtevant to Sonneborn, January 7, 1946). There seemed to be nothing in the setup that could be shown to act like Sonneborn's *Kappa*. In the *Neurospora* work, too, what at first seemed to be a case of cytoplasmic inheritance was given a nuclear basis. There were also problems with the work on yeast reported by the Lindegrens. The results seemed to be inconsistent (Lindegren, 1949).

There were even some apparent limitations with regard to the evidence for cytoplasmic genetic particles in *Paramecium*. One of Sonneborn's students, J. R. Preer (1948b), showed that *Kappa* could actually be seen under the microscope in the cytoplasm of appropriately stained killer *Paramecium,* and Sonneborn (1948a) reported that *Kappa* could be transmitted by infections under certain laboratory conditions. Similarly, the cytoplasmic genetic particle "*Sigma*" controlling CO_2 sensitivity was later shown by L'Héritier to be infectious by the technique of transplantation and injection. It soon became apparent to many biologists outside of Sonneborn's laboratory that *Kappa* were vulgar virus-like particles: a somewhat strange parasite of little significance for physiological genetics. The favorable attitude towards plasmagenes quickly declined in American genetics.

It would be false, however, to assume that discussions of cytoplasmic inheritance and its genetic investigations withdrew to a minor concern in American genetics. By the mid-1940s Sonneborn had left Johns Hopkins University and had set up a major laboratory for genetic investigations of unicellular organisms based on *Paramecium* at Indiana University. As one of the few key geneticists who worked with microorganisms, he and his laboratory at Indiana University attracted graduate students from all over the United States during the 1940s and 1950s (Nanney, 1983). Sonneborn quickly became not only one of the leading authorities on the genetics of microorganisms, but also a vigorous advocate of the biological importance of cytoplasmic inheritance.

"I Always Liked Unorthodoxy"

> For myself there was no particular logic in embarking upon this type of research, except that perhaps I was interested in all aspects of genetics, and the cytoplasm seemed to have been neglected. I always liked unorthodoxy! Sonneborn at that time was perhaps more strongly motivated by the feeling that there was more to genetics than the chromosome theory: at times he seemed to want to prove . . . that the cytoplasm contained some alternative system of genetic determinants, which in some organisms could be as important as, or even more important than, that of the nuclear genes. (G. H. Beale, letter to the author, July 17, 1981)

Together with his collaborators and many students, Sonneborn continued to lead genetic investigations of *Paramecium* away from the mainstream of nuclear, genic studies of physiological genetics. The systematic study of the cytoplasmic factor *Kappa* remained central to investigations at the Indiana laboratory, and Sonneborn continued to promote its possible value as a model mechanism of cellular differentiation in higher organisms. In addition to the case of *Kappa,* beginning in 1948, Sonneborn brought forward what he considered to be "an even more interesting and probably more general system of interactions between genes, cytoplasm and environment" (Sonneborn, 1948a, p. 157).

With the collaboration of Geoffrey Beale who had worked with Darlington at the John Innes Institute in London before the war and who later established a laboratory based on *Paramecium* genetics at the University of Edinburgh, Sonneborn and his co-workers carried out extensive investigations on a new cytoplasmic condition concerned with serological properties of *Paramecium* (Sonneborn, 1948a,c; Sonneborn and Beale, 1949; Beale, 1948). Here again, the inheritance of environmental effects was demonstrated. This time, the external environment, namely temperature, could induce heritable transformations concerning the antigenic type produced. However, unlike the situation in the control of the killer trait, the cytoplasmic property controlling antigenic characteristics was not ascribable to the intrusion of any cytologically visible cytoplasmic particle. The complicated roles of the genes, cytoplasm, and environment in the case of antigenic traits in *Paramecium* differed from that of the killer trait in another important aspect as well. It seemed that the antigenic "plasmagenes" could only be formed when proper nuclear genes were present, not otherwise. Interpretations of the physical basis for the inheritance of this property would again lead Sonneborn into a great deal of controversy throughout the 1950s.

Briefly, the situation may be described as follows: When a culture of *Paramecium* is repeatedly injected into rabbits, the blood of the rabbit produces antibodies which are specifically directed against the antigens of the injected strain. In the presence of the specific antibody, *Paramecia* become rapidly immobilized. Their cilia, which carry the antigen, can no longer beat in a coordinated manner. Sonneborn showed that if a series of independent cultures of *Paramecium* are compared in this way, they frequently exhibit different antigenic traits.

In this manner Sonneborn constructed a series of eight types of *Paramecium* which differed in the kind of antigen carried on their cilia. Each of the types bred true both vegetatively and sexually for many generations, and the antigenic traits were shown to be inherited through the cytoplasm. It was further shown that all of the types could be transformed to other hereditary types by growing them under different conditions of temperature.

If *Kappa* did not bear on the poorly understood *Dauermodifikationen* as Sonneborn had hoped, antigenic traits certainly did. The antigenic traits of *Paramecium* could be induced by environmental conditions and then be manifested temporarily and inherited through the cytoplasm. After a time under the original conditions, the induced trait disappears and the original trait reappears. Jollos, it will be recalled, assumed that the temporary cytoplasmic inheritance was terminated as a result of the ultimate "triumph" of nuclear genes over cytoplasmic components. Sonneborn (1948d) and Beale (1948b) adopted a similar hypothesis and considered the possibility that "plasmagenes" might be gene-initiated after all, as had been suggested by Wright (1941), Darlington (1944), and Spiegelman (1946) to account for non-Mendelian inheritance in other organisms.

In view of the overwhelming importance of genes in the control of hereditary traits and the possible parasitic nature of *Kappa,* Sonneborn reasoned in 1948 that such a system of determination and inheritance was likely to be more general and significant than the system involving the killer trait (Sonneborn, 1948a, pp. 157–160). To Sonneborn, discouraged by the possible ultimately parasitic origin of *Kappa,* the inheritance of serological properties was an encouraging discovery: here, finally, were cytoplasmic traits hardly ascribable to the intrusion of a parasite. In 1948 Sonneborn expressed his beliefs about *Kappa* and the antigenic work clearly in a letter to Ruth Lynch, a former student of Jennings who now worked as Jennings's assistant:

> As to Delbruck, he told me long ago that he believed paramecin and *Kappa* were the same thing and it is quite possible that he is right. There is, however, no evidence for this at present and there is much evidence against it. I have never considered the matter is settled and would still welcome a decisive test. So far as *Kappa* and the killer story is concerned, I consider it essentially finished—at least my part of it is finished. For the last six months I have been devoting myself to the study of antigenic traits, for the gene-plasmagene environment interactions we are discovering seem to me far more important and of far more general significance than anything we ever found out about *Kappa*. There is no possibility of confusing the plasmagenes controlling the antigens with any symbiotic organisms. Further, we now know that the plasmagenes for the antigens are formed directly under the influence of the genes. This alone makes the antigen analysis likely to be more general than the killer analysis in which the genes could not initiate the control of the cytoplasmic factor. (Sonneborn to Lynch, June 12, 1948)

Sonneborn's initial attempt to submit the results on antigenic properties to all-exclusive nuclear control soon faced conflicting experimental evidence. By 1950, he ruled out the idea of gene-initiated plasmagenes to account for this case. Instead he came to support the notion that the antigenic traits were due to plasmagenes that were autonomous in their reproduction, but exhibiting a specificity dictated by the nuclear genes. In this case, then, the cytoplasm and external conditions would call into action one of a series of genes. The predominance, under a given set of conditions, of one type of plasmagene to the exclusion of others could be explained by a sort of intracellular competition and selection. "The competition could be for sources of energy, for simpler substances used in the synthesis of the specific substances, or for position in the cell" (Sonneborn, 1950e, p. 30).

Finally, just when it seemed that Sonneborn and Beale had settled on the notion of gene-specified plasmagenes, the influential physicist Max Delbrück (Sonneborn and Beale, 1949) attacked their interpretation and attempted to replace it with an alternative mechanism based on self-perpetuating metabolic feedback systems, which did not recognize the existence of any cytoplasmic genetic particles. The two conflicting models of plasmagenes on the one hand, and alternative steady states on the other, were perpetuated throughout the 1950s in relation to serological properties in *Paramecium*. The same conflict gradually spread to embrace interpretations of other cases of cytoplasmic inheritance, including mating types in varieties of *Paramecium,* investigated in Sonneborn's laboratory most prominently by D. L. Nanney (see Chapter 7).

However, by 1950 Sonneborn constructed a notion of "plasmagenes" that was significantly different from the gene-initiated plasmagene formulated by Wright and others. In their size, organization, degree of autonomy from the nucleus, and hierarchical position within the cell, his plasmagenes were unlike any cytoplasmic entities previously considered by an American geneticist. Sonneborn did not limit the term to what he considered to be "a prejudice in favor of sub-microscopic units of heredity." It was meant to embrace the gross structures of plastids in plants, centrioles, mitochondria, kinetosomes, *Kappa, Sigma,* as well as the subunits possibly contained within them. The name "plasmagene," he argued (Sonneborn, 1950e, p. 16), emphasized "their common cytoplasmic localization and their common gene-like properties" of self-duplication and mutability.

In Sonneborn's words, plasmagenes were "self-duplicating, mutable, cytoplasmic particles . . . which depend on the nuclear genes for their maintenance or normal functioning, but not for their origin or for their specificity" (Sonneborn, 1950e, p. 31).

> These plasmagenes, in different cases, vary in size from microscopically visible particles down to submicroscopical particles probably of the same order of size as the genes. There are indications that at least some of the larger plasmagenes contain or consist of smaller genetic units. Plasmagenes are probably never independent of the genes, but this does not support the conclusion that they occupy in the cell hierarchy a position inferior to the genes. Genes and cytoplasm are mutually interdependent for their maintenance and normal functioning. Evolution may proceed by parallel but independent mutations of nuclear genes and plasmagenes.

To support his ideas, in 1950 Sonneborn brought forward about twenty cases of what he considered to be reports of cytoplasmic inheritance scattered throughout the genetic literature since the first decade of the century. The work of German geneticists, generally dismissed or ignored by American geneticists, was most important. Support by these investigations, those of Billingham and Medawar in England, and those of L'Héritier and Ephrussi in France, Sonneborn (1950e, p. 11) proclaimed:

> Cytoplasmic inheritance among both plants and animals, and in multicellular as well as unicellular organisms, is thus not an hypothesis, but a fact—one of the capital facts of biology.

Sonneborn attempted to integrate both cytoplasmic and nuclear genetic mechanisms, which, he came to believe, interacted and cooperated to determine and control the hereditary traits of the organism. His belief in the interactional character of nuclear and cytoplasmic heredity was supported by several cases, such as the production of chlorophyll in higher plants and the absence of respiratory enzymes in yeast, where it seemed that the same hereditary trait could appear to be controlled in one study by nuclear genes and in another study by the cytoplasm.

The situation did not seem unlike the paradox which prevailed in past discussions of heredity and environment. Observations by geneticists had shown that the same difference in phenotype could in one comparison result from a difference in genes and in another comparison from a difference in environment. In Sonneborn's view, the particular reaction norm, which denoted different responses under different conditions without change in the responding genetic materials, applied to both the nuclear and the cytoplasmic genetic materials of the cell. The reactions between cytoplasmic and nuclear genetic constituents would be reciprocal:

> The cytoplasmic genetic materials constitute or control part of the conditions to which the genes respond, and the nuclear genetic materials constitute or control part of the conditions to which the cytoplasmic genetic materials respond. The traits that develop are the result of interaction between the two components of the genetic system under the conditions in which they are operating. (Sonneborn, 1951a, pp. 200–201)

Sonneborn (1951b) proposed three primary functions for the cytoplasmic part of the genetic system of the cell as it concerned cellular differentiation. First, the differential distribution of self-duplicating cytoplasmic particles behaving like *Kappa* offered one mechanism of cellular differentiation in higher organisms. Second, the cytoplasm could control which of an alternative series of genes would come into phenotypic expression and be maintained in the course of cell multiplication. The antigen system in *Parmecium* thought to be controlled by interactions of plasmagenes and genes was held to be a model mechanism for this process in cellular differentiation.

On the other hand, neither of these two mechanisms, in Sonneborn's view, touched the "master problem of the control of the *pattern* of cellular changes in time and space during the course of development" (Sonneborn, 1951b, p. 308) As discussed in Chapters 1 and 3, many embryologists and *Plasmon* theorists postulated that the submicroscopic organization or structure of the egg cytoplasm was a hereditary property. It was held to be necessary as representing a "ground plan" for the symmetry and pattern of the multicellular organism. Sonneborn endorsed this view and went one step further, providing genetic reasons for the logical necessity of a particular self-perpetuating molecular pattern, as a sort of ground plan for each cell.

To Sonneborn, the rapid and efficient operation of enzyme systems, with many enzymes participating in a regular sequence, seemed to require a precision of localization in enzyme-bearing particles such as mitochondria. The surfaces of self-duplicating structures such as plastids and mitochondria and afibrous "ground substances would provide a hereditary surface pattern on which enzymes would be absorbed" (Sonneborn, 1951b, p. 310). The actual development of a genetic research program on supramolecular cytoplasmic organization, led by Sonneborn and several of his former students and associates, would not be realized until a decade later (see Chapter 7). But in 1951 Sonneborn stated the issue in the following figurative terms:

> If the nucleus were in complete and exclusive control of heredity, then it would have to be concluded that nuclei, isolated under conditions that permit their multiplication, would be capable of reconstituting cells of the kind from which they were taken. If this did not happen, then it would have to be concluded that the cell, including the cytoplasm, somehow serves as a necessary model for the formation of new cellular material in essentially the same sense as the genes are necessary models for the transformation of new genes. . . .
>
> Perhaps it will be objected that there are some self-duplicating cytoplasmic elements which the nucleus cannot make. Then suppose these too can be cultivated *in vitro*. Is anyone willing to believe that, if all such self-duplicating components of the cell were thrown together in a test tube in the proper proportions with adequate food for their multiplication, a *Chilomonas* cell or any cell at all would result? Although the whole picture is admittedly imaginary, it makes the nature of the problem sharp and clear. If cells cannot be reconstituted in the way suggested, then it seems to me we are forced to admit that the molecular and particulate arrangement of the cellular materials, their organization into a working system, is itself a part of the genetic system of the cell. (Sonneborn, 1951b, pp. 310–311)

Sonneborn perceived the study of the latter proposed cytoplasmic genetic mechanisms to be problems of the future, when the two major divisions of the cell nucleus and cytoplasm—which had been "torn asunder" in the genetic analysis of the past 50 years—would again be reunited in an "integrated, interactional conception" of the genetic and developmental systems of the cell. Only then could the "long-sought" fusion between genetics and embryology be achieved.

The Way to Power

> Sonneborn would seem to be one of the classical workers in the field, not of biochemistry, biophysics or biomathematics but in the straight line field of biological biology. (W. F. Loomis, diary, February 13, 1951)

Sonneborn became a skilled publicist for the biological importance of cytoplasmic heredity. He headed and participated in various symposia on "plasmagenes" in the United States and France and lectured and wrote extensively. In semipopular articles, such as "Partner of the Genes" in *Scientific American* (1950d), he publicized the experimental evidence that some factors that controlled heredity existed outside the nucleus of the cell. In other articles, such as "Beyond the Gene" in *American Scientist* (1949), the complicated, somewhat mysterious, undefined interrelationships among nucleus, cytoplasm, and environment were highlighted.

Their potential resolution and clarification, Sonneborn promised, would be most important not only for problems relating to development but for cancer and aging as well. Investigations of cytoplasmic inheritance now had an applied basis.

Cancer was a constant rubric under which Sonneborn promoted the importance of cytoplasmic inheritance (Sonneborn, 1947a) and obtained funding for his research. As was well-known, cancer was a cellular transformation in the sense that certain cells of the body became altered in their characteristics and reproduced true to type. Investigations into the causes of cellular transformations and of the intracellular mechanisms and physical basis involved in them could provide clues to important aspects of cancer.

Beginning with his successful introduction of unicellular organisms into the domain of genetics, Sonneborn quickly acquired the necessary credibility to be a serious defender of the cytoplasm and a constant challenge to the predominance of the nuclear gene. His continued genetic investigations were widely praised for their elegance, thoroughness, and novelty and earned him the most prestigious awards and positions available to a geneticist. Awards won by Sonneborn between 1945 and 1960 included the Newcombe Cleveland Research Prize of the American Association for the Advancement of Science (1946), an honorary D.Sc., from Johns Hopkins University (1957), and the Kimber Genetics Medal and Prize of the National Academy of Sciences (1959). Indeed, Sonneborn briskly climbed up the institutional hierarchy of American genetics. As early as 1946, at the young age of 41, he was elected to the prestigious National Academy of Sciences. His name appeared on the editorial boards of many leading biological journals ranging from the *Journal of Experimental Zoology* to the *Annual Review of Microbiology*. By the end of the 1940s he had firmly established himself at the center of American biology as president of the American Society of Naturalists and the Genetics Society of America.

His teacher, the aging, distinguished biologist Jennings, played a significant part in assuring that Sonneborn and his work became well recognized by funding agencies as well as by professional honorary organizations. On September 6, 1946, Ruth Lynch wrote Sonneborn concerning Jennings's role in Sonneborn's election to the National Academy of Sciences:

> Did I ever tell you he [Jennings] and Dr. Castle shook hands by mail, over their part in your election? They were so happy to have had a hand in it, and so delighted that you had been elected by such a large vote.

Jennings also informed Rockefeller Foundation officials as early as 1940 that "there was no one else in the same class with Sonneborn" (F. B. Hansen, diary, April 1–18, 1940).

The Rockefeller Foundation, more specifically the natural sciences division, played a leading role in fostering microbial genetics and the integration of biochemical and physical approaches into biology, especially genetics, during the 1930s and 1940s. The natural sciences division was directed by Warren Weaver, a former physicist, whose principal aim was the development of what he, in 1938, called "molecular biology." The officials at the Rockefeller Foundation developed and mastered the art of conducting a large program of relatively modest research grants for individuals and projects. Weaver's program aimed at developing fundamental problems in biology rather than solely developing research that had immediate applied ends (see Kohler, 1976). Its support of microbial and biochemical genetics was to prove to be one of Weaver's most successful initiatives. The Rockefeller Foundation played an instrumental role in fostering the Neurospora school led by Beadle at the California Institute of Technology and biochemical genetics led by Ephrussi in France (see Chapter 6). It was also influential in building up the biology department at Indiana University in the 1940s.

Sonneborn's strategic institutional position, after World War II, was also instrumental in promoting cytoplasmic inheritance and its investigation. In 1939, shortly after he worked out the methods for controlling mating in *Paramecium* and was beginning to subject the organism to some fundamental genetic analyses, Sonneborn decided to reject a job offer as associate professor at Johns Hopkins University. Instead he accepted a two-year appointment at the little-known midwestern state university in Indiana (Sonneborn, 1978, unpublished autobiography, pp. 69–70). Sonneborn's location at Indiana after World War II was a major asset in obtaining funding, recruiting and training researchers in *Paramecium* genetics, and pushing cytoplasmic heredity research towards the center of American genetic investigations.

Sonneborn (1978, unpublished autobiography, pp. 65–71) cited various reasons for leaving the prestigious Johns Hopkins University and the East Coast where genetics had flourished. They included racist, economic, and various other social issues. Johns Hopkins University had become greatly weakened during the Depression and many of its top people had left, some to midwestern universities such as the University of Chicago. Furthermore, the anti-Semitic environment at Johns Hopkins represented a major obstacle which threatened to restrict him from climbing the academic and administrative hierarchy. The president of Johns Hopkins, Isaiah Bowman, a famous geographer who had been important in fixing national boundaries after World War I, made the situation clear for Sonneborn. According to Sonneborn, Bowman frankly told him that he would never be made head of the department, since as a Jew he would be subjected to "irresistible pressures" to take Jews in his department, which according to Bowman would make the non-Jews leave. As Bowman saw it, such an appointment would "ruin the department." It was not just that Sonneborn could not be expected to be promoted in rank, however, but finding a position anywhere in the United States during the late 1930s seemed to present problems for him. F. B. Hansen, one of the officials at the Rockefeller Foundation's natural sciences division, viewed the situation in 1939 as follows:

> Jennings finds small sums from year to year to support Sonneborn who is rated highly as a geneticist but almost impossible to place in a permanent post because of strong Jewish traits. (Hansen, diary, July 1, 1939)

Sonneborn's decision to accept the offer at Indiana University followed the advice of several influential biologists such as Raymond Pearl at Johns Hopkins Medical School and Abraham Flexner at Princeton's Institute for Advanced Study, whose international institutional studies of the universities were well known

(Sonneborn, 1978, unpublished autobiography pp. 69–70). Under President Wells and Dean Fernandus Payne, Indiana University was making a serious effort to build up its biology department in the late 1930s and 1940s. New buildings were under construction and the university was being rebuilt both in physical plant and in staff, with ninety-one new appointments made to the science faculty. Payne, a former geneticist, toured the country seeking the best available scientists. In 1938 they secured the plant geneticist R. E. Cleland as head of the Department of Botany and in 1939 appointed five biologists in zoology between the ages of thirty and thirty-five years, Sonneborn being the oldest of the group.

The directors at Indiana University made their offer so attractive that Sonneborn could not refuse it. Sonneborn was immediately appointed associate professor of zoology. He was offered a good salary and a research assistant and had limited teaching responsibilities, with one semester plus the summer entirely free for research. As Hansen saw it from the perspective of the Rockefeller Foundation, "Sonneborn, who is politically somewhere to the left of center, has an idea that the future of education and research in this country lies with the State Universities rather than with the privately endowed institutions" (Hansen, diary, July 13, 1939).

However, the decision to remain at Indiana was not automatic. In the late 1940s, for example, Sonneborn received an offer of a laboratory at the Carnegie Institution of Washington at Cold Spring Harbor on Long Island, New York. Founded in 1904, the Cold Spring Harbor laboratories successfully competed with other private institutions and the state universities. During the 1940s, the Department of Genetics included various geneticists such as M. Demerec and Barbara McClintock. There was also a very close collaboration between the genetics group at Columbia and the geneticists at Cold Spring Harbor. Joint seminars were held alternately at Columbia and Cold Spring Harbor, and Columbia graduate students spent part of their time working at Cold Spring Harbor. The decision not to accept the offer at Cold Spring Harbor was not an easy one. Cold Spring Harbor offered ample opportunity to carry out research in an intense investigative environment without the burden of teaching and other responsibilities of a campus scientist. The issues to be confronted and Sonneborn's reactions are well documented in the following passage from a letter he wrote to Jennings (October 17, 1944):

> Cleland here has expressed himself very strongly concerning the "moral" issue he thinks is involved. He seems to think I have an obligation to teach and train students! . . . He seems to think one should "enter vigorously into the life of the campus and community" meaning, I think, committee work, church activities and other community affairs. I need hardly tell you how I feel on such matters.
>
> Metz's reaction was more pertinent. He is strongly against my going to CSH. He stressed the importance of training graduate students to expand one's field, the bad prospects for the future that he thinks endowed institutions have, the very narrowing influence of the CSH environment. In support of his general mistrust of the future of CSH, I have heard from several others as well that the present Carnegie Head (Bush) is in favor of diverting Carnegie support from the biological to the physical sciences.

Sonneborn stayed at Indiana University for the rest of his investigative career, building up a leading biological research department. An array of eminent figures in his field, which included H. J. Muller, Salvador Luria, Ralph Cleland, and Marcus Rhoades, joined the department. Indiana University provided favorable conditions for the development of *Paramecium* genetics and cytoplasmic inheritance investigations. Sonneborn trained over thirty-six Ph.D. students, and his research enjoyed ample funding. During the late 1940s when cytoplasmic genetic investigations were beginning to flourish, Sonneborn's research projects on "The Nature of Plasmagenes and Their Relations to Genes and Characteristics in *Paramecium*" were supported with grants from Indiana University, the Rockefeller Foundation, the Jane Coffine Childs Memorial Fund, and the U.S. Public Health Service.

The officials at the Natural Sciences Division of the Rockefeller Foundation played a leading role in both the funding and the organization of the genetics groups at Indiana. In 1940 they awarded $20,000 to Cleland and Sonneborn. They recognized Cleland to be an authority on the cytology and genetics of the genus *Oenothera* and Sonneborn an authority on the genetics of unicells. As early as 1939 when the natural sciences division of the Rockefeller Foundation began to consider funding the "field of environmentally induced modifications on protozoa," Sonneborn was a chief consultant giving appraisal and advice (see F. B. Hansen, diary, July 13–August 31, 1940). In 1940 he had outlined an ambitious research proposal which included:

> 1) Clarification of the normal genetic phenomena in the Protozoa, 2) a study of the role of the environment in the determination of genetic characters, 3) a study of the mechanism of gene action in the development of characters, 4) investigation of the nature of the gene. (F. B. Hansen, diary, January 19, 1940)

At the end of World War II, the Rockefeller Foundation officials played an instrumental part in fortifying the genetics group at Indiana by strongly promoting the addition of the prestigious geneticist, H. J. Muller. Muller was noted as a brilliant and intuitive geneticist. His work on inducing mutations by X-ray and ultraviolet radiation in the late 1920s had opened the doors to extensive study of genic mutations (see Carlson, 1981). Muller's speculations on the nature of the gene and the use of microorganisms as tools for its study were later judged by many as quite prophetic (see Judson, 1979, pp. 47–49; Ephrussi, 1953). The year following his appointment at Indiana, Muller was awarded a Nobel Prize. This award came as no surprise to the Rockefeller Foundation officials. In 1945 M. Demerec had informed Hansen at the Rockefeller Foundation that "when the history of genetics is written 100 or 500 years from now . . . only three names will survive, G. Mendel, T. H. Morgan and H. J. Muller" (Hansen, diary, January 9, 1945). Nonetheless, before his appointment at Indiana, Muller was having difficulty in securing a satisfactory position in American universities. His personality difficulties and his conflicts with Morgan and others were as well-known as his notoriety as a poor teacher.

Hansen was anxious to see a collaboration between Muller and Sonneborn at Indiana. On March 5, 1945, when Sonneborn decided not to leave for Cold Spring Harbor, Hansen quickly wrote to Muller:

> I heard indirectly the other day that Sonneborn has definitely decided to remain in Bloomington. I am very glad for this; since I know it will be mutually avantageous for you and Sonneborn to be in the same department.

Hansen also made his opinion known to both the dean and the president at Indiana University. On February 23, 1945, before Muller's appointment, Hansen wrote to him:

> I am sure I made it clear, both to Payne and Wells, that the Rockefeller Foundation would have no interest at all in helping Indiana over an initial period of years, probably about five, unless they were prepared to go on from there with a permanent appointment for you. So I doubt if you need worry about tenure until age 70 if you receive an offer from them.

The financial support from the Rockefeller Foundation was substantial indeed. In 1945 the foundation donated $95,000 to support Sonneborn, Cleland, and Muller for six years beginning July 1 when Muller arrived, with the understanding that the university would provide $69,500 during the same period. By 1946, when Sonneborn again considered leaving Indiana, this time for Pennsylvania, Hansen (diary, January 28, 1946) stated the socioeconomic issue clearly. Sonneborn was "tied" to Indiana by nontransferable Rockefeller Foundation grants at the level of about $13,500 a year and could not hope for research assistance of this magnitude at Pennsylvania. When the first Rockefeller Foundation grant awarded to the three geneticists expired in 1951, Sonneborn shared with Muller and Cleland an unusually large grant awarded to the university toward a program of research in genetics for the amount of $200,000 over five years.

Certainly, the addition of Muller as a most distinguished and honored geneticist with his Nobel Prize brought luster to Indiana University. However, the hoped-for collaboration between Muller and Sonneborn was not to be realized. At Indiana, Muller was constantly visiting Sonneborn's laboratory, which was just down the hall, asking what he was doing and offering advice. Sonneborn strove to keep Muller at a comfortable distance. Muller's fame as a classical geneticist rested on the nuclear genes as the sole or most important factors controlling heredity and evolution; Sonneborn's work, of course, was not in keeping with Muller's line of genetic inquiry. Muller did his best to "dissuade Sonneborn from the experimental course he was following." (Nanney, unpublished, 1982, p. 11)

Muller and Sonneborn were intellectually incompatible, and Muller did not develop the strong interaction with other geneticists at Indiana that Rockefeller officials had hoped for. Young biologists who would later become leading geneticists and who had been attracted to Indiana and had taken courses with Muller, such as J. D. Watson, D. L. Nanney, and many others, eventually centered their attention around the microbiologist Salvador Luria and Sonneborn, who held informal seminar meetings weekly at his home. There was an active interchange of ideas between Luria's and Sonneborn's laboratories.

In 1950, when Warren Weaver of the Rockefeller Foundation visited Indiana University to review the work of the genetics groups there, he found that Sonneborn was not only investigating novel modes of inheritance in *Paramecium*, but that he was developing a whole school of genetics and sending his followers out around the country. In his diary Weaver (November 17, 1950) wrote:

> This is absolutely first class work if Warren Weaver ever saw any; (and rather in contrast with Muller) there is also the most lively and inspiring teacher-student relation in the group.

Indeed, Weaver was so impressed by the work of Sonneborn that he had the intention to nominate Sonneborn for a Nobel Prize.

By the late 1940s and 1950s, Sonneborn's laboratory was well established, with two full-time research associates and many students and visiting researchers. Sonneborn himself was recognized as one of the few leading authorities on the genetics of microorganisms, certifying the research grant applications of such well-known biologists as Salvador Luria, Sol Spiegelman, J. D. Watson, and many others. On the other hand, outside his laboratory Sonneborn perceived his cytoplasmic genetic research to be reciprocated by overwhelming antagonism and criticism on the part of many leading American geneticists. In complete symmetry to the views of Dunn and others who saw the protests to the nuclear monopoly as an emotional issue, Sonneborn understood the criticisms launched against cytoplasmic inheritance as emotional ones.

From Plasmagenes to Human Serfdom

> You will doubtless find, as I have found, that there are strong emotional responses on the part of American geneticists to the term "plasmagene." (T. M. Sonneborn to D. F. Jones, July 5, 1950)

Edgar Altenburg (1946a,b), George Beadle (1948), H. J. Muller (1951), Jack Schultz (1950), and many others publicly attacked the new genetic evidence for cytoplasmic inheritance and the notions of plasmagenes, claiming that cytoplasmic entities were relatively unimportant compared to nuclear genes, or that they were due largely to the transmission of infectious agents. Sonneborn (1950f, p. 243) characterized the views of his critics in the following politico-economic terms: the cytoplasm was

> the moat that guards the hereditary estate from the inroads of the variable environment, the dual highway across which supplies from the outside are conveyed to the nucleus and products of cellular activity are transported to the outside world; the factory in which these cross-currents of materials interact to yield produce for home consumption and for export. . . . The cytoplasm is the protector of the genes, their purveyor, their workshop, and the display case in which the products of their activity are shown.

Certainly cytoplasmic inheritance continued to represent a challenge to genetic orthodoxy as established by classical genetics, inasmuch as it added another possible form of inheritance to be incorporated into cell theory and the genetics of evolutionary processes. But the idea of cytoplasmic inheritance in the 1940s and 1950s did not represent simply an addition to the already established Mendelian principles. It represented a direct challenge to the prestige and authority of nuclear geneticists, their claims for control over cell function and embryonic development, and the "synthetic theory" of evolution.

The cellular distinction between nucleus and cytoplasm had been manifested socially in the United States by the struggle between embryologists and geneticists for control over the physiological functions of the cell and macroevolutionary phenomena. As late as 1945, the geneticist Edgar Altenburg still felt it necessary to defend the primary role of the nuclear gene from the embryological view when he wrote:

> Some embryologists are of the opinion that the cytoplasm determines important traits and that the genes in the chromosomes are concerned with only minor varietal differences between individuals within a species. (Altenburg, 1945, p. 410)

The next year Altenburg publicly criticized Sonneborn's genetic evidence for cytoplasmic inheritance.

During the 1940s the theory of plasmagenes and the attempts to understand the principles of somatic cell inheritance stood in direct conflict with the predominant research programs and doctrines of biochemical genetics. The *Neurospora* school, led by George Beadle, emerged as a bastion of Mendelism. Within that school there seemed to be little, if any, room for cytoplasmic genetic particles as normal constituents of all cells. The control of the cell by the nucleus was a central doctrine of biochemical genetics as understood by G. W. Beadle and E. L. Tatum in their classical paper of 1941: "From the standpoint of physiological genetics," they said, "the development and functioning of an organism consists essentially of an integrated system of chemical reactions controlled in some manner by [nuclear] genes" (p. 499).

When genetics centered around investigations of gene action, the stakes were precise and the struggle between the cytoplasm and the nucleus concerned a struggle for control over protein specificity. Beadle was quick to seize upon the idea of genetic particles in the cytoplasm and attempted to dispel any threatening notions of plasmagenes with a degree of autonomy or frequency of occurrence comparable to that of nuclear genes. After reviewing the evidence for plasmagenes based on plastid inheritance in plants known to be associated with the essential physiological function of photosynthesis, cytoplasmic inheritance of *Kappa* in *Paramecium,* and CO_2 sensitivity in *Drosophila,* Beadle concluded:

> The examples cited above leave no doubt whatever about the existence of cytoplasmic units capable of self-duplication and having a limited degree of autonomy. In most of these cases the cytoplasmic factors concerned seem clearly to be directed in a specific way by nuclear genes. There is no convincing evidence in any single instance of the functioning of an essential gene being entirely taken over by a so-called plasmagene. On the contrary, the great bulk of available evidence bearing on the question clearly indicates that the primary control of protein specificity resides in nuclear genes. In view of the elaborate mechanisms of mitosis and meiosis, which have evolved and persisted throughout almost the whole of the plant and animal kingdom and which evidently have a great selective advantage, it would be most remarkable if the cytoplasm could compete as a carrier and transmitter of hereditary units in any except a few very special circumstances. (Beadle, 1948, pp. 232–233)

Despite Beadle's claim, there was no direct evidence that nuclear genes controlled protein specificity and all other important physiological processes of the cell beyond the fact that the genetic evidence obtained in *Neurospora* was compatible with that belief. The extensive investigations of biochemical mutants in *Neurospora* led by Beadle and his many followers showed that the loss of the ability to carry out various specific reactions was associated with the loss or alteration of single genes. The idea that genes controlled protein specificity was a necessary corollary of the central dogma of the Morgan school concerning the central control of heredity by the nucleus. Its legitimacy during the 1940s and 1950s was supported through the social power of Mendelian geneticists throughout the first half of the century.

At the purely theoretical level, the whole question of protein synthesis remained in darkness during the 1940s and 1950s, and the control of protein specificity by nuclear genes was questioned by many. The gene itself already had been attributed with the power to direct the formation of enzymes and was still the formal abstract unit whose physical nature had not been directly related to the interpretation of the experimental results. For example, there was no evidence that a gene made a difference in the structure of an enzyme in the sense that one could actually work out the amino acid composition of a protein and show that with a nuclear gene mutation it was changed. This is to say nothing about how genes could control the synthesis of even larger cellular structures.

Until nuclear control over the synthetic processes of the cell was definitely demonstrated, the apologists of the chromosome theory of inheritance could engage only in polemics. However, they did have an arsenal of evidence provided by classical genetics to dismiss the current conception that cytoplasmic components formed an essential part of the genetic contribution of all complex organisms. H. J. Muller summarized the major issues when he attacked the evidence in support of the importance of cytoplasmic inheritance in 1951. Although Muller recognized "chloroplastid" inheritance in plants, he denied the claim that cytoplasmic genes or gene complexes formed an essential part of the genetic constitution of animals. He selected a choice passage from the earliest edition of E. B. Wilson's authoritative text, *The Cell in Development and Inheritance,* as representative of what he considered to be the opinion of "the more progressive group of biologists." The quoted passage concluded with the following statement on the nucleo-cytoplasmic controversy:

> The nucleus cannot operate without a cytoplasmic field in which its peculiar powers may come into play; but this field is created and molded by itself. Both are necessary to *development;* the nucleus alone suffices for the *inheritance* of specific possibilities of development. (in Muller, 1951, p. 77)

In a passage that deserves to be quoted in full, Muller (1951, pp. 82–83) outlined the arguments against the genetic importance of cytoplasmic components clearly:

> In mitigation of the current conception that cytoplasmically located genes or gene-complexes form an essential part of the genetic constitution of animals, the following points should be noted: (1) the extreme rarity with which illustrations of such inheritance have been found in animal material, incontrast to the thousands of Mendelian differences found in them; (2) the dispensability of the cytoplasmically located particles in the cases studied and the absence of evidence of the existence of normal alternative forms of them; (3) the fact that, in these same cases, the agents have

> been proved to be able to pass as infections from one cell to another; and (4) the lack of a fundamental basis for distinguishing between these and cases of undoubtedly parasitic or symbiotic microorganisms or viruses of exogenous derivation. These are points which, taken together, would appear to argue for most or all of these agents in animals having at one time arrived as invaders; for their still constituting, in a sense, an adventitious part of the inheritance, and for their tenure usually being insecure, as compared with that of the native chromosomes. This conclusion is, moreover, reinforced by a consideration of their mode of distribution and aggregation, since it is not only rather precarious but apparently far less suitable than that of the chromosomal genes for the simultaneous retention and the accumulation of numerous different types within the same germ plasm.

Sonneborn would not let the public criticisms of cytoplasmic inheritance go unanswered, and throughout the 1950s he addressed the issue raised by geneticists who attempted to diminish their importance. First Sonneborn clearly recognized that very few cases of cytoplasmic inheritance had been reported by geneticists, especially in animals. The detection of nuclear gene mutations, he argued (Sonneborn, 1950e, pp. 22–23), relied on the phenomenon of diploidy, which had no counterpart in the cytoplasm. Moreover, if plasmagenes controlled fundamental cellular processes that could not be altered without disastrous consequences to the organism, they would be seldom detected. As will be more fully discussed in the next chapter, the argument that technical obstacles to studying extranuclear hereditary factors exaggerated the seemingly dominant role of the nucleus would play a crucial role for those who defended the importance of the cytoplasm. However, even without employing genetic methods, it seemed obvious to Sonneborn that cytologically visible cell structures such as kinetosomes, mitochondria, and centrioles formed much of the essential equipment of the cell.

The criticism of Beadle, Muller, and other geneticists who claimed that cytoplasmic genetic elements lacked a mechanism to transmit themselves safely from one generation to the next also did not hold up, in Sonneborn's opinion. Even *Kappa* particles in *Paramecium,* which were randomly distributed from one cell generation to the next, possessed a mechanism that would assure their continuity. Preer (1948b) showed that when the cellular concentration of *Kappa* dropped, its rate of reproduction rose; when the cellular concentration rose, its rate of reproduction fell. Still other extranuclear structures, such as kinetosomes, Sonneborn (1950e, pp. 20–21) pointed out, remained fixed on the cell cortex of ciliates. There they duplicated, and at fission every animal received its full allotment of kinetosomes. He claimed that other cytoplasmic structures, such as chloroplasts, mitochondria, and centrioles, were as precisely distributed as were chromosomes by mitosis. As already implied, of course, the reproduction of some plasmagenes was not necessarily synchronized with nuclear and cell division. The rate of reproduction of *Kappa* and chloroplasts, for example, could be controlled directly by environmental conditions, such as nutrition, light, and heat.

Finally, there was the common criticism that cytoplasmically located genes or gene-complexes in animals were parasites of extrinsic origin and that they did not, in the words of Muller, "form an essential part of the genetic constitution of animals." Certainly, Sonneborn was willing to admit the possible origin of *Kappa* and *Sigma* as symbionts by 1950. However, he was unwilling to accept the charge that they and other plasmagenes were merely infectious parasites of little genetic significance. Sonneborn, Ruth Dippell, and Preer had shown that *Kappa* was well integrated into the physiology of some strains of *Paramecium*. Dippell had shown that *Kappa* particles, like genes, possessed the ability to mutate, and diverse mutant *Kappa*s could be maintained and multiplied in cells. They clearly possessed the properties of the building blocks required by the process of evolution which Muller had reserved solely for nuclear genes in animals. *Kappa,* like the cytoplasmic particle *Sigma,* was infectious only under laboratory conditions which could scarcely be imitated in nature.

The belief in the parasitic nature of *Kappa* and other hereditary cytoplasmic inclusions was in direct conflict with Sonneborn's belief in the importance and wide incidence of cytoplasmic heredity and his view that the environment could direct hereditary changes. As a particulate mechanism of cytoplasmic heredity, *Kappa* proved to be a useful model in explaining the behavior of chloroplasts in cell heredity and that of other possible cytoplasmic genetic components such as mitochondria, found in all cells. Moreover, as one of the most thoroughly investigated cases of cytoplasmic inheritance, *Kappa* was important in providing a possible mechanism for *Dauermodifikationen* and for cellular differentiation in metazoa. To say that it was a symbiont was much more than a semantic argument. In Sonneborn's view, these statements were designed to deny the genetic importance of the particles and confine them to pathology.

The parasitic nature of *Kappa* and other cytoplasmic inclusions was not only a threat to Sonneborn's belief in cytoplasmic heredity. It also represented a serious assault on the survival of *Paramecium* itself as a useful genetic organism. For the study of the role of the cytoplasm in heredity, *Paramecium* was an exceptionally suitable organismic tool. At conjugation nuclei are exchanged between cells, but the two cytoplasms do not normally mix. As a result, if *Paramecia* differing in both cytoplasms and nuclei conjugate, their progeny will possess identical nuclei but diverse cytoplasms. It was possible, then, to obtain at will various combinations of different nuclei and cytoplasms, and study the role of both. The mutual fertilization between two *Paramecia* made the organism a remarkably suitable tool for investigations of cytoplasmic inheritance.

However, *Paramecium* presented problems for the study of the biochemical processes by which genes affect the phenotype. It will be recalled that *Paramecium* had been selected as a genetic tool in reference to the research program of classical genetics. That is, it was selected because it showed signs of sexual reproduction—conjugation—and if mating could be controlled it satisfied the primary condition of a classical genetic organism, the ability to be cross-bred. In nature *Paramecium* feed on bacteria, and they are cultured in the laboratory on bacterial media (Nanney, 1981). On the other hand, the study of the genetic control of cell physiology after World War II was based on the analysis of biochemical markers which required a defined and preferably synthetic culture medium. Because the media upon which *Parmecium* was grown could not be easily defined, biochemical studies were compromised.

In an attempt to remedy this situation, Sonneborn launched a serious attempt to adapt his organism to the new milieu of physiological genetics (Nanney, 1982, unpublished, pp. 27–30). In the late 1940s he sought out a ciliate nutritionalist to construct for *Paramecium* a minimal medium which would enable one to collect biochemical markers. Auxotrophic mutants were an instrumental part of the biochemical technology exploited and publicized by George Beadle in *Neurospora*. At the advice of E. L. Tatum, who together with Joshua Lederberg pioneered the technology of auxotrophic dissection effectively on *Escherichia coli,* Sonneborn hired a biochemist, Willem van Wagtendunk. He brought van Wagtendunk to Bloomington, set him up in a laboratory, and arranged a tenure-track appointment in the Zoology Department with limited teaching responsibilities for him with the intention of obtaining a quick definition of the nutritional requirements of *Paramecium aurelia*. Unfortunately, von Wagtendunk's efforts to develop a defined medium for *Paramecium* had limited success. To Sonneborn's frustration, it seemed that no combination of simple metabolites would permit *Paramecium* to grow as required. Eventually, and even when the most completely defined medium was concocted, growth was slow and limited. *Paramecium* could not integrate itself well into the new thrust of postwar genetics.

Sonneborn had devoted his research career since the early 1930s to the importance of *Paramecium* as a profitable genetic tool. He had established and standardized methods to be employed in the study of the genetics and general biology of *Paramecium*. He promoted the study of protozoa in general biology and zoology courses throughout the United States (Sonneborn, 1950a,b, 1955). He had built up at Indiana a laboratory based on investigations of *Paramecium* genetics and had recruited, trained, and placed many students in various universities who continued to work on the problems offered by the organism. Nanney (1983, p. 166) has stressed that Sonneborn worked on cytoplasmic inheritance, at least in part, because that is what *Paramecium* had to show him. *Paramecium* had proven to be a successful competitor to the organisms of classical genetics, such as *Drosophila,* maize, and other metazoans.

However, in the decades following World War II when the mainstream of genetic research had shifted to the study of gene action and gene control, *Paramecium* had to compete with various other unicellular organisms such as *Neurospora* and *E. coli,* organisms which had been selected and "domesticated" especially for that purpose. Cytoplasmic inheritance as an unexplained, obscure, and somewhat mystifying phenomenon provided a favorable niche which permitted the existence of *Paramecium* as a prominent organism of genetics. Inasmuch as the genetic importance of *Paramecium* relied on investigations of cytoplasmic inheritance, an assault on cytoplasmic inheritance also represented a threat to the survival of *Paramecium* as a useful genetic organism.

In Sonneborn's view, *Kappa* and *Sigma* were borderline cases; they could not be seen simply as parasites. They "are of genetic significance," Sonneborn argued, "because they are normally transmitted only by heredity." In response to Altenburg, Muller, and others who attempted to dismiss them as insignificant for heredity, Sonneborn (1950e, p. 22) remarked rhetorically:

> One cannot but be impressed by the fact that practically every self-duplicating structure occurring within cells has at one time or another been considered a symbiont

> or parasite. For example, chloroplasts and mitochondria have been interpreted as symbionts. . . . Even the nuclear genes have not been spared.

Indeed, the endosymbiotic origin of chloroplasts and mitochondria had long been discussed throughout the century (see Buchner, 1952). While some cytologists entertained the idea that chloroplasts originated as blue-green algae, I. E. Wallin (1927) claimed to have demonstrated that mitochondria originated as symbiotic bacteria. He further claimed that they provided the chromosomes with new genes necessary for evolution.

Sonneborn rightly perceived Muller's and others' claims that plasmagenes were viruses as an attempt to relegate them to pathology. At the same time, however, as those who were investigating bacteria and their viruses were constructing a new definition of viruses, Darlington (1948, 1951) and Lederberg (1951, 1952) began to develop a notion of "infective" heredity, claiming that viruses themselves were far more intimately and permanently associated with genetic material of their host cells than had ever been imagined. If mitochondria (bacteria) and chloroplasts (blue-green algae) had entered the cell and become integrated in its physiology millions of years ago, viruses seemed to be continuing the processes of aggrandizing the genome today. Darlington and Lederberg attempted to reconcile the attitudes that plasmagenes were symbiotic organisms and that they comprised part of the genetic constitution of the complex organism, playing important roles in development and somatic differentiation.

Darlington led some of the early theorizing on the relations between plasmagenes and viruses for problems of development and cancer. He claimed that "latent plasmagenes" or proviruses could become viruses under certain environmental conditions and be responsible for tumors. On the other hand, infectious viruses could turn over a new way of life as opportunity arose and become integrated into the cell. The *Kappa* particles of *Paramecium,* in Darlington's view, represented examples of the opposite change—from a virus to a plasmagene. This meant that geneticists could not dismiss a plasmagene as "only a virus." But, as Darlington (1951, p. 320) realized, this view contradicted Anglo-American genetic orthodoxy: "We are gradually being drawn to conclude that there is a wider range of cytoplasmic determinants of greater power than our predecessors had dared to suppose." He claimed that plasmagenes offered a "unifying theory of heredity." "Cytoplasmic inheritance," he wrote (1951, p. 331), "will in future enable us to see the relations of heredity, development and infection and thus be the means of establishing genetic principles as the central framework of biology."

Lederberg took a similar view and supported it with studies of transformations and transductions in bacteria. Transformations were cases where hereditary changes in bacteria could be induced by chemically pure nucleic acid. The first celebrated case was that investigated by the immunologist Oswald Avery and his associates (see Avery et al., 1944; McCarty, 1985), based on the experiment of the British pathologist Frederick Griffith (1928). The experiment may be summarized as follows: a number of different strains of *Pneumococcus* bacteria may be distinguished by their virulence or nonvirulence and by whether the outer gelatinous capsule is present or absent. Griffith showed that noncapsulated, nonvirulent cultures of *Pneumococcus* could be induced to the capsulated virulent type by feeding them purified extracts from the capsules of the virulent strain. During the early 1950s, studies directed by Lederberg at the University of Wisconsin, and in France by Harriette Ephrussi-Taylor at the Rothschild Institute for Physico-chemical Biology and by Andre Lwoff at the Pasteur Institute, indicated that similar heritable changes were possible by the transfer of living viruses between different types of bacteria. Lederberg (1952, p. 413) proposed the term "transduction" for this phenomenon.

Bacterial geneticists regarded transduction both as a mechanism of transferring genetic material and as an effective way of investigating the nature of the genetic material. From this perspective, Lederberg (1951, p. 286) argued, infections from extrachromosomal agents such as symbiotic viruses were of great genetic value and were formally indistinguishable from *Pneumococcus* transformations. From the point of view of their evolutionary origin and taxonomy, Lederberg claimed extrachromosomal genetic agents represented a continuum between deleterious parasitic viruses at one extreme and integrated cytoplasmic genes such as plastids at the other (Lederberg, 1951, p. 286). On this basis, he proposed the term "plasmid" "as a generic term for any extra-chromosomal hereditary determinant" (Lederberg, 1952, p. 403).

In 1951, Lederberg attempted to go a step further and erect a second major generalization which would extend the significance of studies of extrachromosomal agents into a sociopolitical arena far beyond the internal or domestic politics of genetics. Infectious extrachromosomal components, he argued, were not only extremely valuable material for experimental study of the nature of genetic material. One could confer "quasi-organismic" status upon the various genetic components in the cell ranging from pathogenic viruses to plasmagenes, and by studying their interaction, form the basis upon which to construct social theory. In other words, geneticists could extend their knowledge claims about intracellular relations to the realm of human social relations. In Lederberg's own words (1951, p. 287):

> The advantage (or drawback) of this unifying view is that it comprehends a continuous spectrum of such genotypic interactions ranging from Ephrussi's granules, lysogenic viruses, and facultative intracellular symbiosis eventually to the least tangible ranges of genotypic (that is, interorganismal) interaction in, for example, human social relations. At each level of interaction pathological deviations can be found, ranging from sick plastids and malignant tumors (on Darlington's theory) to human serfdom.

CHAPTER 5

Boris Ephrussi and the Birth of Genetics in France

> The ability of the genes to vary, and when they vary (mutate) to reproduce themselves in their new form, confers on these cell elements, as Muller has so convincingly pointed out, the properties of the building blocks required by the process of evolution. Thus, the cell robbed of its noblest prerogative, was no longer the ultimate unit of life. This title was now conferred on the genes, subcellular elements of which the cell nucleus contained many thousands and, more precisely, like Noah's Ark, two of each kind. (Boris Ephrussi, 1953, pp. 2–3)

The notion of plasmagenes as self-perpetuating, mutable genetic elements, though often vigorously attacked in the United States, found wide support in France during the decade following World War II. Leading French-speaking biologists including André Lwoff, Jean Brachet, and Philippe L'Héritier brought forward a host of arguments based on diverse observations in support of cytoplasmic inheritance, and challenged what they considered to be the doctrine that Mendelian genes were the sole or principal agents of heredity and evolution. Some of the chief genetic work on cytoplasmic inheritance in France was led by Boris Ephrussi, who soon joined forces with Sonneborn to become a vigorous and skillful defender of the importance of the cytoplasm in heredity.

Ephrussi (1901–1979) is well known in the history of biology for his contributions to physiological genetics. His celebration by biologists and historians is primarily due to his epoch-making efforts in the mid-1930s to combine physiology and genetics in the study of *Drosophila* and for his success in demonstrating the influence exercised by genes on certain chemical reactions of the organism. Ephrussi's influence on the development of physiological and molecular genetics was monumental. He introduced Beadle to the investigation of genic action in the early 1930s. Beadle and Tatum would later, in 1958, be awarded Nobel Prize for their work on *Neurospora,* which led to systematic investigations of the relations of genes to the enzymatic control of metabolic reactions. He also introduced Jacques Monod to the problems of genic action. In 1965 the work of Monod, André Lwoff, and François Jacob on regulatory systems in bacteria would also be recognized with a Nobel Prize (see Chapter 7). It is little known, however, that Ephrussi also played a leading role in the institutional development of French genetics and in fostering research on cytoplasmic inheritance and challenging the predominant role of nuclear genes in heredity and evolution.

Immediately following World War II, Ephrussi was appointed to the first chair of genetics at the Sorbonne and became largely responsible for the development of genetics in the French university curriculum. After establishing an institute for genetic research in France in 1946, Ephrussi organized a comprehensive research program which centered on investigations of genetic regulation based on cytoplasmic inheritance in *Drosophila, Podospora,* and yeast. The research carried out by Ephrussi and his co-workers on yeast concerned the non-Mendelian inheritance of respiratory-deficient *petite* mutations. Continued investigation of this characteristic for more than a decade would ultimately be regarded as providing the first genetic evidence for mitochondrial heredity (see Chapter 7).

The Neo-Lamarckian Hegemony

> The Napoleonic structure was rigidly hierarchical. It was a mixture of an ecclesiastical control of ideas, governmental bureaucracy, and the military style of the emperor. (Terrence Clark, 1973, p. 18)

Ephrussi was born in a suburb of Moscow and was initially introduced to genetics as a university student. Following the Bolshevik Revolution, he left Russia for Rumania, where he tried to follow the Tolstoyan way as a farmer. After a year and a half of this way of life, Ephrussi disavowed himself from agriculture and religion and emigrated to France to return to science at the University of Paris. Though genetics had established itself in Britain and the United States by the 1920s, it had been resisted in France. In direct conflict with the United States and England, classical Mendelian genetics was almost nonexistent in French universities, which remained a bastion of neo-Lamarckism (Boesiger, 1980; Limoges, 1980; Buican, 1984).

Neo-Lamarckism had emerged in France during the 1880s when Darwinian theory, which had previously included the notion of the inheritance of acquired characteristics, was beginning to be identified solely with natural selection (see Chapter 1). Since that time almost all biologists in France not only acknowledged a role for the inheritance of acquired characteristics but used it as a central determinant in an explicitly anti-Darwinian view of evolution. A nationalistic component contributed to the emergence of the neo-Lamarckian viewpoint in France. The defeat of France in the 1870 war with Bismarck's Prussia brought with it a patriotic tendency of French biologists to encourage support of their own theories rather than to accept or develop those of other countries, especially Germany. In the face of the impending neo-Darwinian views, French biologists upheld Lamarck, attributing to him notions that were quite alien to those conveyed in his writings (Boesiger, 1980; Limoges, 1980).

World War I helped to further imbed Lamarckism in French patriotism. In fact,

Lamarckism in France came to represent a form of nationalism which was posited in direct opposition to Germany and German militarism with Bismarck. Traces of this anti-German sentiment can be detected in much of the French biological literature between the two World Wars. When Maurice Caullery, who held the chair of evolutionary biology at the Sorbonne, evaluated the contributions that diverse countries had made to the development of biology, he wrote:

> It is necessary after having escaped the peril of Germanic political hegemony, not to lose sight of another danger which threatens us, in the name of unjustified pretensions, that of the intellectual hegemony of Germany. (Caullery, 1922, p. 23, my translation)

It is worth mentioning in passing that German biological descriptions of cellular processes were not without their military metaphors of machine guns and armies. In 1885, the leading neo-Darwinian Weismann (p. 195) described the action of the nucleus during development in the following metaphorical terms:

> The development of the nucleo-plasm during ontogeny may be to some extent compared to an army composed of corps, which are made up of divisions, and these of brigades, and so on. The whole army may be taken to represent the nucleoplasm of the germ-cell: the earliest cell-division . . . may be represented by the separation of the two corps, similarly formed but with different duties; and the following cell-divisions by the successive detachment of divisions, brigades, regiments, battalions, companies, etc.; and as the groups became simpler so does their sphere of action become limited.

The Mendelian-chromosome theory as formulated by the Morgan school stood in virtual conflict with some of the central presuppositions of the neo-Lamarckian approach to heredity. Neo-Lamarckian biologists such as Yves Delage viewed heredity as an epigenetic process (Fischer, 1979). The adult characteristics of the organism were seen to result from a series of complex and integrated processes. Both intraorganismic and extraorganismic environmental circumstances contributed to this process. As discussed in Chapter 1, from the epigenetic perspective of the organism as a whole, any particulate theory of "determiners" seemed to be wrong and naive.

For Delage (1903, p. 806) the germ plasm of the egg contained only two essential factors, a relatively simple chemical composition and an arrangement of its parts. Delage searched for the causes of variations and their transmissibility in problems of ontogenesis. Lamarckism was a "somationist" theory of evolution which presupposed that characteristics acquired by the soma could have a specific influence on the germ. There could be no fundamental separation of the somatoplasm from the germ plasm. For example, Delage thought that an amputation of a gland could have a correlative effect and become hereditary. He attached great importance to nutrition, which he considered to have a morphogenetic action by adding to the chemical composition of cells, and modifying the nature of the substances and their arrangement. In his view, the Irish, English, and Arabs obtained their "racial" characteristics from differences in their diets (Delage, 1903, p. 836).

The views of Delage were representative of the thinking of many French biologists until 1915. It was only after Caullery visited Morgan's laboratory in 1916 that he became an advocate and a popularizer of the chromosome theory (Limoges, 1980, p. 326). Though Caullery accepted the basic principles of Mendelian heredity, he remained a Lamarckian and continued to resist the general significance of genes and natural selection for macroevolution.

Many leading French biologists embraced unknowable vitalistic forces to account for the progressive orthogenetic development and harmonious functioning of complex organs and structures. The texts of philosophers such as Henri Bergson (1907), who claimed the existence of an *élan vital,* were widely read by French biologists prior to World War II (Boesiger, 1980). As Limoges (1980, p. 327) has argued, "The spiritual overtones of even the most rationalist of French Philosophers created an intellectual environment uncongenial to a Darwinian approach." These spiritual overtones can be found in the writings of some French biologists throughout the twentieth century, including Pierre-Paul Grassé and Lucien Cuénot. Cuénot himself had made significant contributions to Mendelian genetics prior to World War I (Limoges, 1976). However, by the 1940s he emerged as one of the most authoritative representatives of the vitalist tradition in France. Cuénot (1941) envisaged the germ cells to possess a "teleological power of invention," a sort of intelligence which had immanent power equivalent to the intentionality of a sculptor who carved a statue. Cuénot wrote in direct conflict with the views of Etienne Rabau at the Sorbonne, the leading French proponent of mechanistic tradition, who denied any notion of progress in evolution (see Lwoff, 1944, pp. 234–237).

Mendelian genetics with its particulate gene, largely immune from extranuclear influences, represented a challenge and a threat to Lamarckism. It had to be disproved or rationalized in order for evolution to conform to Lamarckian principles. Faced with Mendelian genetics by the 1920s and 1930s, leading French biologists trivialized the significance of genes, claiming that Mendelian characters were of little, if any, importance for evolution. Many maintained similar views to embryologists, claiming that Mendelism was concerned only with "superficial" particularities, not with the most essential properties of the organism, those which decide the life or death of the embryo or the young individual. Félix Le Dantec, for example, compared Mendelian characters to the thirty-six vests of a circus clown, which having been removed, one after the other, leave no less a complete man (see Caullery, 1935, p. 70). Others such as Jean Rostand (1928) argued, like embryologists, that although genes were responsible for the constitution and differences of the individual, they were concerned only with the transmission of characteristics that did not exceed the framework of the species. This view was similar to that held by Ludwig Plate and other neo-Lamarckians in Germany (see Chapter 3) and was maintained by Caullery as late as 1935:

> The properties of the characters to which Mendelism applies limit themselves, in an almost absolute way, to variations which do not extend beyond the framework of the species. . . . The fundamental constitutional elements corresponding to the family, order, or subkingdom remain outside of Mendelian analysis, and these are those, however, which are the very essence of heredity. (Caullery, 1935, p. 263, my translation)

It is not enough to know that the majority of French biologists were neo-Lamarckians and therefore were reluctant to develop Mendelian genetics. One also has to know how they maintained their authority in French biology. New chairs devoted to genetics, necessary for the training of students and assistants, and new institutes devoted to genetic analysis were not easily established in France. As we shall see in greater detail in this chapter (see also Chapter 6) when discussing the institutionalization of genetics in France after World War II, the centralization of the French university system represented a most serious obstacle to the development of genetics. French science, which had flourished in the nineteenth century, became stifled by its bureaucratic and centralized structure. Renovated under the Third Republic, the French university system possessed rigid structure which was constructed by a Napoleonic tradition of French administration. This reality represented severe constraints for the differentiation and development of scientific research in France. The description that Clark (1973) gave for the nineteenth century rang true for French geneticists in the twentieth century.

From the nineteenth to the twentieth century, the procedure for university innovation in France was to convey the ideas to the government Ministry of Education, which legislated on the examinations, hours of each class, and method of teaching throughout the state-owned university system. This form of control, more strict than that which existed in Germany (see Chapter 3), may be contrasted with that of the United States, for example, where there was no central authority to lay down policy for the whole country and where the state universities competed not only among themselves, but also with the private universities. There geneticists could multiply very quickly and easily. In France, however, where almost all universities were state-owned, piecemeal change was formally resisted. In contrast to the situation in Germany, recognition of a new discipline in France entailed much more than a budgetary commitment for a new chair. It necessitated provisions for a national system of examinations in the subject and a staff to prepare students for the new examinations in many, if not all, of the universities.

Finally, mobility in the French university system was largely vertical. There was Paris at the center, and at the peak of the institutional hierarchy, the University of Paris. The centralization of formal authority made it essential for people on the periphery to establish communication channels with the center if they were to have some control over their individual destinies. The Ministry of Education constrained the types of professors appointed to faculty chairs and the types of courses they would offer, but the Latin Quarter had influence on the Ministry and Parliament; Sorbonne professors advised the Ministry about examinations and promotions in their particular field.

In effect, the centralization of the university system left the control of scientific research in the hands of a few individual chair holders who fostered the development of their own interests and beliefs. Such a system tended to induce intellectual continuity and conformity. Exemplifying this continuity and conformity was the chair of evolutionary biology at the University of Paris, which was created for Alfred Giard in 1887, who was succeeded by Maurice Caullery in 1908, and at the time Ephrussi received his chair of genetics in 1945, by Pierre-Paul Grassé. All were supporters of neo-Lamarckism, and the latter two were also hostile to

Morganist genetic doctrines, which included the central evolutionary role of natural selection.

It was only after World War II that geneticists became professors at the Sorbonne, when the situation in France was quite unique. The same year Ephrussi was given his chair of genetics, Georges Teissier received his chair, not because he was favored by a powerful professor but as a reward for his activities in the resistance movement during the war (Boesiger, 1980, p. 319). The following year L'Héritier was given a chair at the University of Paris. Teissier and L'Héritier, both trained as mathematicians, made significant contributions to the theory of natural selection in the 1930s. They are known today mainly for their technological advance in the study of *Drosophila* populations with the invention of "population cages," which provided a method for the experimental study of evolution. Ephrussi's appointment was pushed not by biologists but by mathematicians and physicists.

It is striking that Mendelian genetics found little support from French biology prior to World War II, and then, when a discipline of genetics finally emerged, it did not conform to Mendelian orthodoxy. The opposition to Mendelism before and after the war, however, had different causes. Prior to World War II, the development of genetics in France was stifled primarily because of neo-Lamarckism, which precluded a Mendelian approach to heredity, and because of the centralized structure of the university system, which represented a severe obstacle to the development of a new discipline of genetics. On the other hand, genetic investigations led by Ephrussi after World War II, which challenged the hegemonic position of Mendelian genetics, classical neo-Darwinian assumptions, and the "evolutionary synthesis" were not based on classical neo-Lamarckian presuppositions. In fact, as will be discussed more fully in the next chapter, genetic research in France emerged in a constant struggle against the traditional forms of French neo-Lamarckism.

From Embryology to Physiological Genetics

> Ever since I took up the genetic tool, the problem of embryonic differentiation has been in the back of my mind. All along, one of my persistent claims has been that changes during development do lie within the province of genetics. (Boris Ephrussi, 1958, p. 35)

Ephrussi was trained in embryology during the 1920s, and the course from embryological investigations to the development of cytoplasmic genetics was indirect. First, it required an evolution from embryology to genetics, which in turn, required recognition of the importance of intrinsic factors for embryonic differentiation. Second, it recessitated the construction of a genetic approach to developmental problems. It also involved the chance isolation of a non-Mendelian mutation and a theoretical disposition capable of recognizing its possible importance.

The skepticism of French neo-Lamarckians about the possibility of building a truly Lamarckian evolutionary theory led to an attitude which Limoges (1980, p. 325) has referred to as "theoretical agnosticism." French biologists who held the traditional chairs at the universities generally were reluctant to develop any of the major theoretical trends in biology. This resulted in a remarkably descriptive character of French biology during the 1920s, 1930s, and 1940s. In the 1920s, for example, the *licence ès science naturelle* at the Sorbonne was composed of *certificats* of zoology, botany, and geology or mineralogy. Although for the *certificat* of botany some plant physiology was taught, one could become a *licencie ès sciences* ignoring almost all biochemistry and genetics (Lwoff, 1981, p. 6).

However, there had been a tradition at the Sorbonne that the thirty or so students in zoology complete a *stage* at the Marine Biological Station at Roscoff. It was there that Ephrussi began his career as an embryologist, a career which ultimately led him to genetics and to the study of genic action. The station at Roscoff was a privileged place where young researchers could meet eminent personalities working in diverse disciplines. "These meetings," as André Lwoff (1981, p. 6) recalls, "played a decisive role in the career of a number of scientists.

Ephrussi met, among others, such well-known French biologists as Edouard Chatton, Marcel Prenant, and Georges Teissier, who headed the station during the 1920s. Roscoff was also frequented by the leading representatives of experimental embryology of Belgium and Sweden. As discussed in the previous chapters, many experimental embryologists found insurmountable obstacles in reconciling the atomistic nature of Mendelian genes with the holistic and orderly aspects of epigenetic development, with its morphogenetic fields and gradients. Many felt obliged to exclude genes from playing an important part in primary morphogenesis (gastrulation, cleavage, or segmentation, and organ initiation), though they readily conceded that genes intervene in the final details of the developing individual.

These views are very apparent in the writings of the Belgian Embryologist Albert Brachet, whose text of 1917, *L'Oeuf et les facteurs d'ontogénèse* had a great influence on Ephrussi and other French-speaking biologists. Brachet, who, unlike the majority of his French colleagues, was not a Lamarckian, distinguished between what he referred to as *hérédité générale,* which had its seat, if not exclusively, at least principally, in the cytoplasm of the egg, and *hérédité spéciale,* or Mendelism, which added the finishing touches to the organism, such as eye color, wing shape, etc. (Brachet, 1917, pp. 176–177).

Similar views were held by the Belgian canon Victor Grégoire, a botanist at the Catholic University of Louvain. Although he is best known for his early cytological work on chromosome structure and behavior, Grégoire could not accept the idea that gene mutations had any importance for evolution. "It is the protoplasm," he wrote,

> which develops itself and differentiates itself, and it is it, which at all stages of ontogenetic evolution, is the seat of the capacities which determine the course of development and differentiation. The mission of the chromosomes, during ontogenesis, is to furnish the protoplasm with certain substances which the protoplasm itself uses to accomplish its normal functioning. . . . The chromosomes have nothing to

> do with the governing of protoplasmic work; they are its instruments. (Grégoire, 1927, p. 870, my translation)

Even for Emile Guyénot, a Swiss student of Caullery, who by 1918 had become convinced of the Mendelian principles and adopted a neo-Darwinian viewpoint, the cytoplasm of the egg could not be excluded in heredity. In his text on heredity, which was the first textbook on classical genetics written in French, Guyénot (1924, pp. 414–418) offered an evaluation of the hereditary roles of the cytoplasm and the nucleus. Guyénot recognized the importance of Brachet's distinction between the role of the cytoplasm and that of the nucleus in heredity. It seemed to him to have a relatively incontestable value for a great number of hereditary processes. However, since it was possible that the nucleus also played a role in the establishment of the cytoplasmic localizations which constituted the "plan" of the future organism, he could not accept Brachet's distinction as being "universal" and "fundamental." Instead, he defended Mendelism from the criticisms that it was concerned only with accessory or superficial traits and argued that Mendelian factors were concerned with some considerably general modifications dealing with metabolism, fertility, and vitality of the organism, and that a great number of Mendelian factors were lethal. Nothing, he argued, was opposed to the idea that even the fundamental structures of the species depended not only on the cytoplasm, but on the nucleus as well. However, unlike Morgan in the United States, who in 1926 dogmatically asserted that the cytoplasm could be ignored genetically, Guyénot encouraged its investigation.

Ephrussi completed his doctoral degree based on two projects at Roscoff, concerned with the chemistry of embryonic development and tissue culture. The first project was carried out under the direction of the influential biochemist Louis Rapkine. It was Rapkine who introduced Ephrussi, Lwoff, and Monod to biochemistry and encouraged them to describe living processes in biochemical detail. Ephrussi's work under Rapkine concerned with the effect of temperature on the first stages of development and an analysis of the modifications of the chemical composition of the sea urchin egg during development (Ephrussi, 1933).

His second project, on growth and regeneration in tissue cultures (Ephrussi, 1932), was done under the direction of the celebrated biologist Emmanuel Fauré-Fremiet, who held the chair of comparative embryology at the Collège de France. Before his retirement in 1955, Fauré-Fremiet authored some 500 papers and books primarily in the research domains of cytology, developmental biology, and ciliate protozoology, where he made his most lasting contributions (see Corliss, 1972).

Fauré-Fremiet's cytological investigations did not center on the behavior and structure of chromosomes, and he excluded genetic investigations from his various scientific activities. The primary direction in all of his cytological research was elucidating the underlying ultrastructural organization of the cell. In an attempt to support his belief in the Lamarckian notion of orthogenetic evolution, he investigated organismic pattern and sought evidence of affinity between organisms through comparisons of the arrangement and distribution of intraciliary structures.

Fauré-Fremiet's views on cellular organization and its relations to embryology had a lasting impact on Ephrussi. Beginning in 1907, Fauré-Fremiet conducted extensive investigations on the behavior and physiology of mitochondria in ciliates. As mentioned in Chapter 1, he, along with many other cytologists, had attributed to these cytoplasmic granules a large part in hereditary phenomena and a significant role in the processes of differentiation and fertilization. Some cytologists had associated them with hypothetical "organ-forming substances." Following the work of Fauré-Fremiet, Ephrussi (1925) investigated the possibility that mitochondria might be the basis of reciprocal differences found in crosses between two different species of *Drosophila* reported by the Morgan school which seemed to him to have a cytoplasmic basis. However, definitive genetic evidence indicating mitochondrial heredity was lacking.

Ephrussi's collaboration with Rapkine was responsible for convincing him of the value of a chemical understanding of development. But it was his work on tissue culture that led him to the conviction that the great problems of development could not be solved outside a genetic context. During the early 1920s, when a great gap persisted between genetics and embryology, cellular differentiation was thought to be primarily an epigenetic phenomenon due to the reactions of essentially similar cellular protoplasms to different local environmental conditions. On the other hand, Ephrussi's work on tissue culture seemed to contradict this belief. This work showed clearly that at least some of the differences among cells of one organism persisted when they were taken out of the body and permitted to grow in a test tube. These studies persuaded Ephrussi that ". . . development of cellular specificities during ontogenesis was based, above all, on the action of intrinsic factors, therefore, in all probability, on chromosomal genes." (Ephrussi, "Notice sur les titres et travaux," unpublished, undated, p. 4)

It was this work which led him to orient his studies around investigations of the mechanism of genic action, which at that time was only the object of speculation. Ephrussi's primary interest in genetics, then, was to build up "the chain of reactions connecting the gene with the character, this chain being important not only as an eventual indicator of the nature of the gene, but also having a bearing on the general problem of differentiation" (Ephrussi, 1938, p. 6). In the 1930s *Drosophila* was the only convenient material, in view of the number of hereditary characters described in that organism. In order to familiarize himself with the genetics of *Drosophila,* he spent the academic year 1934–1935 on a Rockefeller Foundation Fellowship in Morgan's laboratory, which had moved from Columbia to Cal Tech.

When Ephrussi arrived in the United States the problem of cellular differentiation had been loosely excluded from American genetics. The extent to which genetics and embryology remained apart is illustrated by a conversation between Ephrussi and Morgan in the summer of 1934 at Woods Hole. Morgan, like Ephrussi, was trained originally as an embryologist, and his book *Embryology and Genetics* had just come off the press. After Ephrussi told him how interested he was in reading the book, Morgan gave Ephrussi a copy on the promise that he would give him his frank opinion about it. As Ephrussi (1958, p. 36) recollected several years later:

> I accepted, and a few days later went in to report. I said I found the book very interesting, but I thought that the title was misleading because he did not try to bridge the gap between embryology and genetics as he had promised in the title. Morgan looked at me with a smile and said, "You think the title is misleading! What is the title?" "Embryology and Genetics," I said. "Well," he asked, "is not there some embryology and some genetics?"

Morgan's evasive answer to Ephrussi's criticism bears witness to the extent to which Ephrussi became polarized on the gap between embryology and Mendelism. Cellular differentiation in the face of an apparent genomic equivalence represented a paradox for genetics. The most common assumption was that differentiation was under cytoplasmic control. Ephrussi (1953, p. 4) phrased the dilemma for geneticists clearly:

> *Unless development involves a rather unlikely process of orderly and directed gene mutation, the differential must have its seat in the cytoplasm.*

During the 1930s many geneticists wanted to know what a gene was and how it acted. Knowledge of the nature of the gene would lead to knowledge of how it worked. And reciprocally, knowledge of the mechanism of genic action would give some indication of its structure. These two possibilities indicated two principal ways of approaching the gene. First, one could try a direct approach and attempt to investigate the gene from the "gene end" of the chain of reactions connecting the gene with the character. In the 1930s such studies were carried out in relation with the various particular problems, such as studies of the differences in the effects of the same gene in different positions on the chromosome, studies of the process of mutation, studies of the structure of the salivary gland chromosomes, etc.

On the other hand, studies of the developmental effects of genes, which interested Ephrussi, represented a second way, starting at the "character end" of the postulated chain of reaction. When Ephrussi was working with Sturtevant it appeared to him that genes in which manifestations are not autonomous (that is, those characters which are gene-dependent in development) should provide access to phenomena anterior to the differentiation of characters. And eventually, "with a little luck," he could describe the events in biochemical terms. Only with "non-autonomous" characters could the most efficient methods of experimental embryology be used, such as grafting (Ephrussi, 1938, p. 6).

When Ephrussi returned to France from Morgan's laboratory in 1935, he was followed by George Beadle. Unlike Ephrussi, Beadle was trained not in biochemistry or embryology but in plant genetics. Together, Ephrussi and Beadle constructed a technique for transplanting eye disks from *Drosophila* larvae of one genotypic constitution to the body of larvae of a different genotypic constitution. By using eye colors as examples of characters controlled by genes and by applying transplantation methods, it was possible to begin to reconstruct the chain of reactions leading to pigment formation. This technique was applied to a series of eye color mutations and revealed reciprocal influences which suggested the existence of two diffusible and specific gene-controlled substances intervening in the formation of eye pigment. In other words, there were two successive links in the formation of eye color (Beadle and Ephrussi, 1936). Unfortunately, a description of the events in biochemical terms was difficult to obtain.

Although the work of Ephrussi and Beadle was representative of its beginnings, physiological genetics would not develop into normative practice with well-defined methods and standards until unicellular organisms were incorporated into the domain of genetics. There were difficulties in the use of *Drosophila* and other organisms used by classical geneticists for investigating the genetic control of the development of an organism. First, investigations which attempted to determine the physiological and biochemical basis of already known hereditary traits were confined to a study of nonlethal heritable characters. Beadle and Tatum (1941, p. 499), who helped to establish the pattern of research characteristic of biochemical genetics after World War II, claimed this to be the most serious limitation. As they realized, as long as geneticists confined themselves to the study of such characters the possibility existed that genes controlled only "superficial" characters.

A second difficulty, not unrelated to the first, and recognized by Beadle and Tatum, was that the hereditary characters established by classical genetics were characters with visible manifestations which distinguished individuals of a species. Many such characters involved morphological variations which were likely to be based on systems of biochemical reactions so complex as to make analysis exceedingly difficult. For these two reasons, then, it was necessary to reverse the procedure for investigating the general problem of genic action. Instead of attempting to work out the chemical basis for known genetic characters, physiological genetics, led by Beadle and Tatum, set out to determine "if and how genes control known biochemical reactions" (Beadle and Tatum, 1941, p. 500).

The program of physiological genetics required an experimental material that was accessible to both cross-breeding and biochemical analysis. The organisms studied by classical geneticists were not suitable for a detailed biochemical description. As discussed in the last chapter, the domestication of an organism for physiological genetics required that not only its sex life, but its growth as well, be brought under meticulous control. Physiological genetics was based on the study of biochemical markers which required a defined and preferably synthetic culture medium. This material was provided by microorganisms, especially fungi and bacteria.

The organism employed by biologists conditions the range of questions which can be posed and constrains the nature of the answer received. Beadle and Tatum chose the fungus *Neurospora*. In an attempt to bridge the gap between the synthetic processes of the cell and the Mendelian gene, they proposed that a gene acts by determining the specificity of a particular enzyme and thereby controls enzymatic synthesis and other chemical reactions in the organism. This view found support among many American geneticists who extended the work on *Neurospora,* such as H. K. Mitchell, N. Horowitz, and D. Bonner at the California Institute of Technology, and others at other laboratories. On the other hand, after World War II Ephrussi chose yeast.

The Competitive Strategy of French Genetics

> You will be amused to hear that Pasteur undertook the study of yeasts for patriotic reasons. "I was inspired in these investigations by our misfortunes," says Pasteur in the introduction to his classic book, *Etudes sur la Bière*. "I undertook them directly after the 1870 war and continued them relentlessly ever since, with the resolution of carrying them far enough to stamp with a lasting progress an industry in which Germany is superior to us." I am afraid that in spite of Pasteur's efforts, German beer remained much better than French beer. Meanwhile, however, Pasteur's studies on yeast have laid the foundation of modern biochemistry. (Boris Ephrussi, 1953, p. 13)

When the German army occupied Paris, Ephrussi fled to America where he took up a post at Johns Hopkins University. In 1944, however, before the Germans were out of Paris, he was in England active in *les forces françaises libres* and was ready to fly to Paris at the moment of liberation. Shortly following his return to France, Ephrussi was appointed to the first chair of genetics at the Sorbonne and headed a Parisian laboratory of genetics of the Rothschild Institute of Physico-Chemical Biology.

Throughout the 1940s and 1950s Ephrussi attempted to establish a major center for genetic research in France. In 1946, when Teissier was director of the *Centre National de la Recherche Scientifique* (*C.N.R.S.*), an *Institut de Génétique* was created. Established at the beginning of World War II, under the Ministry of Education, the C.N.R.S. quicky became the leading scientific institution providing both funding and buildings for scientific research in France. The Institute of Genetics was constructed with Ephrussi elected as its director with a *Conseil de Direction* represented by P. Auger, L'Héritier, Rapkine, and Teissier. With the death of Rapkine, Ephrussi had asked for Lwoff to be nominated to the council. Lwoff was head of the Department of Microbial Physiology at the *Institut Pasteur* where Monod was working on "adaptive enzyme" formation in bacteria. The Institute of Genetics were structured around three primary goals: (1) to recruit and to train researchers in the domain of genetics, in particular in those domains within the research framework of the institute (in 1946, when L'Héritier was appointed professor at the Sorbonne, Ephrussi invited him and his collaborators to install themselves in his laboratory at the Rothschild Institute); (2) to prepare the construction of a building which would be its future center and to acquire the financial means to effect this project; (3) to carry out research in both plant and animal genetics.

The objectives of the institute were divided into three *services:* (1) *Service de Génétique Formelle,* directed by L'Héritier, was concerned primarily with genetic analysis with *Drosophila;* (2) *Service de Génétique Évolutive,* led by Teissier, was based on statistical analysis of populations; (3) *Service de Génétique Physiologique,* which was the largest division, was headed by Ephrussi. The two primary aims of physiological genetics were to understand the heterocatalytic function of the genes—the process by which genes affect the phenotype—and the autocatalytic function—the process by which the genes are duplicated. Ephrussi was largely concerned with the former problem.

However, neither classical genetics nor studies of the action of chromosomal genes were highly represented at the Institute of Genetics. French geneticists were largely preoccupied with investigations of somatic cell differentiation, adaptive enzyme formation, and cytoplasmic inheritance. That genetics in France at the Institute of Genetics had come to be centered around investigations of cytoplasmic heredity, or what Ephrussi called in 1949 "*la génétique transcendante,*" had a social determinant as well. Ephrussi was persuaded that working in teams, and limiting the number of problems for genetic investigations to a very few, was the most effective strategy to be competitive with geneticists in the United States and England. In 1949 when reporting on the activity of the Institute of Genetics, he wrote:

> I am convinced that, at this time, working in teams is the most effective form of scientific experimental research. The formation of teams, the establishment of a tradition of working in collaboration, appears to me to be a particularly important objective in a country where individualism is pushed often to an ill-fated degree, and in a domain retarded as French genetics which finds itself naturally exposed to a strong foreign competition. Since the beginning, I have therefore sought to stimulate by all possible means collaboration and teams. I think we have succeeded to a great extent, and if you are soon struck by the small number of research subjects with which we are occupied, it's because most of the researchers are working around the same problems. I think that, in total, we will gain by it. (Ephrussi, 1949, "Rapport sur l'activité de Génétique," unpublished, p. 4, my translation)

In Ephrussi's *Service de Génétique Physiologique* his wife, Harriette Ephrussi Taylor, worked on new specifically induced transformations in *Pneumococcus,* while he and his associates Hélène Hottinguer, Anne-Marie Chimènes, and Piotr Slominski began to work on the "problem of induction" and "adaptive enzyme formation" in yeast. At the time they began their work, the genetic literature contained many novel results indicating strange new relationships among the nucleus, cytoplasm, and environment in the control of hereditary traits. *Dauermodifikationen* and Sonneborn's work on *Paramecium* were widely publicized. The specific hereditary effects of *Pneumococcus* transformation, and genetic resistance to antibiotics in bacteria investigated by Salvador Luria and Max Delbrück (1943), seemed to many to challenge genetic and evolutionary orthodoxy. All this work led to the possibility that environmentally adaptive changes occurred in microorganisms. They smacked of Lamarckism.

Yeast in particular seemed to exhibit a very great genetic "plasticity." Mutations in yeast seemed to have been "provoked" by cold, heat, various chemical products, and radiation (see Ephrussi, 1949). Yeast offered other advantages. Ever since Pasteur's day, it had been frequently used for biochemical studies and was one of the biochemically best-known organisms of the 1940s. Moreover, the mode of origin of the mutations and the genetic environmental control of "adaptive" enzyme formation in yeast remained unclear. Ephrussi's central problem was to establish their seat in the cell and whether the hereditary variations in yeast arose by selection of mutations preexisting in the cultures or whether the mutations were induced by environmental agents. In bacteria, Delbrück and Luria were the most authoritative proponents of the former view. Ephrussi, as we shall see, became one of the leading spokespersons for the latter process operating in yeast.

When Ephrussi and his co-workers began their studies, yeast genetics was a turbulent domain. Several geneticists who worked on yeast associated their results with those of Sonneborn and were openly challenging some of the basic principles and doctrines of classical genetics. In 1940 O. Winge and his collaborator O. Lausten at the Carlsberg Laboratories in Copenhagen had reported the non-Mendelian inheritance of reduced fertility in yeast, which they interpreted to have a cytoplasmic basis, perhaps in mitochondria. But the work concerning cytoplasmic genetic elements in yeast did not begin to flourish until 1945, when Sol Spiegelman and Carl and Gertrude Lindegren reported their results on "adaptive fermentation." As discussed in the last chapter, these were cases in which the addition of a certain compound (the sugar melobose, for example) to the substrate on which microorganisms were cultured elicited the formation of a specific enzyme capable of utilizing this substance. The environmentally induced adaptive genetic changes were inherited as long as the substance on which the enzyme acted was present in the medium. The Lindegrens and Spiegelman initially interpreted the phenomenon in terms of cytoplasmic plasmagenes.

Similar phenomena were reported in bacteria. Jacques Monod (1947) at the Pasteur Institute wrote the first extensive theoretical and experimental account of the phenomenon of enzyme adaptation. Monod had been introduced to genetics by Ephrussi, who in 1936 took him to Morgan's laboratory at Cal Tech. When they returned to France, Ephrussi wanted Monod to work with him to follow up the work which he had begun with Beadle. However, Monod refused to collaborate with Ephrussi, whom he saw as an "extreme authoritarian and disciplinarian," qualities which did not appeal to him, at least when he was a student (see Judson, 1979, pp. 358–363). In 1943 Monod joined the Communist Party and the *Franc-Tireurs,* the armed resistance movement. After the war he resigned from the party over Lysenkoism (see Chapter 6). In 1946 he began to work systematically on adaptive enzyme formation, which he had begun as a doctoral student before the war. As will be discussed in Chapter 7, the work of Monod, Jacob, and collaborators, by 1961, would provide a central model for nuclear gene regulation and would be recognized later with a Nobel Prize.

In 1947, however, the interpretation of adaptive enzyme formation remained a subject of speculation. Monod (1947, p. 224) began his review of the problem with the familiar difficulty of somatic cell differentiation in the face of genomic equivalence, or, as Monod put it, how cells with identical genomes could acquire "the property of manufacturing molecules with new, or at least different, specific patterns or configurations." Monod did recognize the possibility that the loss of cytoplasmic plasmagenes during cell division could account for irreversible cellular differentiation. However, he generally dismissed the plasmagene hypothesis of self-reproducing enzyme-forming gene replicas or proteins to account for irreversible cellular differentiation. He claimed that "no authenticated cases of *cytoplasmic* inheritance of an enzyme have been reported, whereas the mere existence of Mendelian genetics makes it obvious that *purely nuclear* inheritance of enzymatical properties must be considered as an absolute rule" (Monod, 1947,

p. 274). However, Monod realized the difficulty of a nuclear basis for permanent or irreversible differentiation: "The main reason for doubting this possibility is that the complete autonomy, the 'randomness' and the rarity of gene mutations, do not seem to afford any explanation of the apparently orderly processes of ontogeny" (Monod, 1947, p. 284). In his view the only recourse was simply to deny the orderliness of ontogeny for later development and claim that "automatic" differentiations were confined to the early stages when the embryo consisted of a few cells and when it remained possible that the early differentiations were reversible. Selection of spontaneous gene mutations could not be overlooked when accounting for later differentiations (Monod, 1947, p. 285). At the same time, Monod recognized that one could not deny that enzymatic adaptation did occur, that its mechanism was unknown, and that it could not be accounted for by any selection hypothesis.

By 1946, when Ephrussi and co-workers were beginning to carry out their first experiments on yeast, the research on cytoplasmic genetic elements in microorganisms was in a state of turmoil. First, the authority on yeast genetics, O. Winge in Copenhagen, totally rejected the interpretation of the Lindegrens and Spiegelman on the basis of his own extensive studies of the inheritance of enzymatic capacities in yeast. He maintained that the results could be explained without any recourse to the notion of self-perpetuating entities in the cytoplasm (Lindegren, 1949). Lindegren's highly speculative "cytogene theory" was attacked by Sonneborn at Cold Spring Harbor in 1946 (Sonneborn, in Lindegren and Lindegren, 1946)

Perhaps more dramatically, the early results were inconsistent. Several efforts to repeat them in Lindegren's laboratory had failed (Lindegren, 1946). The situation was best described by Spiegelman when attempting to encourage Ephrussi to put yeast genetics at the center of research in his laboratory at the Rothschild Institute in Paris:

> I should like to express the fervent hope that your laboratory is prosecuting yeast genetics with some vigor. I say this because since I have returned I have had the opportunity to go over the Lindegren data, in the raw, of the past year and a half quite carefully. The whole business is in a pitifully confusing state. What makes it particularly disturbing is that constancy and reproducibility of results is apparently difficult to attain. Crosses which at one time give beautifully clear-cut Mendelian segregations with respect to a given enzyme will go completely haywire when redone at a subsequent time. Other heterozygotes which have consistently yielded non-Mendelian segregation ratios will suddenly and unpredictably clear up and yield only 1:1 ratios. It is quite evident that some as yet unknown factor (or factors) are working to disturb normal segregations of enzymatic characters. (Spiegelman to Ephrussi, November 16, 1946)

On the other hand, the notion of plasmagenes as self-perpetuating, mutable, cytoplasmic genetic particles, as indicated by non-Mendelian inheritance in *Paramecium,* seemed to be highly plausible. Ephrussi, for example, realized well, and from the very beginning, the great potential significance of Sonneborn's genetic work. Following the appearance of the first series of papers in *Kappa,* he wrote to Sonneborn (November 1943):

> I read yesterday your two papers with the greatest interest and admiration, and I still think, as I told you here, that it is a magnificent piece of work—whatever the final interpretation will be; and that, if your present interpretation will be the final one, this work may later appear as the most important development in genetics made after the establishment of the fundamental principles.
>
> I spent the evening yesterday in thinking over your interpretation and all possible objections. I must confess that I did not proceed altogether scientifically in the sense that sympathetic as I am with your interpretation, I was definitely attempting to refute possible objections rather than to objectively evaluate the possibility of a different interpretation.

Ephrussi's research strategy in yeast involved first studying the problem of induction of mutations by chemical means, more precisely by acriflavine. Acriflavine was a well-known bacteriocide whose action was attributed to its ability to combine with nucleic acids, which were believed to be constituents of all genetic elements. Therefore, Ephrussi and his co-workers hoped they might also be able to say something about the unsolved mechanism of induction (Ephrussi *et al.,* 1949a,b,c). Their results were striking.

When yeast was grown on a medium containing acriflavine, slow-growing, respiratory-deficient colonies were produced at a frequency approaching 100%. Ephrussi called the mutants *petites*. Moreover, the effect of acriflavine was semi-specific; it always led to the same genetic mutation, which was inherited in a non-Mendelian fashion. The mutation was irreversible and led to a permanent loss of respiratory ability. Biochemical studies by Slonimski led to the conclusion that the slow growth of *petite* colonies was the result of a loss of certain respiratory enzymes. The synthesis of the respiratory enzymes appeared to be under the joint control of nuclear genes and cytoplasmic particles. The cytoplasmic particles were independent of the genome in terms of their reproduction but dependent on it for their formation (see Ephrussi, 1950, p. 58).

The mechanism underlying adaptive enzyme formation in yeast remained at the center of genetic controversy throughout the late 1940s and early 1950s. Since the environment (presence of a sugar in the medium) induces a specific response (production of enzyme) which gradually disappears in the course of a few generations if the cells are removed from the medium, some geneticists considered these cases as examples of *Dauermodifikationen* (see review by Rhoades, 1955, pp. 53–54). The central question was whether they represented an environmentally induced change in the genotype or a change in the phenotype.

By subjecting *petite* mutations to a complex series of tests in response to environmental influences (anaerobic conditions), Ephrussi and his co-workers were able to produce gradual adaptive reversible changes in enzymatic constitution. In Ephrussi's opinion (1950, pp. 65–66), the changes he observed were phenotypic in character. Contrary to genotypic characters governed by autocatalytic particles, he claimed they were governed by "reversible chemical equilibria." Only their amplitude was limited by the nuclear and cytoplasmic *genetic* constitution of the cell.

The essential point for Ephrussi was that his work on respiratory deficiency indicated that the enzymatic constitution of the cell was not determined in a mechanistic way by genes. It was controlled by genes, the cytoplasm, and the environment, offering a variety of ways for an organism to adapt to its environment. Ephrussi seldom published his views on evolutionary processes, but when he did, as he freely admitted, they had a teleological flavor. However, Ephrussi denied the neo-Lamarckian view which assumed that the adaptive features of organisms were perfectly fitted to their environment:

> The admirable perfection of the adaptations of organisms and of their parts to the functions they perform has detracted attention from the fact that adaptedness does not consist of perfect fit, but capacity to fit or adapt in a *variety* of ways: only in this sense is adaptedness a guarantee of further survival and evolutionary progress, for too perfect a fit is fatal to the species if not to the individual.
>
> This, I think, sets phylogeny and ontogeny in the correct perspective. It is the genotype which bears the marks of past experience of the species and defines the range of possible fits. What fit is actually chosen, what phenotype is actually evolved, is determined by the ever renewed individual history. (Ephrussi, 1950, pp. 45–46)

The plurality of mechanisms which Ephrussi saw to be operating in yeast fit this view perfectly. In 1950 he concluded:

> The respiratory function of yeast is shown, by the study of its diverse types of variation, to depend on a complex system of enzymes controlled by genes, by cytoplasmic particles and by external factors. The yeast cell is thus endowed with mechanisms enabling it at every instant to respond to the demands of both long range evolution and of the contingencies of every-day life by adequate, either reversible or irreversible, variations.
>
> Evolution, in its pursuit of adaptation, uses both strategic and tactical arms. (Ephrussi, 1950, p. 67)

However, Ephrussi saw the greatest theoretical significance of his result on *petites,* not from the evolutionary perspective of an organism, but from the developmental perspective of a differentiating cell in a complex multicellular organism.

During the 1940s and 1950s, when direct genetic analysis of somatic cells was lacking, the functional equivalence of nuclei of different somatic cells remained a highly plausible hypothesis. Crosses between somatic cells, nuclear transplantation from one somatic cell to another, or grafting of fragments of cytoplasm could provide the definitive information. Such experiments, however, had to await the development of adequate technical devices. In the meantime, the closest approximation to the evidence geneticists required to understand the mechanism of cellular differentiation was provided by the study of lower, single-celled organisms which propagate by vegetative reproduction and possess no isolated germ line.

Microbial genetics provided a new methodology for the study of somatic cell lines. It provided several new techniques for genetic analysis of somatic cell heredity: transformation, transduction, and mitotic recombination. It also provided a number of new mechanisms of inherited variations, involving the interplay of the nucleus, cytoplasm, and enviornment, none of which could be rejected on *a priori* grounds as possible agents of somatic variation in higher forms. Sonneborn's *Kappa* particles in *Paramecium* represented a textbook case for the behavior of cytoplasmic genetic particles. Genetic and cytological work on plastids, which many cytologists related to mitochondria, seemed to indicate that they too possessed genetic continuity. The case of antigenic characters in *Paramecium* and Ephrussi's case of *petite* mutations resulting in a permanent inability of cells to respire also indicated the existence of cytoplasmic genetic elements. To those who reported the phenomena, these results seemed to confirm that the cytoplasm was the seat of the causes of the irreversible changes involved in differentiation.

"Still Threatened with Vertigo"

> The *Drosophila* school . . . presented the world with a model of heredity constructed entirely on chromosomes. This model, which was at the same time mechanistic and dogmatic, did not fail to shock embryologists as well as geneticists in Europe (C.D. Darlington, 1949b, p. 123, my translation).

> The mechanist is intimately convinced that a precise knowledge of the chemical constitution, structure, and properties of the various organelles of a cell will solve biological problems. This will come in a few centuries. For the time being, the biologist has to face such concepts as orienting forces or morphogenetic fields. Owing to the scarcity of chemical data and to the complexity of life, and despite the progress of biochemistry, the biologist is still threatened with vertigo. (André Lwoff, 1950, p. 93)

In 1948 the C.N.R.S., supported by the Rockefeller Foundation, held an international symposium, hosted by André Lwoff and entitled "*Unités biologiques douées de continuité génétique*." At this meeting Lwoff, L'Héritier, Brachet, and Ephrussi brought together diverse evidence indicating cytoplasmic genetic elements and attempted to show their importance for a general theory of somatic heredity. Their work was incorporated with that on chloroplasts in higher plants, presented by Rhoades, and with that on the role of genes, plasmagenes, and the environment in the determination of antigenic traits in *Paramecium,* presented by Beale. Sonneborn and Muller had been invited but neither could attend the meeting. Darlington presented a theoretical introduction to the problem of cytoplasmic genetic particles primarily concerned with the similar properties of plasmagenes, viruses, and genes. This time, equipped with arguments and evidence based on observations from an array of disciplines, including cytology, embryology, protozoology, and microbial genetics, Lwoff, L'Héritier, Brachet, and Ephrussi reconsidered the distinction between "superficial" and "fundamental" characteristics corresponding to the nucleus and cytoplasm, respectively.

In 1937, in the course of their experiments on selection in mixed populations, L'Héritier and Teissier had stumbled upon a "hereditary physiological anomaly" which involved the non-Mendelian inheritance of CO_2 sensitivity in *Drosophila*. Normal resistant *Drosophila* are not permanently injured when exposed to an atmosphere of pure CO_2 for a short period of time. However, CO_2-sensitive flies die. L'Héritier considered this phenomenon to be so important that he placed it at the center of his theoretical and experimental investigations throughout the remainder of his investigative career. The work was interrupted by the war and the sensitive stock was lost at the time of the German invasion of France in 1940. However, at the beginning of the war H. J. Muller carried some of the stock to America and had it preserved at Cold Spring Harbor (L'Héritier, 1948, p. 326). Thanks to Muller's efforts, the work on CO_2 sensitivity resumed again in France in 1942 under poor working conditions.

L'Héritier and his co-workers soon reported a series of striking results associated with CO_2 sensitivity. First, the hereditary agent, which he called a "genoid," could migrate from somatic cells to germ cells and was greatly affected by temperature. By raising the temperature during different stages of development one could "cure" sensitivity in both somatic cells and germ cells permanently or temporarily. Furthermore, with the advice and technical assistance of Ephrussi, L'Héritier and F. Hugon de Scoeux (1947) found that the genoid could also be acquired by the implantation of an organ of a sensitive individual into a resistant fly or by injection of the supernatant of ground and centrifuged sensitive animals. Perhaps even more strikingly, the sensitivity acquired by implantation or injection could be transmitted to the germ cells and be carried by the gametes to the next generation (see L'Héritier, 1948).

The environmentally (temperature) induced adaptive changes from sensitive to resistant occurring in both somatic and germ cells raised the question of the inheritance of acquired characteristics, which remained at the center of discussion among many French biologists. L'Héritier (1948, p. 341) himself acknowledged this interpretation:

> When a partial or complete gametic cure happens to coincide with a somatic cure, this may be looked upon as a case of inheritance of an acquired character. It occurs regularly with permanently cured flies which never breed any sensitive offspring. . . . It is known from transplantation experiments that in females the germ cells can acquire the genoids from the somatic cells and *vice versa*.

However, L'Héritier denied the Lamarckian idea that the germ cells could be continuously and progressively altered in response to adaptive changes in the somatic cells, as was commonly assumed in France. "In my opinion," he continued,

> the acquired character inheritances have no deep biological significance, in the sense that they do not involve any direct and constant action of the somatic cells on the germ cells. The inheritance of the acquired cure seems to be rather the result of the identity of responses of both kinds of cell to the same external condition, the temperature. The reciprocal infection of somatic and germ cells does not seem to agree any better with the progressive germinal response to somatic adaptive changes, which is assumed to take place by Lamarckian-minded biologists. (L'Héritier, 1948, p. 341)

Although L'Héritier opposed classical Lamarckian interpretations, he also opposed what he (1955, p. 494) called the proponents of *le néodarwinisme classique*" (who included H. J. Muller, Dobzhansky, Fisher, and by that time Ernst Mayr and many others), who, he claimed, refused to acknowledge any genetic specificity in the cytoplasm and any role for cytoplasmic entities in evolution. In fact, throughout the 1940s and most of the 1950s, L'Héritier maintained a distinction between "special" and "general" heredity and claimed that macroevolution could have a cytoplasmic basis. L'Héritier (1955, p. 494) wrote:

> Experimental analysis of genetic differences is necessarily limited to the comparison of neighboring organisms on the systematic ladder and we have no direct knowledge of the nature of the determinants responsible for the differences of organization which separate the higher groups. To admit that they are all chromosomal is only an extrapolation, based on the relative rarity of cases of cytoplasmic heredity. Its legitimacy is perhaps debatable. (my translation)

It was not until the rise of molecular biology (see Chapter 7) that L'Héritier changed his view and claimed that there was hardly any more room for doubt that the major distinction between "special" and "general" heredity was "unreal" (see L'Héritier, 1964, p. 12).

In the meantime, L'Héritier remained a staunch defender of the primary role of diverse mechanisms of cytoplasmic inheritance in cellular differentiation, morphogenesis, and evolution. Indeed, as discussed in the previous chapter, the ability of CO_2 sensitivity to be transmitted artificially led to the question of whether the agent responsible was a normal genetic constituent of universal significance or a parasite like *Rickettsia*. L'Héritier, like Darlington and Lederberg, opposed the interpretation of Muller and his followers, who attempted to dismiss various cytoplasmic particles simply as infectious parasites. He argued that the genoid had properties similar to that of *Kappa* and plastids in plants and it was not contagious under normal conditions. It was inherited according to precise and elaborate rules which allowed prediction of the outcome of crosses with nearly the same accuracy as when Mendelian characters were dealt with (L'Héritier, 1948, p. 345). Moreover, even if it was an intimate virus, it would have major hereditary and evolutionary consequences (L'Héritier, 1955). By 1965, when L'Héritier fully acknowledged the genoid as a virus, he insisted that hereditary nonpathological viruses played significant roles as agents of speciation and claimed that they could be detected only by genetic experiments and then with a great deal of difficulty (see L'Héritier, 1970, p. 206). During the late 1940s, however, the virus-genoid question remained open and L'Heritier (1948, p. 346) refused to classify the genoid in any definite category, since "its classification is doomed to remain so much a question of definition and personal feelings."

Lwoff offered one of the most elaborate and novel theoretical treatments of plasmagenes as agents of development. First in 1948, and then in 1950 in a small book which since has become a classic, *Problems of Morphogenesis in Ciliates: The Kinetosomes in Development, Reproduction and Evolution,* Lwoff offered a far-reaching theory of the kinetosomes in cytoplasmic heredity. His arguments were based upon investigations of the natural history and life cycle of ciliates which he and his teacher, the well-known protozoologist Edouard Chatton, carried out during the 1920s and early 1930s. Their early work was primarily descriptive and was not concerned with the problems of heredity and development.

In the late 1940s and early 1950s, though, Lwoff argued forcefully to ally protozoological studies based on the evolution of kinetosomes during the ciliate life cycle with the embryologists' insistence that development was primarily cytoplasmic. Among other cytoplasmic bodies, kinetosomes seemed to be particularly important in development. They were located on the cell surface at the base of the cilia, and the importance of the cell surface, or cortex, in particular had been emphasized by many embryologists (Lwoff, 1950, pp. 83–87).

The role of the cell surface in the spatial organization of the cell had been emphasized on conceptual grounds as an important ingredient in Paul Weiss's (1947) concept of "molecular ecology," according to which the cell is viewed generally as an organized mixed population of molecules and molecular groups. Lwoff found it easy to apply Weiss's concept in a wholesale way to ciliate development. To Weiss the spatial localization of parts of a cell resulted from the "response of organized elements to fields of organized (i.e., non-random) physical and chemical conditions" (Weiss, 1947, p. 255). Weiss summarized his concept of "molecular ecology" in a series of ten propositions. All of them centered on the notion that the spatial organization of the content of the cell and its constituent particulate elements required "a primordial system of spatially organized 'conditions' to set the frame for the later differential settlement of different members of the dispersed molecular populations" (Weiss, 1947, p. 253). The guidance mechanism would operate in the solid cell surfaces and interfaces in the cell which would favor the absorption of a given assortment of molecular species. Throughout development the surface populations of each cell, responding to outside influences, would acquire a unique role in determining the subsequent course of cellular events. In 1948 Fauré-Fremiet and H. Mugard reported that very important cortical changes took place during the development of some organisms. The importance of the cell surface in localization and morphogenesis, as Lwoff noted, also had been promoted most forcefully in France during the 1930s by the American embryologist E. E. Just.

Just received his Ph.D. in 1916 at the University of Chicago under the direction of F. R. Lillie (Manning, 1983; Gilbert, 1985). His first paper, in 1912, was a highly acclaimed study in which he showed that the plane of symmetry of development is determined by the polar bodies and the point of entrance of the spermatozoon in the egg of the annelid *Nereis*. This was followed by about fifty papers in the next twenty-five years dealing with fertilization, experimental parthenogenesis in marine eggs, and the action of the cell surface during development. A Black American, Just was affiliated with the Department of Zoology at Howard University, Washington, D.C., 1912 until his death in 1941. Founded in 1869, Howard University became the first American institution of higher learning for Black citizens. Just's work was well recognized by such influential physiologists and embryologists as Jacques Loeb, Ross Harrison, and, of course, Lillie. As early as 1915 he received the first award of the Spingarn Medal for his embryological investigations and his efforts to improve medical education at Howard and other Black universities in the United States. The Spingarn Medal was presented annually to "the man or woman of African descent who shall have made the highest achievement during the preceding year, or years, in any honorable field of human endeavor" (Lillie, 1942).

In spite of his highly acclaimed research, as Manning (1983) has argued, Just's scientific career was a constant struggle for opportunity to carry out his investigations. In America he was condemned by race to remain attached to a Black institution of low academic status and financial support which was unable to provide him with the facilities and full opportunity to do his research. At Howard he was burdened with heavy teaching and administrative responsibilities and he was unable to receive an appointment in one of the large universities or research institutes in the United States. Just spent the last ten years or so of his career largely in exile, working in various European laboratories: in Germany at the *Kaiser-Wilhelm-Institut für Biologie* in Berlin, in Italy at the Naples Zoological Station, and in France at the Sorbonne and the marine biological station at Roscoff.

In the late 1930s, when he wrote his well-known text *The Biology of the Cell Surface* (1939), Just was stationed in France at Roscoff. The central focus of his book was on the importance of the cortex or "ectoplasm" of eggs in the initiation and the further course of development. Just conceived the behavior of the ectoplasm to be a major factor in differentiation. The behavior of the chromosomes in the nucleus, in his view, was too "rigidly mechanical for them to be responsible as primary agents of heredity" (Just, 1932, p. 73). He compared their action to "puppets in a puppet show," their activities being controlled by the reactions of the cell cortex. He believed that the first effects of the environment manifested themselves on the cortex and ultimately these reactions would affect the whole cell system (Just, 1932, p. 74). According to his model, the cortex played a crucial role in controlling the action of genes in heredity:

> As the boundary, the living mobile limit of the cell, the ectoplasm controls the integration between the living cell and all else external to it. . . . It is keyed to the outside world as no other part of the cell. It stands guard over the peculiar form of the living substance, is buffer against the attacks of the surroundings and the means of communication with it. (Just, 1939, p. 366)

To Lwoff, the kinetosome represented a model example of a visible, self-duplicating, cytoplasmic particle, or "plasmagene." Like chloroplasts, killer particles, and the genoid, kinetosomes were also sensitive to certain variations of environmental conditions. The metabolism of the host, light, and temperature, he argued, could cause variations in the relative multiplication of the particles and in certain cases lead to their disappearance (Lwoff, 1950, p. 84). He also suggested the possibility that kinetosomes were responsible for the inheritance of antigenic characters in *Paramecium* investigated by Sonneborn and Beale. (Lwoff, 1949)

Kinetosomes were endowed with some very special properties which made them particularly fundamental in the morphogenesis of protozoa and metazoa. According to Lwoff, the kinetosomes not only possessed properties of growth and division and were responsible for the production of cilia. They also had another property, what embryologists called "prospective potencies": they were organized into different systems and structures according to their position in the cell. The fate of a kinetosome was determined by its immediate cytoplasmic environment. The environment, in turn, varied from place to place within the cell at any one time, and at the same place in different stages of the life cycle of the cell. Based upon morphological observations and viewing kinetosomes as a model for visible plasmagenes, Lwoff postulated that "one plasmagene may possess many potencies and turn out different organelles according to its position and to the phase of the life cycle" (Lwoff, 1949, p. 32). Kinetosomes could develop into various morphological bodies, though usually they developed into cilia.

As a protozoologist, Lwoff acknowledged that the cytoplasm represented a very differentiated system, with its cortex, its mitochondria, its kinetosomes, and its chloroplasts, which he claimed were all endowed with genetic continuity. As Lwoff (1950, p. 2) stated:

> Cytoplasm is not just a collection of enzymes or a plastic and complaisant receptor passively submitted to the dictatorship of genes, but certainly contains self-reproducing bodies endowed with specificity.

However, only rare cases of cytoplasmic inheritance were known by 1950, and Lwoff realized that many geneticists regarded the role of the cytoplasm to be minimal in heredity. In spite of the paucity of examples, Lwoff defended the integrity of cytoplasmic organelles and, like L'Héritier, suggested that they were in control of fundamental characteristics which would be difficult to detect by typical genetic procedures.

Brachet and Ephrussi also supported the idea that the cytoplasm controlled fundamental characteristics, and, like Lwoff, they opposed the predominant roles assigned to Mendelian genes in heredity and development. Based on evidence from biochemical embryology, Brachet (1949) postulated that microsomes, which he associated with protein synthesis, were autonomous cytoplasmic organelles representing "plasmagenes." He had been investigating these small spherical bodies at the University of Brussels during World War II, when he established that they contained RNA. He and his collaborators had shown that microsomes existed in all cells and they correlated their abundance with the intensity of protein synthesis in various types of tissues. In 1942 he published a full account of his pathbreaking findings in a fifty-one-page article in the Parisian journal *Archives de biologie*. His early work was paralleled with that of Torbjorn Caspersson in Stockholm, which corroborated his findings.

Brachet and Caspersson localized RNA in the cytoplasm, where protein synthesis occurred, while DNA seemed to be confined to the nucleus. RNA was placed, for the first time, somewhere between the nuclear gene and the protein. However, Brachet did not see a logical necessity to attribute protein synthesis in the cytoplasm directly to DNA in the nucleus, the seat of the Mendelian gene. Instead, he postulated that RNA-containing particles, microsomes, were the direct agents of protein synthesis. He claimed that nuclear genes played only an indirect role in protein synthesis by controlling the multiplication of microsomes (Brachet, 1950, pp. 200–201). Work on the tobacco mosaic virus showed that RNA had genetic continuity, and Brachet thought that microsomes could be little viruses replicating in the cells of higher organisms like *Kappa* in *Paramecium* and the genoid in *Drosophila*.

The results of the research on nucleic acids, which had been scattered through several languages and erratically circulated during the war, were brought together in 1946 at a symposium on nucleic acids held in Cambridge by the Society for Experimental Biology. When summarizing his work, Brachet wrote about the cytoplasmic RNA, concentrated in particles, as follows:

> The results clearly demonstrate that the ribonucleoprotein granules are pre-existing structures in the living cell, where they exert important physiological functions. Another point of interest is the following: when granules are isolated from red blood cells, they are found to contain a small amount of hemoglobin which cannot be eliminated by repeated washings. In the same way, pancreatic granules contain insulin. . . . These facts point towards the following hypothesis: ribonucleoprotein granules might well be the agents of protein synthesis in the cell. (Brachet, 1946, pp. 214–215)

After World War II, the idea that microsomes might represent autonomous cytoplasmic organelles or "plasmagenes" seemed plausible indeed, especially when taken in connection with the new genetic and protozoological investigations indicating cytoplasmic heredity. As Brachet recalls:

> I was impressed by the work of Sonneborn. I was impressed by the work of L'Héritier on the genoid in *Drosophila*. I was impressed, perhaps not by the work, but by the ideas of Lwoff on the ciliates and the formation of new cilia. And I had somehow the impression that in the egg there must be self-reproducing cytoplasmic units. (interview, December 10, 1981)

As late as 1958, Brachet still insisted that microsomes would provide respiratory enzymes, would be the agents of protein synthesis, and consequently, be responsible for embryonic differentiation in all organisms. Brachet used the genetic investigations of Sonneborn as a centerpiece in his argument for cytoplasmic control of protein specificity. After summarizing his views on microsomes as plasmagenes in 1950, he wrote:

> This way of seeing things connects in a striking manner with the point of view adopted by Sonneborn based on genetic evidence in protists. Sonneborn admits, in effect, that cytoplasmic heredity is due to cytoplasmic particles endowed with genetic continuity, susceptible of undergoing mutations and dependent on nuclear genes for their maintenance or their functioning, but not dependent from the point of view of their origin or their specificity. The two points of view are ready to be united, if one admits the identity of plasmagenes and microsomes. (Brachet, 1950, p. 201, my translation)

During the early 1950s, the idea that cytoplasmic genetic elements represented the primary basis of somatic cell heredity found biochemical genetic support from the intensive investigations of Ephrussi and Slonimski and co-workers. Biochemical studies of the *petite* mutants in yeast indicated that the respiratory deficiency was due to a loss of the ability to synthesize a whole series of respiratory enzymes, which were reported by other biochemists to be carried by mitochondria (Slonimski and Ephrussi, 1949; Ephrussi, 1951).

The nature and behavior of mitochondria provided a plausible model for cellular differentiation. Biochemical investigations of mitochondria in rat liver and diverse cytological observations suggested that the mitochondria of a cell represented a heterogeneous population. Like *Kappa,* plastids, and other cytoplasmic bodies, the differential segregation of mitochondria under environmental influences during cell division provided a cytoplasmic interpretation of how cell lines with identical nuclear constitution could become differentiated in metazoa. In 1951 Ephrussi put his case as follows:

> If this view is accepted, we have then a consistent interpretation of how the two cell lines of identical nuclear constitution are differentiated from each other in a way which satisfies the requirements of the lasting differentiation of metazoan cells. At the cell level, the differentiation of the two lines of yeast appears as a mutation; in terms of intracellular mechanisms, however, if the ensemble of the population of intracellular units is taken into account, it is to be regarded rather as the result of a segregation. This segregation, accidental in the case of spontaneous mutation, can be directed by an environmental factor, which causes an irreversible restriction of the potencies of newly formed cells without apparently affecting the totipotency of the generating cells. It may be rash to see in this relationship an exact equivalent of the relations existing in the metazoan organism between the cells of the germ line and those of somatic tissues, but such a parallel does indeed suggest itself. (Ephrussi, 1951, p. 254)

The cytoplasmic basis of cell heredity indicated to Ephrussi that protein synthesis might also be under extranuclear control. With the rise of microbial genetics, numerous examples of biochemically deficient mutants had been accumulated. But most biochemical geneticists in the United States, led by the work of George Beadle, interpreted them formally in terms of nuclear gene modifications of enzyme specificity. The idea that genes control specific reactions either by acting directly as enzymes or by determining the specificity of enzymes was pervasive in genetics from its earliest beginnings. The basic gene-enzyme relation goes back to the first decade of the century and was popularized in various genetic texts. By the 1930s, many geneticists, from those analyzing pigments in guinea pigs to those investigating the chemical effects of genes on flower colors, were interpreting their results in terms of genes controlling the character of enzymes.

In direct conflict with the views of the *Neurospora* school led by Beadle in the United States, Ephrussi, like Brachet, challenged the assumption that the nuclear genes were in direct control of the immediate synthetic processes of the cell throughout most of the 1950s. A number of possible alternative interpretations existed. The role of cytoplasmic genetic particles containing RNA in protein synthesis was suggested by the work of Brachet and others. The possibility also existed that proteins were formed by "copying" protein templates and that the control of the basic structure of cytoplasmic proteins was a function of cytoplasmic elements, with DNA in the nucleus merely determining the final folded-up configuration (Haurowitz, 1950). Ephrussi interpreted the results in yeast indicating cytoplasmic control of the synthesis of respiratory enzymes to be consistent with the view that plasmagenes themselves were proteins. Following one of the suggestions of Wright (see Chapter 4), Ephrussi claimed that in somatic cells the plasmagene proteins would combine with prosthetic groups emanating from the nucleus to form molecules that multiplied in the cytoplasm thereafter.

Ephrussi argued that there was no decisive proof for nuclear control over protein specificity. Applying this theory to the case of non-Mendelian inheritance in yeast, he wrote in 1951:

> There is no evidence . . . for the ultimate nuclear origin of the postulated cytoplasmic units of yeast. Indeed, the very existence of the so-called vegetative mutants is against the initiation of these elements by the nucleus. No valid argument can be presented however against an interpretation postulating for example that, just as the functional activity of these units is awakened by something produced by the genes, so the capacity of reproduction is conferred upon them by, say, ribonucleic acid emanating from the nucleus. Such an hypothesis could be supplemented by the further assumption that the purely cytoplasmic component is protein, for there is no real proof, I believe, of genic control of protein synthesis, and indeed some arguments against it. No doubt, many would prefer to refer to such a protein as a 'starter' rather than as a self-reproducing unit. (Ephrussi, 1951, p. 260)

Position Effect

> I suppose you have heard of Poky, the presumably cytoplasmic case in Neurospora discovered by the Mitchells. It fills me with joy: it's a dream to have this thing happen in the fortress of all exclusive Mendelism! (Boris Ephrussi to T. M. Sonneborn, May 24, 1952)

As discussed in the last chapter, during the early 1940s and 1950s cytoplasmic heredity continued to represent a challenge to genetic orthodoxy. There could be no room for plasmagenes as general constituents in the economy of the cell for those American geneticists imbued in the Morganist tradition. The arguments for the existence of plasmagenes formulated on nongenetic lines by Lwoff and Brachet were ignored by most geneticists. They did not conform to the methods and explanatory standards of microbial genetics and therefore were not recognized as hereditary phenomena by those who upheld the dominant role of the Mendelian gene. However, they were taken into consideration by Ephrussi and Sonneborn, who challenged the technical capacity of Mendelian analysis and who attempted to expand the realm of hereditary phenomena.

The genetic results in yeast reported by Ephrussi and co-workers were considered. During the early 1950s the validity of cytoplasmic heredity relied on its recognition by Ephrussi's chief competitors in biochemical genetics. The *Neurospora* school led by Beadle and his associates was the dominant group and, as discussed in the previous chapter, they upheld Morgan and Muller's dictum that nuclear genes played the exclusive or primary role in heredity. As for other cases of non-Mendelian inheritance, the influentials at the California Institute of Technology sought and found alternative interpretations for the results in yeast. They were interpreted formally in terms of accepted theory. The most common opinion in the United States followed the suggestion of A. H. Sturtevant that Ephrussi's results could be explained by a Mendelian phenomenon called "position effect": a change in a gene activity due to a change in its position in the chromosome.

Some *Neurospora* geneticists only began to consider the evidence for cytoplasmic heredity when a member of their school reported an example. In 1952 at the California Institute of Technology there was unexpectedly reported a case of non-Mendelian inheritance in *Neurospora* called "poky," which resembled *petite* both biochemically and genetically (Mitchell and Mitchell, 1952). A brief glance at some of the discussion surrounding this case underlines the social process involved in validating knowledge claims. The case of non-Mendelian inheritance in *Neurospora* was exactly the sort of thing Ephrussi had been waiting for. In a letter to one of the authors of the work, H. K. Mitchell, he wrote (March 29, 1952):

> A rumor has reached me yesterday according to which you have discovered a case of cytoplasmic heredity in *Neurospora:* you will easily imagine how much this (by me) long awaited news item excites my curiosity, especially if I add that it was transmitted in Western Union style and that it contained a reference to "a new plasmagene doing something to the cytochrome system." Will you consider it as an act of arrogance if I ask you to drop me a few lines about the facts which are the basis of the rumor? I am sure you are not cruel enough to refuse such a favor to a man who credits himself with modest contributions to both cytoplasmic heredity and your marriage.

To some members of Beadle's school the case of non-Mendelian inheritance reported in *Neurospora* actually verified the case in yeast reported in France. Norman Horowitz, for example, who was also situated at Cal Tech, changed his opinion of the case of non-Mendelian inheritance in yeast. He wrote to Ephrussi (June 10, 1953):

> Incidentally, I don't remember whether I mentioned it to you, but I think that the *poky* results eliminate whatever doubts they may have been concerning the cytoplasmic inheritance of *petite*.

Ephrussi recognized the role played by social elements in the verification of scientific findings. In an attempt to defend his priority he quickly wrote back to Horowitz (June 26, 1953) and protested against the lack of credibility given to his own results:

> . . . seeing the stand you are taking now that "the *poky* results eliminate whatever doubts there may have been concerning the cytoplasmic inheritance of *petite*." I hope you will take this opportunity to explain what system of logic has come to dominate the cartesian one so as to enable you to remove the doubtful elements of the results obtained with yeast by working on *Neurospora*. I am afraid, dear Norm, that it will be hard for us to agree, for, contrary to you, I am beginning to believe that Sturt [A. H. Sturtevant] is right in invoking position effect in this business: the credibility of the same results obtained in Paris and Pasadena is obviously very different.

Despite his rhetorical response to Horowitz's claim, Ephrussi considered "a case of cytoplasmic heredity discovered in *Neurospora*" to be "an event of major importance" (Ephrussi to Sonneborn, May 24, 1952). Ephrussi had cast cytoplasmic inheritance at the center of his investigations in France, and by the early 1950s he had considerable vested interest in its biological importance. Taking the long-awaited results in *Neurospora* into consideration with those reported by Sonneborn, himself, and others in microorganisms, it seemed to him that cytoplasmic heredity would lead ultimately to the downfall of Mendelian supremacy in hereditary phenomena. To Ephrussi the results reproduced in *Neurospora* not only verified his case of cytoplasmic heredity in yeast but legitimated the general occurrence of cytoplasmic heredity in all organisms.

Despite Ephrussi's excitement, the real significance of poky nevertheless remained a subject of contention. Sonneborn, for instance, had been convinced of the importance of cytoplasmic heredity for some time. To him the results in *Neurospora* were not seen to be a discovery at all, but merely another isolated example of cytoplasmic inheritance. After all, Sonneborn's work had convinced him of the importance of cytoplasmic inheritance for some time. Moreover, unlike Ephrussi, who had reported only one example in yeast, Sonneborn had accumulated several examples of cytoplasmic heredity in *Paramecium*. It was this kind of result, in his opinion, that would change the views of most American geneticists. As he wrote back to Ephrussi (November 17, 1952), the verification of the existence of cytoplasmic heredity required, not another case of non-Mendelian inheritance in another organism, but just the opposite. It required several cases of non-Mendelian inheritance in one organism:

> The "poky" story did not excite me because I am not surprised to hear of such results in any organism and because isolated cases like this will do no more for the *general* question of cytoplasmic inheritance than the relatively few isolated cases already known. Any major change in current thinking must come, I think, from accumulated results on a particular organism in which perhaps it may be shown that the cytoplasmic role in heredity may be broadly and widely integrated with the nuclear role.

This flurry of method talk over the significance of *petite* in yeast and poky in *Neurospora* highlights the social negotiations which take place in the process of scientific discovery. We cannot understand the significance of the results outside of the power relations in the field. Each participant in this controversy struggled to occupy the dominant position by attempting to impose the greatest significance on the scientific values most closely related to him personally or institutionally. However, Ephrussi's hope for a major revolution in genetics, thought to be signaled by the report of poky, was not realized, and poky remained just another isolated genetic anomaly.

"A Common Language and a Common Ideology"

> The present knowledge of the biochemical constitution of the cell was achieved largely by the use of destructive methods. Trained in the tradition of the theory of solutions, many a biochemist tends, even today, to regard the cell as a "bag of enzymes." However, everyone realizes now that the biochemical processes studied *in vitro* may have only a remote resemblance to the events actually occurring in the living cell. (Boris Ephrussi, 1953, p. 108)

Throughout the 1950s Ephrussi attempted to defind the recurrent claims of evidence for cytoplasmic inheritance and the theoretical need for postulating its existence. As discussed in the previous chapters, many geneticists traced developmental differentiation to the action of nuclear genes alone. They claimed that the initial cytoplasmic organization of the uncleaved egg was a consequence of prior gene activity. However, relatively little investigation and theoretical activity had been directed toward a synthesis between genetics and embryology. On the other hand, Ephrussi was a member of a small group of geneticists who were unwilling to concede that the methods of analytic abstraction, which had demonstrated that the nuclear genes intervene in every activity of the cell and the organism, had excluded the possibility of an equally pervasive participation of cytoplasmic materials with genetic properties.

In 1953 Ephrussi published a small book in which he attempted to synthesize the genetic work on microorganisms carried out on yeast, *Paramecium,* and *Podospora* with the experimental work on nongenetic lines, chiefly that of experimental embryology, generally ignored by geneticists. The book, based on a series of lectures delivered at the Birmingham Medical School and suggested by Peter Medawar, was entitled *Nucleo-Cytoplasmic Relations in Micro-Organisms*. Its primary aim was to defend the thesis that the cytoplasm was endowed with genetic properties which interacted with nuclear genes and that this provided geneticists with a basis for interpreting the phenomenon of embryonic differentiation.

Ephrussi began with the well-known argument that reproduction of diverse tissue cells true to type in tissue culture implied a cytoplasmic basis of these genetic differences on the cellular level, for all the diverse cell types in a metazoan body were thought to be alike in the kinds of genes they contained. He then reviewed a number of examples in microorganisms in which cells with the same gene constitution manifested persistent diversities owing to cytoplasmic differences. By way of summary, he offered a series of models for the possible processes by which metazoan cells could develop along diverse lines and produce both reversible and irreversible changes in the face of nuclear equivalence:

> The non-living environment can induce changes of the concentration of Kappa particles and of antigenic type in Paramecia, and the loss of cytoplasmic particles in yeasts. The contact with the living environment (s strains) induces the change from S^s to S in *Podospora*. The reverse cytoplasmic change in this organism is apparently due to the interaction of certain nuclear genes. Lastly, we find that nucleus and cytoplasm affect each other's activity. The cytoplasmic particles of yeast are activated by a nuclear gene. In turn, in Paramecia, definite cytoplasmic states permit the expression of definite nuclear genes. (Ephrussi, 1953, p. 100)

Cytoplasmic variations of both persistent and irreversible character could be initiated in several ways, and several mechanisms based on purely cytoplasmic features or on nucleo-cytoplasmic relations could account for cellular differentiation. In Ephrussi's view, the choice of appropriate mechanisms was outside his disciplinary boundaries: "It will be the task of the embryologist to choose among them" (Ephrussi, 1953, p. 101).

Despite the enormous potential importance of these cytoplasmic mechanisms, none of them could account for the principal problem of development. They could only be *instruments* of somatic cell differentiation; they were secondary to another cytoplasmic property which played the leading part in development. It will be recalled that many embryologists who protested against the dominant position attributed to the particulate nuclear gene maintained that the single-celled or complex organism was an integrated and organized unit which could be broken down only for analytic purposes. Embryologists who defended the integrity of the "organism as a whole" viewed development as an orderly process which follows a "plan" engraved in the cytoplasm of the undivided egg. As Ephrussi (1953, p. 101) wrote of it:

> Sometimes it is indicated by the visible distribution of cytoplasmic materials, sometimes it can be revealed only by experiment. But at this stage the "ground plan" is neither complete, nor necessarily definite: it will be gradually refined and fixed in the course of subsequent development.

To Ephrussi, the "ground plan" in the cytoplasm of the cell dictated when and where the *instruments* of differentiation came into action. The primary cause of differentiation, then, resided in the initial anistropy of the egg, expressed by its polarity and symmetry. Following the views of Fauré-Fremiet, Harrison, Lillie, and others, Ephrussi (1953, p. 101) maintained that "*the fundamental anisotropy of the egg cytoplasm itself has a genetic basis.*" Hence, he argued, the "*fundamental problem of genetics in relation to development becomes that of the origin of the specific molecular pattern of the cytoplasm which confers to the egg its vectorial properties*" (Ephrussi, 1953, p. 104).

However, even the solution to this problem did not represent the end of the problem of development. The fact that each of the parts of an egg is capable of giving rise to a complete organism, and yet does not when left in its natural position, proved to embryologists that the embryo is an integrated unit. The developing embryo has the properties of a "supracellular continuum." The "ground plan," potentially contained in each of the cells, is superseded by that of the "embryo as a whole," of which the individual cells are now only the subordinate, "executive agents." Cell boundaries appeared to be no obstacle to the all-pervading integrative forces, the nature of which Ephrussi 1953 (p. 104) considered to be the "key to biological organization." Similar views, he remarked, were upheld by the British embryologist V. B. Wigglesworth, who, like many other embryologists, maintained that genes controlled only the trivial details of the organism. In consideration of the integrated character of organisms, Wigglesworth wrote in 1945:

> The essential organism is something apart from the cells which support it. It exists before the cells dispose themselves and define its form. The cells and their nuclei, as the vehicles of the genes, play great part in controlling the details of the form the organism will take; but the framework which marks the main outlines of that form, which says that the organism shall be a vertebrate, an amphibian, a frog, or an insect, a dipteron, a *Drosophila,* which defines the head and the tail, the main regions of the body and the limbs—this framework exists before the cells. (cited in Ephrussi, 1953, pp. 111–112)

Like the relations of the cells to the "organism as a whole," Ephrussi argued, the autonomy of intracellular constituents could be relative. The cell, he maintained, was an integrated unit which could be broken down only for purposes of analysis. Ephrussi supported his position with quotations from various writers who viewed the cell essentially as a "protoplasmic crystal" in which a myriad of protein molecules are associated in a definite ultramicroscopic pattern characteristic of the particular type of cell. The celebrated British biochemist F. G. Hopkins (1932, p. 333) summarized this view as follows:

> A cell has a history; its structure is inherited, it grows, divides, and, as in the embryo of higher animals, the products of division differentiate on complex lines. Living cells, moreover, transmit all that is involved in their complex lines. I am far from maintaining that these fundamental properties may not depend upon organization at levels above any chemical level; to understand them may even call for different methods of thought; I do not pretend to know. But, if there be a hierarchy of levels, we must recognize each one, and the physical and chemical level which, I would again say, may be the level of self-maintenance, must always have a place in any ultimate complete description.

Ephrussi also supported the conception that the integrated properties of cells represent a hereditary property with arguments drawn from the writings of Sonneborn, who, as we have seen, also maintained that the molecular arrangement of the cellular materials, their organization into a working system, is itself a part of the genetic system of the cell.

However, the claim that the organization of the cell represented a hereditary property stood in direct conflict with the dominant doctrines of genetics and demanded that the technical capacity of Mendelian genetics be brought into question. The autonomy of the nuclear genes and their primary role in heredity was a doctrine based on conclusions from previous genetic experiments. In order to challenge this doctrine one had to question in technical terms the experiments on which it was based, matters normally ignored by defenders of the nuclear monopoly.

As discussed earlier, the genetic notion of heredity had become endowed with connotations which restricted its usage to phenomena studied by certain techniques. In this instance, Ephrussi employed arguments focusing on the prevailing analytic techniques in an attempt to distinguish what occurs *in vitro* from what possibly occurs *in vivo*. Genetics was based on differences between organisms, not on similarities; and cross-breeding analysis presented a picture of genes as discrete units acting autonomously from the rest of the cell and dictating its activities. Ephrussi (1953, p. 108) raised the issue of technique dictating theory:

> . . . that the method of genetics, although it involves no "bloodshed," is as analytical in its essence. Indeed, the "resolving power" of this method is amazing. It provides us with a picture of the cell's nuclear constitution with unequalled "definition." But, so long as the basis of genetics is the study of differences it cannot be expected to give us an undistorted picture of the cell as a whole. The integrative character of the cell, which is its fundamental property, is bound to escape our notice most of the time.

Ephrussi knew well that by limiting the scope of Mendelian genetics he might appear as a "nonprogressive" biologist, perhaps of the nature of P.-P. Grassé and other neo-Lamarckians, against whom he himself had struggled in France (see Chapter 6). As he wrote to Pontecorvo at the University of Glasgow:

> All along in preparing the manuscript I knew that many of my readers, as a result of reading my book, will classify me as belonging to what you so aptly call the "obscurantist biological right." I hope however that *you* will not make this mistake, and that you will see how near the points I tried to make are to your own idea of the necessity to introduce into genetics the 'space element.' I regret very much that I did not expand on the pattern problem from the point of view of enzyme organization. (Ephrussi to Pontecorvo, August 25, 1952)

In the 1950s, an acknowledgment of the hereditary properties of the cell as a whole entailed a confrontation not only with the methods of Mendelian genetics, but also with the biochemical perspective of the cell. When geneticists turned decisively to the study of the mechanics of genic expression with the aid of Warren Weaver's program for the management of science at the Rockefeller Foundation, many physicists and biochemists immigrated into biology. They brought with them the methods, theories, explanatory standards, objectives, and doctrines of their disciplines. Due to the rapid development of the biochemical and molecular approaches in genetics following World War II, this period is usually marked as representing the origins of molecular biology. (On the role of the Rockefeller Foundation in fostering the development of molecular biology, see, for example, Kohler, 1976, 1978; Abir-Am, 1982, 1984; Fuerst, 1984; Bartels, 1984; Olby, 1984; Yoxen, 1984). Ephrussi vigorously opposed the purely biochemical view of the cell, which conflicted with his integrative conceptions of heredity.

The struggle between physiological geneticists and the emigrants from the physical sciences was represented at the social level as well. For example, when, in 1956, Ephrussi recommended a corn geneticist who turned to yeast after the war for a position of "physiological genetics" at Berkeley, the issue centered around a disciplinary dichotomy between genetics and biochemistry:

> My suggestion may, at first sight, surprise you. I have little doubt that very few of our fellow geneticists would classify Roman as a physiological geneticist. Indeed, he does not belong to the type of geneticist generally referred to as physiological or biochemical. However, I wish to submit that most of the latter category, using genetic material as a tool, in fact contribute to the progress of biochemistry, while what is desirable from our point of view is to make biochemical information bear on genetic problems, in order to contribute to the progress of genetics.
>
> The fact that Professor Roman's background is genetical rather than biochemical appears to me in the present connection a further advantage to your Department. In so far as I know this Department's tradition is chiefly one of 'formal genetics' and it will be easier for your group to find a common language and a common ideology with Professor Roman than with an emigrant from biochemistry. (Ephrussi to J. A. Jenkins, December 4, 1956)

Ephrussi was certainly not the only geneticist to object to the biochemical approach, which threatened to dominate biology, or at least genetics, after World War II. The situation of Sol Spiegelman, as he saw it at Washington University in St. Louis in 1946, provides a vivid testimony to the social and intellectual conflicts facing geneticists with the emergence of "big science" in the United States:

> Affairs at this university, as far as the biological sciences are concerned, have deteriorated considerably over the summer. The great American 'team spirit' has hit this place with a vengeance under the influence of our new president, 'Atomic Scientist, A. H. Compton.' In the biological sciences they have decided to back the 'team' headed by Carl Cori, our university's candidate for the Nobel Prize. Cori is a great biochemist but he is no biologist and has very little understanding or sympathy with biological programs. If you can't isolate it and crystallize it you aren't doing science as far as he is concerned. Since, in terms of money, our program is the biggest thing outside of Cori's department, you can guess what might result. While we were away for the summer, without our knowing it, they almost organized us out of existence. When Kamen and I returned, we put up quite a fight and managed to save the program. Cori and I, because of our diametrically opposed views about the nature of biological research, never got along very well in any case, and this affair has not made things any pleasanter. Kamen, who was in Cori's department, has at his own request been shifted to a regular chemistry department. So we have a truce in which we are free to carry out our program but it is obviously no long term solution. Kamen and Steinbach, who was also involved in our project, as well as myself, are completely soured on this place. . . . Biological research as we understand it, will be stifled here for some time to come and no amount of brilliant biochemistry on cell-free extracts can substitute for it. As a result everybody is leaving. Hershey, by the way, is probably leaving also. (Spiegelman to Ephrussi, November 16, 1946)

By the early 1950s, then, biologists who supported the view of "the cell as a whole" and who claimed that the cytoplasmic properties controlled the "fundamental" characteristics of the organism faced a great deal of resistance. However, they did not understand the "incorrect" views of their opponents in terms of a lack of adherence to a formal scientific methodology. Instead, they accounted for conflicting views in terms of divergent techniques and objectives in the field and in terms of scientific indoctrination. In the following section we will see in a more detailed examination how accounts based on the technique-ladenness of observations, scientific indoctrination, and power relations in science can be used by scientists in an attempt to delegitimate opposing and dominant views in the field.

Accounting for Mendelian Error

> We must not forget that Mendelian genetics is a very resourceful science. As Morgan once said: "Give me six pairs of genes and I will explain the results of any cross." (Boris Ephrussi, 1950, p. 55)

The postulation of cytoplasmic heredity represented a major effort to fill the gap between the problems of development and heredity. While this interpretation had great merit as a theoretical model in accounting for cellular differentiation and morphogenesis, it also had some striking weaknesses by the 1950s. First it had been greeted with a great deal of hostility, and secondly, relatively little genetic evidence had been obtained for cytoplasmic inheritance as a general phenomenon. During the 1950s, advocates of cytoplasmic heredity raised various technical issues, and drew them to the center of a complex sociopolitical understanding of the nature of scientific development and scientific activity, in an attempt to counteract these 2 major weaknesses of their genetic theory. In effect, Sonneborn and Ephrussi were attempting to make a new and major scientific fact out of what many others considered to be a few genetic anomalies by placing them in a theoretical construction irreducible to purely genetic evidence. Within this context, arguments based on interpretations of the development of science, the social nature of scientific activity, and the technique-ladenness of observations played instrumental roles.

Certainly, if the relative frequencies of known examples of the two types of inheritance were an index of the relative frequencies of their occurrence in nature, then the cytoplasm would clearly play a minor role in heredity. On the other hand, Sonneborn and Ephrussi reasoned that such a measure did not provide a reliable index of their occurrence in nature. It merely represented the extent to which cytoplasmic inheritance could be, and had been, investigated by geneticists. There was a technical incommensurability. One could not compare the results produced in favor of nuclear inheritance with those of cytoplasmic inheritance. Sonneborn (1951a, p. 199) summarized the major issues concisely:

> In the first place, there is a simple, familiar and highly successful method—the Mendelian method—for studying genic inheritance; no method comparable in simplicity, power, and reliability is available for the study of cytoplasmic inheritance. This results in considerable selection in the examples of heredity studied and reported in the literature: the simpler cases that yield readily to a familiar methodology are preferentially attacked and reported. Complex or less standard cases tend to be put aside or interpreted formally in terms of accepted theory. Secondly, genetic methods are designed chiefly for analysis of differences between individuals that can interbreed, hence, properties common to all individuals of an interbreeding group remain largely unanalyzable by the ordinary methods of genetics except under unusually favorable conditions.

This quotation contains several independent and important suggestions that were articulated throughout the 1950s by defenders of the cytoplasm, and which deserve specific attention. First and most generally, the facility of Mendelian analysis was seen to be largely responsible for deceiving geneticists and for leading many geneticists away from what actually occurs in the cell. But although technical opportunism could help explain the attention focused on nuclear inheritance, this alone could not account for the attitude of most American geneticists towards alternative forms of inheritance. Since, for advocates of cytoplasmic heredity, the attitude of nucleocentric geneticists could not be understood in rational terms (that is, the terms they claimed they themselves were using to arrive at an opposite conclusion), they turned to historical interpretations of the nature of scientific development.

They postulated that throughout the twentieth century, Mendelian geneticists had in part been advocating a cause and necessarily had to ward off alternative theories in their scientific struggle. This strategy would no longer be necessary once the chromosome theory became accepted and the presumed limits of its possibilities were realized. Within this context, the argument for the final recognition of cytoplasmic heredity in the 1950s was held to rest not only on the intrinsic properties of nature, but also on the natural order of social processes. Thus, Ephrussi (1951, p. 242) wrote:

> The rapid rise of genetics, resulting in the ever widening recognition of the universal character of Mendelian mechanisms, has indubitably overshadowed both the recurrent claims of evidence for the occurrence of extra-nuclear heredity, and the theoretical need for postulating its existence. No doubt, to be effective, revolutionary ideas, whether political or scientific, have for a time to disguise themselves in simplified forms and pretend to the monopoly of virtues. Moreover, since science proceeds by successive approximations, a prevailing theory has of necessity to explain all the facts it can account for, before a new theory can be allowed in. It is only natural therefore that the complete exploration of Mendelian principles should have delayed the recognition of cytoplasmic heredity.

Continuing his essentially political understanding of the nature of scientific development, Ephrussi (1951, p. 260) concluded. "'The gene as a basis of life' is only a revolutionary slogan which must not over-shadow the concept of 'the cell as a whole.'"

Another major question raised by cytoplasmic geneticists concerned the actual ability of Mendelian procedures to detect cytoplasmic inheritance when it occurred. This problem was addressed in Sonneborn's second suggestion that genetic analysis dealt mainly with differences between individuals of an interbreeding group, not with *similarities* or properties *common* to members of an interbreeding group. Implicit in this suggestion was the idea that non-Mendelian cytoplasmic heredity might be concerned especially with properties that distinguished higher taxonomic groups. This view had two consequences which could elevate the importance of cytoplasmic heredity and its researchers. On the one hand, "macro-evolution" has recourse to cytoplasmic as well as nuclear gene mutations, and on the other hand, cytoplasmic heredity is particularly concerned with fundamental cellular functions. It was also noted that if the latter were true, as Lwoff and Ephrussi (1953, p. 118) noted, cytoplasmic mutations would most of the time be incompatible with survival and would therefore escape detection.

In effect, this latter theoretical move allowed those who protested against the primary control of nuclear genes to turn the lack of reported genetic evidence for cytoplasmic heredity into an argument for the cytoplasm's crucial role in heredity. Sonneborn held a similar opinion. When referring to the views of many embryologists and physiologists, he wrote:

> Long ago it was suggested that these so-called *fundamental* traits were controlled by the cytoplasmic part of the genetic system, the nuclear genes modifying only relatively minor superficial individual variations. This view, in the extreme form just set forth, is certainly untenable. . . . Yet it is also remarkable that among the traits and properties known to be controlled by cytoplasmic genetic materials, a large proportion are undoubtedly very fundamental. For example, chloroplasts and chlorophyll production are common to nearly all plants except fungi; centrioles and the fibres they form occur in all higher plants and animals; kinetosomes and blepharoplasts, and the cilia and flagella that arise from them are common to great groups of unicellular organisms. The respiratory enzymes cytochrome oxidase and succinic

> dehydrogenase, known to be inherited through the cytoplasm in yeast, are probably common to all kinds of organisms. Perhaps other traits of comparable fundamental importance may prove to be similarly inherited. (Sonneborn, 1951a, p. 199)

Indeed, as Sonneborn argued, in some instances the loss of cytoplasmic elements endowed with genetic continuity was in fact shown to be incompatible with survival, such as the loss of chloroplasts in *Euglena* when grown in the dark. The *petite* mutations in yeast represented another illuminating example. Here a fundamental function, respiration, could be abolished. However, the actual detection of this phenomenon raised the issue of the role of the organism itself as a technological tool for investigating various types of biological mechanisms. In this case of respiratory deficiency it was argued that the mutation could only be detected because the organism in question possessed a remarkably efficient alternative energy-yielding metabolic pathway. As Ephrussi (1953, p. 119) argued:

> It is to this circumstance hardly realized with similar perfection in many other sexually breeding organisms, that we owe the possibility of its detailed study. The cytoplasmic control of fundamental functions may escape detection precisely for this reason.

By these collective arguments Ephrussi attempted to revitalize the notion that the nuclear gene mutations may be concerned only with "superficial" characteristics:

> The emphasis placed in the above statements to the term "fundamental" will probably be strongly objected to by many a reader as reminiscent of the old distinction between "general" and "special" heredity (cf. Brachet) and the controversies as to whether gene mutations concern both "superficial" and "fundamental" characteristics, rather than the former alone. I think that the question is today in need of serious reconsideration, and that it should not be answered by metaphors or by the usual counter-questions. (Ephrussi, 1953, p. 119)

The distinction between these hierarchies of organization was crucial when considering the role of cytoplasmic heredity in macroevolution, and equally important in understanding the possible reasons for the scarcity of reported cases of cytoplasmic heredity in higher organisms. It could be argued, for example, that what was fundamental to the cell is of necessity fundamental to the species and to the complex organism also, but the proposition could not be interted. Fundamental characters and cytoplasmic heredity, therefore, could be detected definitively only at the cellular level with the use of microorganisms.

The functional nature of the distinction between "fundamental" and "superficial" characters in the early 1950s becomes clearer within the context of a polemical discussion between Ephrussi and the evolutionary geneticist T. Dobzhansky. In the second edition of his celebrated text *Genetics and the Origin of Species,* Dobzhansky denied the meaningfulness of a distinction between "fundamental" and "superficial" characters:

> It has been contended, for instance, that mutations, meaning gene mutations, involve only "superficial" characteristics, leaving the "fundamental" ones unaffected. Those making such assertions have wisely refrained from revealing their criteria for the discrimination between superficial and fundamental traits.
>
> The presence of one pair of wings and one pair of balancers, as opposed to two

> pair of wings, is one of the most striking distinguishing marks of the order of flies (*Diptera*). One may ask then, is the appearance of four-winged *Drosophila* a fundamental or a superficial change? Is a mutation which diverts the embryonic development to a wrong course and thus causes death fundamental or superficial? Those who would like to see a mutant fly without an alimentary canal, or with the location of the heart and the nerve chain exchanged, overlook the fact that such a mutant could not survive and hence could never be detected. (cited in Ephrussi, 1953, p. 119)

Ephrussi directly confronted Dobzhansky's remarks and suggested that he was confusing the issues by failing to recognize the different meanings of "fundamental" when applied to the species, to the individual organism, and to the cell. In response to Dobzhansky's first question, he argued that the two-winged condition of a *Drosophila* species is nonfundamental to the individual and its constituent cells, since four-winged flies survive. His answer to the second question was that a mutation which diverts embryonic development to a wrong course and thus causes death may or may not be fundamental. Lethality of embryos, Ephrussi argued, is often due to the breakdown of the correlation which assures the harmonious development of the parts of the embryo. That is, the breakdown itself is at the organismic level and may be insignificant to the individual cell. On the other hand, Ephrussi maintained that changes at the cellular level due to cytoplasmic heredity could indeed be the cause of large developmental disturbances, which in turn could account for the scarcity of detected cases of cytoplasmic heredity in higher organisms:

> The fact that a "fly without an alimentary canal, or with the location of the heart and the nerve chain exchanged" could not survive is apparently more readily remembered than the possibility that this very fact is the cause of the apparent rarity of cytoplasmic heredity (Ephrussi, 1953, p. 120).

Whatever theoretical merit we may want to give to Ephrussi's arguments for the importance of cytoplasmic inheritance and for a major change in the genetic conception of heredity, we have to keep in mind that it is not the strength of ideas in themselves that decides the outcome of scientific controversy. A victory in genetics could be achieved only on the terrain of the technical procedures of that discipline. Cytoplasmic geneticists simply lacked the necessary technical capacity and institutional power to effect a revolution in genetics. Indeed, the struggle for authority is not simply a matter of *receiving recognition* through "valuable contributions to scientific knowledge." It also involves the power to *grant recognition* and define what is "valuable." Nucleocentric geneticists had been able to perpetuate a system of norms that best suited their particular interests and assured the continuous growth of the assembled data that could be treated by their technical procedures. Holistic or synthetic conceptions of heredity were largely ignored. It is impossible to understand the legitimacy of the nuclear monopoly of the cell during this period outside the social struggles in the field of production, where participants upheld those values which gave them the competitive advantage.

As we have seen, it has been difficult for us to decide whether scientific knowledge claims are accepted because those who uphold them have an interest in truth or whether they simply uphold the truth which suits their interest. This problem was recognized by Ephrussi and Sonneborn, and left them with the paradox of how certain and objective knowledge could ever be attained by a community of scientists. Ephrussi could only allude to the Baconian statement that "truth is the daughter of time," not of authority, which in his opinion was responsible for the theory of primary nuclear control. As he wrote in the preface of his small book on nucleo-cytoplasmic relations:

> I am . . . aware that many of the suggested deductions may not stand the test of time. I know also that very few of my fellow geneticists would subscribe to the views I have expressed, but I am encouraged by the thought that "we cannot determine the truth of a hypothesis by counting the number of people who believe it" and that "a hypothesis does not cease to be a hypothesis when a lot of people believe it." (Ephrussi, 1953, p. vi)

Throughout the period since the emergence of molecular biology the conflict over the significance of nuclear and extranuclear heredity has continued. Advocates of cytoplasmic heredity continued to use social accounting to help explain the attitudes of their competitors. On the one hand, historical interpretations have been employed to describe the collective behavior of scientists, and the technical advantage of one research program over another has been used to explain how the balance was tipped in favor of Mendelian inheritance. On the other hand, the individual behavior of those scientists who focused on the study of traits which could be easily investigated by Mendelian methods was explained by an enquiry into the nature of laboratory activity, the motivations of competitors, and publishing practices. During the late 1950s, D. L. Nanney, a former student of Sonneborn, went further and raised the question as to whether social forces acting upon individual scientists biased the evidence against cytoplasmic heredity:

> We are all aware that many considerations—both conscious and unconscious—are involved in the selection of traits to study and are cognizant of the non-randomness of the published accounts. Certain types of traits are more likely to be examined and certain types of results are more likely to be reported. One may argue, moreover, that these biases systematically prejudice judgement in favor of chromosomal inheritance (Nanney, 1957, pp. 137–138).

These biases, Nanney argued, began with a decision as to which traits should be studied. He maintained, like Sonneborn and Ephrussi, that from the beginning of genetics attention was focused on differences which could be readily distinguished between individuals, not on similarities. However, even after the study of a trait is initiated, a further source of bias may appear. Unlike the clear operational approach to Mendelian characters, the establishment of a cytoplasmic basis was a difficult and tedious task, since many possible vagaries of chromosomal behavior had to be eliminated. Thus Nanney (1957, p. 138) maintained:

> It is perhaps only natural that investigations of 'messy' characteristics are discontinued before publication and that investigators move on to traits more readily analyzed.

If this was, in fact, a correct evaluation of the practice of Mendelian geneticists, the role of the chromosomes in heredity could be easily overemphasized. Most American geneticists, in Nanney's view were, for whatever reasons, disregarding the anomalies which contradicted the prevailing Mendelian theory. They were ignoring the legendary exhortation of William Bateson who, in 1908, advised an audience of science students:

> Treasure your exceptions! . . . Keep them always uncovered and in sight. Exceptions are like the rough brickwork of a growing building which tells that there is more to come and shows where the next construction will be. (Bateson, 1912, p. 21)

"On the other hand," Nanney argued (1957, pp. 138–139), "one would like to believe that the rule regarding the cherishing of exceptions is not only an ideal but a widespread practice." In this light nuclear geneticists were simply skimming off the easy work which yielded quick rewards and leaving the difficult problems to others. Nanney (1957, p. 139) concluded his account of possible sources of bias in the reported accounts with the claim that "a completely satisfactory answer" to the significance of cytoplasmic inheritance would necessarily entail "*an analysis of the pressures to which scientists are subjected and the motivations of the investigator*" (my emphasis). But here he drew a disciplinary distinction and claimed that such analysis was beyond his competency.

When appreciating Nanney's remarks it is important to keep in mind that it is one thing to be alert to the potential significance of exceptions and another issue entirely, when one is actually confronted with an apparent exception, to know how much significance to attach to it. Though some exceptions prove to be important to new theories, others prove simply trivial, or even worse, seriously misleading. Certainly, in the struggle for credibility in science, a researcher has to weigh the time and risk involved in investigating (investing in) the apparent exceptions to standard rules against the yield in recognition (symbolic profit). Moreover, when a result seems to contradict prevailing rules it may be difficult to accept as not being due to experimental error, especially when the contending rules are supported by great names. Those who had strongly supported the nuclear monopoly continued to be celebrated in America. A Nobel prize had been awarded to Morgan in 1933 and to Muller in 1946, and the *Neurospora* school headed by Beadle was widely acclaimed in America as pioneer in microbial genetics and for bridging the gap between gene and character, genetics and biochemistry. To many geneticists trained in these schools, the risks may have simply outweighed the perceived profits.

Distinctions based on scientific *indoctrination* were also used by individual geneticists to understand the conflicting interpretations of genetic results. When Carl Lindegren broke from the Morgan school and began to support the inheritance of acquired characteristics and cytoplasmic heredity, he (1949, p. 27) explicitly referred to the bias due to indoctrination:

> Recently in America there has been a recognition of the phenomena discovered by the German workers and much discussion of cytoplasmic inheritance especially by

> Sonneborn, and it is noteworthy that his early training did not involve genetic indoctrination as a member of the Morgan school.
>
> In yeast genetics we encountered many examples of non-Mendelian phenomena in our early work. I interpreted them as involving the hereditary transmission of autonomous entities. This was a rather difficult thing for me to do because I had been thoroughly indoctrinated in gene theory by my long association with Dr. Morgan and other members of his famous staff.

This is not to say, however, that similar social forces were not operating in cytoplasmic genetic camps, or that those who challenged genetic orthodoxy did not have high hopes of success or that they were disinterested. Indeed the development of the research on cytoplasmic inheritance can also be understood as a strategy (objectively at least) directed toward the maximization of scientific recognition and prestige. With the intense competition which was triggered off with the genetic analysis of nuclear genes, one can easily understand the departure of a fraction of the researchers towards other objects, such as the study of the role of the cytoplasm in heredity and the principles of somatic cell differentiation, where competition was less intense. The possibility that cytoplasmic genetic factors were at least as important, if not more so, than nuclear genes in evolution, and that they held the key to understanding ontogenetic development, increased the potential profits to be gained.

CHAPTER 6

The Cold War in Genetics

> By the way, you and Muller will be amused to know that I am being controverted in this country for my views on Lysenko by people who say, "Professor Sonneborn has been given a prize of $1000.00 for proving that Lysenko is right." (C. D. Darlington to T. M. Sonneborn, October 7, 1948)

During the late 1940s and 1950s, while cytoplasmic geneticists struggled to gain recognition for their claims, they found themselves entangled in the most notorious of movements. Seen to be in support of Lysenkoism, the evidence for cytoplasmic inheritance came to be not only in violation of Mendelian "laws" but aligned with an ensemble of concepts and activities which, in the view of many, threatened scientific activity itself. Cytoplasmic geneticists in the West found themselves and their work to be allied with such issues as Soviet political ideology, the breakdown of scientific professional freedom, political control over the beliefs of scientists and the direction of research, as well as anti-American subversive activities in an ever-increasing Cold War.

During the 1940s biologists in the Soviet Union, particularly geneticists and cytologists, became seriously threatened by so-called "Michurinists" led by T. D. Lysenko, president of the Lenin All-Union Academy of Agricultural Sciences (L.A.A.A.S.) and his chief theoretician I. I. Prezent. Supported by the Central Committee of the Communist Party, Lysenko and his followers denounced Mendelian-Morganist genetics, calling it abstract, idealist, Fascist, racist, and incompatible with Soviet science and dialectical materialism.

The controversy between the non-Mendelian agronomists led by Lysenko who were concerned with improving Soviet agricultural production, and academic Mendelians led by N. I. Vavilov who had failed to show the applicability of genes to agricultural production in the Soviet Union, raged throughout the late 1920s and 1930s. While academic geneticists such as Vavilov, who was ultimately arrested in the late 1930s and died in prison, upheld the exclusive role of the nuclear genes in heredity and evolution, Lysenkoists emphasized cytoplasmic and environmental factors and argued for the cell as a whole as the basis of heredity. As Lysenko stated in 1937: "The hereditary basis does not lie in some special self-reproducing substance. The hereditary base is the cell, which develops and becomes an organism. In this cell different organelles have different significance, but there is not a single bit that is not subject to evolutionary development" (quoted in Medvedev, 1971, p. 27). Lysenkoists advocated a physiological conception of heredity, and, like embryologists, they insisted that heredity was a process resulting from the life history of the organism and its interaction with the environment. All parts of the cell and organism formed an interacting hereditary system. In the end, they went further and recklessly discarded the whole of neo-Mendelian genetics. Lysenkoists claimed that their views were based on practical experience and on the philosophical system of dialectical materialism. Basing their claims on the general principle that two entities which come into contact must modify one another, they argued that the hereditary nature of organisms must be affected by the environment in which the individuals develop. From this they developed a somewhat Lamarckian theory of the inheritance of acquired characteristics.

In effect, the dispute was settled in 1948 when the Communist Party and government organs which directed scientific research in the Soviet Union, of which the principal one was the Academy of Science, pledged support for the strengthening and development of the Michurinist trend in Soviet biology and agriculture. Eminent Soviet geneticists, cytologists, and evolutionists who held opposing views to those of Lysenko and his followers in the Soviet Union were subjected to condemnation and suppression as dangerous, bourgeois, reactionary fascists, regardless of their political views, removed from their positions, deprived of their laboratories, and some imprisoned. Under the legislation of the minister of education, all anti-Lysenko doctrines were to be systematically rooted out of the schools, universities, and libraries.

Lysenkoists attempted to improve Soviet agricultural production by creating plants with desired hereditary traits by three different procedures: (1) Grafting: by uniting the tissues of plants of different varieties, Lysenkoists claimed that they could produce seeds that gave rise to plants with mixed characteristics. (2) "Vernalization": plants were exposed to altered environmental conditions at definite moments of their development. When this was repeated for a few generations, and in each generation plants were selected that most nearly conformed to the type one wished to obtain, Lysenkoists claimed it was possible to obtain desired hereditary alterations. (3) Cross-breeding varieties sharply differing in habit or origin. This method was simply to cross different breeds and rear the offspring and later generations under conditions best suited to the variety one wished to obtain. In each generation one would select for further breeding those individuals which thrived the best under desired conditions and which possessed desired traits.

Lysenkoists persistently attacked the germ plasm-somatoplasm distinction to discredit the Western genetic doctrine that selection was the only means of directing hereditary change. The germ plasm-somatoplasm distinction, as Lysenko (1948, p. 19) pointed out, was perpetuated in the writings of Morgan and some other American geneticists when arguing for the theory of nuclear control of heredity. As a zoologist, Morgan was thinking of the fact that shortly after the egg of an animal is fertilized by the sperm, cells are set aside from which sex organs are formed. However, as many Western geneticists realized, Morgan made a serious error in making the isolation of the germ track an essential component of the theory of nuclear control of heredity.

Although the concept of a distinct germ plasm may have been useful in explaining many genetic results and aided in the planning of genetic experiments in the higher animals, many biologists recognized that it had undoubted limitations (see, for example, Sinnot, Dunn, and Dobzhansky, 1950, p. 26). It did not apply at all in single-celled organisms, where no practical distinction between soma and germ can be made. Even in higher animals it was possible that the gametes might arise from tissue which was not fundamentally different from that which produces other parts of the body and that the gonads were not completely insulated from environmental forces which effect changes in the body. Perhaps most importantly, in plants the sex cells are not separate from the soma. Many or all parts of the plant body may give rise to germ cells or to new individuals. Lysenko was primarily a botanist and was well acquainted with the fact that a vegetative cutting, grown by simply putting a twig into the soil, could produce flowers and seeds. Hence the soma can produce new individuals without the intervention of a sexual process.

However, by the end of the 1940s and early 1950s it became increasingly evident to many Western biologists, both Communists and non-Communists, that Lysenkoism as supported by the Soviet government did not recognize and attempted to override all those procedures which had come to be understood as "scientific." Western geneticists, both socialists and nonsocialists, requested Lysenkoists to provide detailed descriptions of their technical procedures so that their alleged revolutionary findings could be independently tested by others. They asked Lysenkoists to repeat their experiments with the controls demanded by their critics and to publish the results with full data so that others could analyze them. Lysenkoists responded largely with arguments based on Soviet ideology and dialectical philosophy.

Lysenkoists maintained their power and official governmental status after Stalin's death in 1953, until Khrushchev was ousted in 1965. In the meantime, as Adams (1977–78) has shown, Mendelian geneticists and molecular biologists went "underground" working unofficially in physical science institutions carefully isolated administratively from Lysenkoism. Throughout this period Lysenkoists failed to make any breakthroughs in Soviet agriculture. Instead, it is widely recognized today that they set Soviet genetics back a generation.

Lysenkoism has been subjected to a great deal of historical analysis and many important contributions have been made towards an understanding of how Lysenkoism arose and maintained its power. The books of Medvedev (1969) and Joravsky (1970) show how dogmatism, authoritarianism, and abuse of state power can help create and sustain erroneous theories. Lysenkoism is often used as a lesson about what can happen when untrained politicians interfere in scientific matters, or as a weapon against Marxist intellectuals (see, for example, Buican, 1984). For Marxist commentators Lysenkoism is seen as a lesson about how Marxist principles can be distorted and abused, and they investigate it with the hope of bringing the insights of Marxism into the practice of science (Lecourt, 1976; Lewontin and Levins, 1976). Too frequently, however, writers have viewed Lysenkoism as coming to hegemonic power between 1948 and 1952 by challenging the Mendelian-chromosome theory and selection theory with "outmoded," "pseudoscientific" ideas based on the inheritance of acquired characteristics which they claim had long since been "disproved."

When taking this attitude these writers have followed established precedent. Their views become indistinguishable from those of the orthodox Mendelian geneticists and population geneticists who opposed Lysenkoist and Lamarckian ideas. They have overlooked the fact that many leading Western geneticists saw a crisis in classical genetic doctrines and a major revolution that would overthrow classical conceptions of the gene and its primary role in heredity and evolution. The conception of the "cell as a whole," protests against deterministic, autonomous genes, and the belief that the cytoplasm might play the primary role in heredity and evolution were pervasive in Western science throughout the 1930s, 1940s, and most of the 1950s. As we have seen, these views were upheld by leading embryologists, physiologists, protozoologists, and geneticists in the United States, Germany, and France. Cytoplasmic heredity rose to such a prominent place in the 1940s and 1950s that Goldschmidt (1958, p. 240), who was always on the outside of whatever side was in, and who continued to criticize the evidence for cytoplasmic genetic elements, could write:

> Cytoplasmic heredity via plasmagenes became so popular that in textbooks and symposia plasmagenes were presented as established facts, and geneticists who remained cautious were considered fossils. Some geneticists and even embryologists ceased speaking of cytoplasm in genetic discussions, but used the term "plasmagenes" for whatever property of the cytoplasm they discussed.

Indeed, one of the leading British physiological geneticists, C. H. Waddington, also realized that developmental biology and genetics were in a state of crisis and required new philosophical concepts such as dialectical materialism. Waddington maintained close relations with his colleagues on the political left in Britain, including J. D. Bernal and Joseph Needham (see Werskey, 1978, p. 226). Like Ephrussi and Sonneborn, he recognized that the cell was an integrated unit which could be broken only for analytic purposes and that one could only "speak of autoreproduction of the integrated unit which is the cell, but it would be more correct not to speak of autoreproducing particles" (Waddington, 1953, pp. 784–785). However, Waddington opposed the notion of plasmagenes as the basis of irreversible cell differentiation, claiming that "they are nothing but our old and discredited friends the 'organ-forming substances' appearing under a new name and with their hair curled in the latest fashion" (Waddington, 1953, p. 784). He attempted to replace cases of cytoplasmic inheritance with self-perpetuating regulatory systems or what he named "epigenetic momentum."

Waddington had developed his views on the relation of genetics and embryology in his text *Organisers and Genes* (1940) where he discussed the importance of dialectical materialism for understanding ontogenetic development and the emergence of biological organization and morphogenetic fields. He acknowledged the importance of the field concept as a spatial unity in accounting for the formation of a developmental pattern. A field, he claimed, quoting from Julian Huxley, represented "an organised whole with certain unitary activities, which must be studied as a unit, not merely as a summative resultant of its parts and their activities" (Waddington, 1940, p. 135). He also accepted the "existence of different levels of organisation as a fact of nature." However, Waddington denied the value of the field concept and biological organization as explanatory concepts. Instead, he claimed that one could account for higher levels of organization in terms of elementary molecules without invoking other vital principles, as had been entertained by some embryologists, including Spemann (see also Baitsell, 1940). In Waddington's opinion, higher levels of organization depended solely on properties which the isolated parts possessed but which could not be expressed until the elements entered into certain relations with each other. For example, he believed that one could explain the cytoplasmic organization of eggs by supposing that they contained only "some orientated protein microstructure similar to that of a liquid crystal." The "existence of the fibre level of organisation," he wrote, "is not accounted for in terms of elementary molecules plus some entity of a higher level, such as an overriding field" (Waddington, 1940, p. 145).

It was here in the difficult problem of emergence that Waddington found the greatest use for dialectical materialism despite its "misuse" by those Soviet biologists who claimed the inheritance of acquired characteristics:

> The developmental side of biology—embryology, genetics and evolution—seems to be reaching a point where radically new types of thinking are called for. In such circumstances it would be very unwise to despise the newer philosophies such as dialectical materialism, which are framed particularly in relation to progressive changes, even if they have sometimes led people astray. (Waddington, 1940, p. 148)

Our understanding of the Lysenkoist-Morganist controversy is highly asymmetrical. Lysenkoists, with their belief in the inheritance of acquired characteristics, are seen as pseudoscientific and ideological, while orthodox genetics and Morganist doctrines appear as politically neutral, value-free, and the result of "free" scientific inquiry. The Soviet government backed the procedures and theories of Lysenkoists as the best strategy for improving Soviet agricultural production. In France, as we have seen, Ephrussi had embraced investigations of cytoplasmic inheritance, problems of directed hereditary alterations, and genetic regulation as the best strategy to be competitive with nucleocentric American geneticists, and Sonneborn did the same in his laboratory at Indiana.

Yet, because the inheritance of acquired characteristics became official Soviet doctrine in a Cold War game, it is important to know to what extent selection theory and gene theory were official American doctrine during the same period. As we have seen, those leading Western biologists who upheld the importance of their work on cytoplasmic inheritance during the late 1940s and 1950s claimed that the theory of exclusive nuclear control was also in part based on dogmatism, *ad hoc* formalistic arguments, and different social strategies, tactics, and modes of control. Since Lysenkoists were willing to suppress their opponents by direct political intervention during the late 1940s and 1950s, was this also true for geneticists in the United States who claimed the exclusive role of nuclear genes and undirected mutations in evolution? Did they use Lysenkoism as a weapon to attack their Western competitors who opposed classical neo-Darwinian doctrines, and how did cytoplasmic geneticists themselves and the Rockefeller Foundation officials respond to Lysenkoism? With these questions in mind, I will first review some of the major genetic evidence which tended to support Lysenkoist views.

Propaganda in America

> The work on Kappa in *Paramecium* and other plasmagenes shows that acquired characteristics can be inherited if the characters fall in a certain sub-division of the non-Mendelian category. (T. M. Sonneborn, 1950a, p. 535)

First, Lysenko's central technical work on vernalization was substantial and has often been overlooked by modern commentators (Rolls-Hansen, 1985). In fact, vernalization was similar to the technique developed by Goldschmidt and Jollos. As discussed in Chapter 3, during the 1930s Jollos reported that when fruit flies were exposed to a high temperature in the larval stage, hereditary alterations occurred. Goldschmidt coined the word "phenocopy" to describe a change which was not a gene change but which had the same appearance as a gene mutation. Jollos's work on heat resistance and other induced traits in *Paramecium* (*Dauermodifikationen*), where the environmentally directed adoptive hereditary cytoplasmic changes gradually faded away, were highly publicized and confirmed by many investigators. His own tragic end in the hands of American geneticists has been discussed earlier. However, Goldschmidt (1958, p. 203) continued to see "the remarkable features of vernalization in cereals" to be a threat to Mendelian orthodoxy:

> By appropriate treatment (e.g., with cold) at the proper time the physiological alternative can be induced; for example, winter wheat can be made to behave like summer wheat. This may be described as induced self-perpetuating cytoplasmic change. Self-perpetuation may mean . . . the presence of changed self-perpetuating bodies like mitochondria, but it may also mean a strictly alternative chemical property of the cytoplasm (a "steady state), which remains until forced again into its original condition.

Among the most publicized Lamarckian experiments of the 1920s were those of Michael Guyer at the University of Wisconsin, who had claimed that certain genes could be altered by immunological influences. Guyer (1924) reported that when antibodies to lens protein were given to or induced in a pregnant female rabbit, they had an effect on newborn rabbits and that this condition was heritable. Guyer's experiments were never taken up by others, though in 1944 there was reported evidence of a mutation in *Neurospora* presumably induced by an antibody (see Irwin, 1951, p. 212).

There was also the celebrated experiment of Griffith of the late 1920s who showed that live *Pneumococcus* of one strain acquired some of the characteristics of dead bacteria of another strain when both strains were injected into a live animal. As discussed in Chapter 4, this phenomenon could be seen as a means of identifying genetic material as nucleic acid. However, it could also be seen as a case of directed hereditary alterations since the heredity of one strain of bacteria was transformed by exposure to a specific environment, that is, killed bacteria of the other strain. As Lewontin and Levins (1976, p. 47) point out, it was widely quoted by Lysenkoists as a case of the inheritance of acquired characteristics. Even non-Lysenkoists viewed the phenomenon in this way. For example, when Darlington and Mather (1949, p. 211) reviewed these cases, they remarked: "Thus at the molecular level we can . . . produce a pseudo-Lamarckian effect. We can control heredity from the outside and control it . . . constructively."

There was a major upsurge in genetic interest in the inheritance of acquired characteristics in the 1940s when microorganisms were domesticated for genetic use. Bacterial resistance to antibiotics and other environmental agents had a common Lamarckian interpretation before the work of Delbrück and Luria, who, based on their statistical studies, showed that in many cases the environmental agent only selected for preexisting resistant organisms. At the same time, the many cases of substrate-induced enzyme formation in yeast and bacteria resisted this interpretation. "Adaptive enzyme" formation could not be explained in terms of the classical conception of the nature of genes (see Chapter 7), and many geneticists proposed a cytoplasmic genetic basis in plasmagenes. Indeed, the genetic evidence for plasmagenes was most illustrative in showing that hereditary changes could be directed by environmental agents in microorganisms and that traits acquired by the soma in higher organisms could be transmitted to the germ cells.

As discussed in Chapter 4, nucleocentric geneticists such as H. J. Muller and his followers attempted to dismiss cases of migratory plasmagenes such as *Sigma* and *Kappa* and other cases of plasmagenes as being due to simple infection, claiming they were of little genetic and evolutionary significance. L'Héritier, Sonneborn, Darlington, Lederberg, Ephrussi, and others opposed this interpretation, claiming that they could be of great evolutionary and developmental significance. There was the classical case of chloroplasts in *Euglena* as investigated by Charlotte Ternitz (1912), André Lwoff (Lwoff and Dusi, 1935; Lwoff, 1944), and others. When *Euglena mesnili,* which normally possesses about one hundred chloroplasts, is put in the dark the number of chloroplasts diminishes to about one or two after some months. The effects of use and disuse are inherited. Moreover, individuals without plastids, which are able to reproduce, may be formed at a mitotic division, though they grow slowly and have never been kept for many generations. *Kappa* in *Paramecium* responded to environmental influences in a similar way.

Then there was the case of the so-called genoid, *Sigma* in the cytoplasm of *Drosophila,* as investigated by Philippe L'Héritier, which was known to be transmitted by the egg as well as by the sperm. L'Héritier had demonstrated that the genoid could migrate could migrate from the body cells to the germ cells which then pass it on to later generations. (L'Héritier, 1948). Plasmagenes which could migrate from soma to germ cells provided a possible mechanism for the inheritance of acquired characteristics.

Another widely discussed case of cytoplasmic inheritance involving the inheritance of acquired characteristics concerned antigens in *Paramecium*. By choosing appropriate environmental conditions, Sonneborn and his collaborators, it will be recalled, demonstrated that in *Paramecium* it was possible to direct transformations to one particular antigenic type among the eight possible antigenic types produced. Moreover, when an immobilization antiserum was used in high enough concentrations to kill *Paramecium,* the transformed organisms were shown to be completely resistant to this agent. The transformation responding to the environmental agent was considered to be adaptive. During the late 1940s and 1950s the cytoplasmic mechanism involved in the inheritance of antigenic types was still being discussed. The interpretations considered involved either plasmagenes, metabolic steady states, or a mechanism of inheritance at the supramolecular level involving cytoplasmic fields and cytologically visible cell structures (see Chapter 7).

Ephrussi reported another case illustrating a close relationship between the environment, cytoplasm, and heredity. If a well-known mutagenic dye called acriflavine is added to a culture, there is no effect unless cells divide. However, irreversible mutation occurs at most cell divisions, resulting in more than 99% of the population being made up of slow-growing, small, respiratory-deficient cells which Ephrussi claimed to be due to cytoplasmic genetic elements perhaps located in mitochondria. Similar cases of non-Mendelian inheritance of acquired characters (including antibiotic resistance) were reported throughout the 1950s (see Chapter 7).

At University College London, J. B. S. Haldane also frequently cited cytoplasmic inheritance phenomena as illustrations of the inheritance of acquired characteristics; Haldane, a Socialist who broke with the Communist Party over Lysenkoism, stated that changes by grafting were possible and sometimes hereditary, but most cases, he believed, could not be transformed by grafting. He also recognized that a variety of chemical and physical agents could alter heredity, and on the basis of investigations concerning cytoplasmic inheritance argued that direct transformations were possible in protozoa. To Haldane (1954, p. 86) the behavior of the chloroplasts of *Euglena mesnili* in the light and in the dark represented "a perfect Lamarckian example of irreversible heritable loss of a function through disease." Haldane also believed that direct changes might also be possible to produce in higher organisms as well, but certainly not as easily as Lysenko believed. He was, however, skeptical of the scope of such phenomena and added:

> It is apparently vain to hope that the existence of such a series of organisms will prevent dogmatic assertions both as to the non-existence of this phenomenon and as to its universality. (Haldane, 1954, pp. 86–87)

Darlington (1958, p. 211), who was not a Communist, also considered the behavior of "plasmagenes" to be illustrative of a Lamarckian principle in evolution, though he did not believe Lamarckian mechanisms were the cause of what he considered to be forward evolutionary change:

> In one evolutionary respect we must note an important distinction between genetic particles in the cytoplasm and in the nucleus. In the cytoplasm they often respond directly to changes in the environment. Particles that are not needed in the dark are lost in the dark. An adaptive change occurs. The effect of disuse is inherited. When this happens in microbes where each cell division gives rise to a new organism we

> have the appearance of a Lamarckian principle in evolution. We may accept this description with the proviso that it is evolution backwards and it is evolution in a subordinate particle, a cytoplasmic particle. The exception helps us to understand the predominant role of the nucleus, the peculiar functions of the chromosomes and of DNA in evolution as a process of forward change.

While as we shall see, many Western Marxists refused to dismiss Mendelism altogether, Lysenko and his more rigid Soviet supporters were unwilling to make allowances. As Medvedev has stressed, "All theoretical constructs of the sect were directed toward the single goal of 'disproving' the chromosome theory of heredity." I. I. Prezent stated at the L.A.A.A.S. session in 1948:

> Cytogenetics is collapsing. It is not for nothing that Morganists invented offhand, in addition to genes, "plasmagenes," "plastidogenes," and similar terms to draw a veil over the theoretical and practical rout of Morganism. . . . Mendelism-Morganism has fully exposed its gaping emptiness; it is rotting also from within and nothing can save it now. (quoted in Medvedev, 1971, p. 145)

Lysenkoists did not refer to the work on nonparticulate inheritance of *Plasmon* theorists in Germany. They did, however, single out the writings of Sonneborn in support of their views. I. I. Prezent (1947, p. 120) in his text *Agrobiologia* referred to the early experimental work and theorizing of Sonneborn in the following manner:

> There are some faithful Morganists who try to deny the facts that overthrow Morganism. They keep in store, as ready answers, to all experimental data that disprove their theory, either the general excuse of "impurity" of original stock, or just one word: mutation.
>
> Sonneborn himself who is so reserved when it comes to discarding Morganism completely, made the reasonable statement: "such interpretations are, however, purely formal and, strange as it may seem, there is no genetical phenomenon that could not be "explained," with due allowances, by mutations. To interpret all the above described phenomena in this manner, it would be necessary to resort frequently to very improbable suppositions."

The experience of the Belgian embryologist Jean Brachet gives vivid testimony to the way in which Lysenko himself viewed the research on cytoplasmic inheritance (Brachet interview, December 10, 1981). In 1949 Brachet, then a member of the Communist Party, was sent by the party to meet with Lysenko to try to evaluate the situation in the Soviet Union. In an attempt to explain the results on grafting which had been reported by Soviet biologists, Brachet suggested the idea that self-replicating virus-like genetic particles in the cytoplasm or "plasmagenes" could replicate and invade the flowers from the somatic cells. He also offered an experimental test which he thought could prove or disprove the idea. The test involved repeating the grafting experiments and inserting a membrane between the two parts of the graft combination which would prevent viruses from passing but would allow smaller nutrients to pass.

Brachet recalls that for Lysenko the idea of doing such experiments showed that Brachet came from a capitalist country and was out of the question for two reasons. Lysenko claimed that there were no plant viruses in the Soviet Union.

Second, such an experiment, in Lysenko's view, was due to pure curiosity of scientists who are not working for the people—and so it was a waste of time. It did not matter how it worked. Brachet was supposed to speak on Russian radio to say that he was convinced. Instead, he left for Brussels where he denounced Lysenkoism at a lecture to a large audience organized by the Communist Party. He was subsequently given special permission to withdraw from the party.

By the late 1940s, the Lysenkoist-Morganist controversy was discussed in weekly and daily journals in many Western countries. In the United States the work of Sonneborn, in particular, was singled out in support of Lysenkoist views. On September 6, 1948, the national weekly magazine *Newsweek,* in an article entitled "Party Line Genetics," ran the following item related to Sonneborn's work, the inheritance of acquired characteristics, and Lysenkoism:

> The irony is that "bourgeois science" in the United States has lately come around, through free inquiry, toward some reconciliation of the opposed schools of heredity and environment. By breeding microscopic paramecia, Prof. Tracy M. Sonneborn of Indiana University has shown that environmental factors can change the nature of this single-celled animal, and that the changes are inherited. The findings have important implications for the theories of embryology and cancer cause, while knocking out some underpinnings for the orthodox theory that all heredity is controlled by genes in the cell nucleus.
>
> Thus Professor Sonneborn's researches undermined, by experiment, the same classical theory of genetics that the Soviet savants of Lysenko's group have been attacking by dialectic argument. But no one has investigated him for subversive or un-American activity. Instead, he and his Indiana co-workers were awarded the annual $1,000 prize of the American Association for the Advancement of Science for 1946. Sonneborn's article on "Genes, Cytoplasm, and Environment in Paramecium" is featured in the current (September) Scientific Monthly, and at the AAAS centennial celebration in Washington on Sept. 14 he is scheduled for one of the principal addresses.

In 1950 the *New York Post* assigned an agent to enroll in the Jefferson School of Social Sciences to report on the Communist Party's current teaching practices. In one of a series of articles on the "little red campus," it was reported that in a lecture on "Michurinism versus Formal Genetics," the instructor, Bernard Friedman, M.A. in embryology and genetics from Cornell University, and a Communist, said that all "capitalist science" was in a state of crisis to which Marxism was the only solution. The article also suggested that Sonneborn, who had asked Friedman for copies of some Russian experiments Friedman had translated, was sympathetic to Lysenkoism and was "trying to maintain his scientific attitude" despite the fact that he was at the same faculty at Indiana University with the "notorious" Professor Muller. Muller, whose contribution to the gene theory was equaled perhaps only by that of Morgan, and who had spent several years at the Soviet Academy of Science in Moscow, had resigned from the academy in protest over the Lysenko issue. He quickly emerged as one of the most authoritative and vocal critics of Lysenko's impact on Soviet science.

In his letter of resignation from the academy in 1948, Muller called Lysenko a charlatan and accused the leaders of the academy of misusing their positions to destroy science for narrow political purposes. In face of the charge that Western genetics was racist and Fascist, Muller was one of the first to point out that the inheritance of acquired characteristics would reach equally as Fascistic conclusions as Nazi science, since it would imply (and did admittedly imply according to some members of the academy) that "the economically less advanced peoples and classes of the world have become actually inferior in their heredity" (quoted in Zirkle, 1949, p. 563).

Friedman, on the other hand, publicized some of Muller's biological determinist views:

> Just to show you how unstable Muller is and the depths of mire to which he would sink, he wrote a book several years ago in which he made the outrageous and ridiculous suggestion that Lenin's sperm should be used to fertilize as many Russian women as possible (February 13, 1950, p. 35).

Like other Western Communists, Friedman believed that Lysenko did not deny that chromosomal genes existed but had demonstrated the existence of other genetic particles that were subject to direct environmental control (Friedman, April 18, 1950). He further claimed that Sonneborn's work was in obvious support of Lysenko and that it was too bad that he did not come out and admit it (Sonneborn to R. C. Cook, December 10, 1949).

The place in which Sonneborn found his work during the Lysenko controversy is perhaps best illustrated by the following statement, which ran in the column next to Friedman's in *The Daily Compass* under the headline: "Is There Any Scientific Basis for the Lysenko Theory?":

> Tracy M. Sonneborn, professor of zoology at Indiana University, supported the theory of the inheritance of acquired characteristics in a paper read before the American Association for the Advancement of Science two years ago.
>
> These and other chromosomes in the wind lead one to suppose that the theories of Trofim Lysenko, which are also the official genetic theories of the Soviet Union, may very well have some basis in fact.
>
> The argument has long since ceased to be a scientific one, however. Not to accept Lysenko, lock, stock and gene, in the U.S.S.R. is close to heresy. To admit in the U.S. today to the possibility of some basis of fact in the Lysenko approach is tantamount to having subversive thoughts. And, unfortunately, science can never flourish without completely free inquiry, so that a proper evaluation of Lysenko will have to wait for the expiration of the cold war. (Boutell, 1950, p. 8)

As a central figure in American genetics being quoted in support of Lysenkoism and struggling to get his own ideas of cytoplasmic inheritance accepted in a Mendelian milieu growing more and more hostile in a Cold War, Sonneborn quickly became defensive. He was actually confronted from time to time by American biologists who asserted that his researches did in fact support Lysenkoism and challenged him to explain that away (Sonneborn, 1978, unpublished autobiography, p. 24). Sonneborn strongly resented having his research cited in support of the ideas of Lysenkoists, who, he charged, appealed to authority, dismissed their opponents, and deprived them of means of conducting their research, and, too often, of their lives. As early as 1948, Sonneborn was already busy attempting to get support from the Intelligence Objectives Agency to smuggle leading geneticists whose lives he believed to be in danger out of the Russian zone of influence (Sonneborn to Captain Weir, December 6, 1948).

He was also busy getting biographical information about Bernard Friedman. Sonneborn had no Marxist sympathies. As a student at Johns Hopkins University, his encounters with Marxists had led him to draw a decisive moralist position. It was then, according to his own account (1978, unpublished autobiography, pp. 21–22), that

> it first became clear to me that a cardinal principle of Marxist tactics was that anything—lies, misrepresentation, or more . . . was approved if the end or object was "good." That alone was quite enough for me. Marxists could not be trusted and Marxism was evil at the core. From that time on, I was a strongly convinced anti-Marxist and never suffered the fate of the many intellectuals who were attracted to it.

Sonneborn was, however, very sympathetic to Lamarckian ideas and throughout his career he was always "on the lookout" for "the inheritance of environmental effects" (Sonneborn to the author, June 2, 1980). As he wrote in his autobiography (1978, unpublished autobiography, p. 24) about his relationship to Lamarckism and the Lysenko affair:

> When Lysenko rose to prominence, I was thoroughly set against him on ideological grounds. Whether this colored my interpretations of my research I cannot say. I don't think it did.

When in 1948 Darlington wrote Sonneborn and nonchalantly mentioned the rumors in England that his work supported Lysenkoism, Sonneborn was not amused. Sonneborn quickly wrote back to him rebuking him for the casualness with which he treated the Lysenko issue. He, and Muller, who had persistently criticized his views on cytoplasmic inheritance, were now determined to cooperate in any way they could:

> Muller and I are going to prepare a joint statement, as representatives of the Human Genetics Society and the Genetics Society respectively. This will deal with the general problem of Communist genetics and we hope it will get wide publicity. And I have still other plans for controlling the metastasizing cancer of Communist "genetics" propaganda. You see Muller and I are definitely not amused by the situation, we are deeply alarmed. (Sonneborn to Darlington, October 13, 1948)

Realizing the seriousness of their intention to enter the politico-scientific debate, Darlington wrote back (October 16, 1948) assuring Sonneborn of his support:

> I hope that you and Muller will go ahead in the most aggressive way. We have been on the defensive too long. Perhaps you could let me have anything you are publishing for quotation in our papers, both *Discovery* and *Nature* are entering the battle in a week or two and will need all the support they can get from genuine scientists.

Sonneborn was resolved to do all he could to correct what he considered to be "the vicious rumor" that his work supported Lysenkoism. As he wrote to Dobzhansky (October 18, 1948):

> You will be interested to see a translation of an article by I. I. Prezent. . . . This article confirms my suspicion that the Communist devils are quoting from my scriptures as if I were a supporter of Lysenko. Darlington writes me that the English Communists quote me in support of Lysenko; and Newsweek did the same in this country. So you see I have a very personal interest in typing to set matters straight.

The following month after the first story appeared in *Newsweek,* Sonneborn published a corrected version of his work, allying himself with nuclear geneticists and stressing that his research was built upon and supported "classical genetics." He further gave a broadcast on the Voice of America in an attempt to clarify the situation. The broadcast was translated into several languages, including Russian, and was beamed to the Soviet Union, among other places (Sonneborn to Darlington, October 8, 1948).

Sonneborn's powerful position as president of the Genetics Society of America and president of the American Society of Naturalists in 1949 allowed him to be extremely active in leading a counteroffensive against Lysenkoism. Sonneborn, Ralph Cleland (president of the Governing Board of the American Institute of Biological Sciences [AIBS]), H. J Muller (president of the Human Genetics Society), and the Russian émigré and population geneticist T. Dobzhansky prepared an official statement to clarify the nature of "the current policy of extermination being carried on by the government of the U.S.S.R. against genetics and to condemn this policy in the name of American biologists." They hoped the AIBS would sponsor and publish it unauthored (Sonneborn to Glass, November 23, 1948). Their primary strategy was to dismiss any scientific basis in Lysenkoist views and disclaim them in a wholesale way. In effect, they mirrored Lysenkoist dogmatism. The inheritance of acquired characteristics was nonnegotiable. They summed up their position in the following propositions:

> 1. The contentions of Lysenko and his supporters have no basis in scientific fact. The Lysenko position is not science, it is a superstition put forward by politics.
> 2. Modern genetics researchers do not support the official Communist views on heredity, and any attempts on the part of Russian proponents of the Lysenko doctrines to bolster their case by citations from the works or conclusions of western scientists are gross distortions of the meaning and intent of these scientists.
> 3. We condemn the actions of the Soviet government in presuming to banish a firmly established science from its schools, publishing houses and research laboratories, and in prosecuting scientists because their field of inquiry is distasteful to the government. (Governing Board of the AIBS, 1949)

The AIBS was established in 1916 by the National Academy of Sciences and organized with the cooperation of the National Scientific and Technical Societies of the United States. Before making the statement public, Cleland wanted approval of the executive committees of the AIBS, the American Society of Naturalist, the Genetics Society of America, the Human Genetics Society, and the Society of Evolution (Cleland to Irwin, November 6, 1948).

However, the attempt to obtain unanimous support for the statement failed. First the Evolution Society expressed its lack of interest in the statement by not responding for several months. Other members of the executive committees voted against publishing an official statement on the grounds that it would only serve to strengthen Lysenko's hand by proving that Mendelism is official United States doctrine and thereby worsening the relations between political authorities and Russian scientists. Still others agreed to go ahead with it, except that it should be restricted to a plea for "freedom of science," and that point 1 was too dogmatic. Fritz Went, at the California Institute of Technology, was among the most prominent biologists who took this view. Went was trained in the Netherlands as a plant physiologist and had a distinguished career in plant geography and ecology. Like Brachet, Haldane, and others, Went believed that characteristics could in principle be transferred by asexual hybridization of plants. Though he could not condone the activities of Lysenkoists, he wrote a long letter to Cleland "blaming the situation in Russia on the intolerance of the Russian Mendelians of all non-Mendelian geneticists" (Sonneborn to Dobzhansky, January 11, 1949). Despite these protests, the view that the official statement did at least represent the great majority of American biologists prevailed, and it was published in *Science* (1949, 110, pp. 124–125).

In the meantime, Sonneborn had established a Committee to Counteract Antigenetics Propaganda of the Genetics Society. He asked Bentley Glass at Johns Hopkins University to be chairman of the committee. "The antigenetics propaganda of the Communists is getting serious in all Western countries," he wrote Glass (November 23, 1948),

> and there is a real need for measures to counteract it. The opinion is widespread, among even intelligent non-biologists, that the issue between the Lysenko group and Western geneticists is one between rival and equally scientific hypotheses. It is necessary to get across to the public that the issue is really between politics and science and that the official Communist writings on the subject are misquoting and misinterpreting the works of Western geneticists.

The committee was composed of a somewhat tainted group of three other members: Muller, Dobzhansky, and R. C. Cook, who edited the *Journal of Heredity*, published by the American Genetic Association, which still carried the subtitle "Eugenics—Heredity—Breeding." Dobzhansky himself, who was in Brazil lecturing and collecting *Drosophilia* (which he hoped would prove interestingly different from the Californian ones) questioned the choice of members. "There is nothing I can refuse to counteract the Lysenko propaganda," he wrote Sonneborn (December 29, 1948):

> But let my first act as a member of said Committee to be a suggestion to the president of the Society. The Committee of Glass-Muller-Cook-Dobzhansky is a bit colored. As you perhaps know, no lesser authority as Minister of Education of U.S.S.R. has publically called me untranslatable but highly uncomplimentary names, and these names have been repeated on radio Moscow. "Pravda" certified that I am "open enemy of Soviet people." Now Muller has recently deserved some similar labels. Cook is editor of a journal that has, at least in the past, published some stuff which is classified racistic.

Dobzhansky suggested that in order to make the committee more convincing to American intellectuals, Sonneborn should appoint two biologists who could not possibly be suspected of such crimes. Sonneborn opposed the idea. He wanted a small group that could act in unison the need arose. He shunned the possibility of adding biologists whose views might not be in agreement. "There is one danger that must be avoided," he wrote to Dobzhansky (January 11, 1949), "and that is tying the hands of the Committee so that unified action cannot be taken. This also would result in not convincing 'American intellectual circles.'"

It must be emphasized that Lysenkoism was not simply a movement which threatened the legitimacy and authority of genetics. It involved much broader issues concerning science policy—state control versus collegiate control, "academic freedom," and "basic science." From the very beginning when a committee was formed, "freedom of scientific inquiry" from political dictation and from Communists was a major issue. The most extreme sentiment of the group was represented by the views of R. C. Cook:

> This member of the Committee is convinced that (1) the commies are convinced (and with pretty good grounds) that they have in Lysenkoism a powerful ideological weapon; and (2) they are going to push it to the limit. And furthermore, there is a great deal of long-haired sucker-bait around that eagerly grabs at just this sort of thing. The opposition is *not going* to let the issue subside. We can stalk with what dignity we can muster back into the wreckage of the Ivory Tower of pre-1939. But we won't be fooling anybody but ourselves if we do that. Like the atomic physicists, we geneticists are right in the middle of the biggest battle of modern times for freedom of the human mind. (Cook to Irwin, December 4, 1949)

As Sonneborn conceived it, the committee was formed to promote united action in public matters of concern to the Genetics Society. However, the committee spoke solely as a committee of the society; it was not supposed to speak officially for the views of the Genetics Society as a whole unless the matter was submitted to a general referendum. It was therefore limited to the activities of its individual members. In 1949 Cook, Dobzhansky, Muller, and Glass each published two or three articles on Lysenkoism in journals ranging from *Science* to the *New York Herald Tribune*. The committee was also active in securing suitable publications of informative articles written by others on Soviet science, the relation of politics to science, and the importance of "scientific freedom."

A major crisis occurred, however, when a young, nontenured chemist (a Communist) was dismissed from the faculty of Oregon State College after having written a letter to *Chemical and Engineering News* in support of some Lysenkoist ideas. The committee was unable to agree upon a statement to be made public. The critical issue centered on the notion of "scientific freedom" itself. Did it mean preventing Communists from teaching unorthodox ideas by political or administrative dictation, or did it mean the freedom of the professor to teach whatever he or she judged fit to teach. (In effect, the discussion was a mirror image of the modern creationist-evolutionist dispute in the United States.) The controversy climaxed at a meeting of the Genetics Society held in New York on December 29, 1949, when attempts were made to extend the committee's mandate for the next year and permit it to speak for the society as a whole on matters affecting the "freedom of science." Bentley Glass set up the issues clearly in a letter to Sonneborn (December 17, 1949) several days before the meeting:

> The largest issue involved, that of the freedom of science from politics is so vital that I for one do not believe that our reaction to the danger of infiltration by Communists into our educational institutions should be allowed to obscure the main issue. In short, I do not believe that we can use political dictation over science and academic freedom even to exclude Communists. The dangers inherent in a recourse to such methods are so great that we must find other ways to handle the immediate danger.
>
> . . . I want it to be clearly understood that I am definitely in favor of a continuation of the Committee to Counteract Anti-Genetics Propaganda, although . . . I think the Committee ought to contain a representation of those members of our Society who are possibly less extreme or less emotional in their reactions to the issue. I do not believe we should hide in the ivory tower, but neither do I believe that we will secure our ends by descending to name-calling. That is why I think that the whole question of the existence of the Committee, its powers and limitations, composition and duties should be fully defined by the entire Society in our annual meeting.

The approval of a committee to speak on behalf of the society as a whole was not obtained and the new committee remained largely inactive. Carl Lindegren was one of the most outspoken of those who protested against it. It will be recalled that, based on his own genetic work on yeast, Lindegren came to oppose the predominant role of nuclear genes and natural selection in heredity and evolution. Lindegren stated that he did not want anyone to speak for him as a member of the society and that if such a committee was formed he would resign from the society. As Lindegren later recalled, "It seemed wrong that a Committee founded for the purpose of suppressing freedom to teach whatever the teacher saw fit should call itself a Committee for 'Scientific Freedom' when it was exactly the opposite" (Lindegren, 1966, p. 8).

Muller, who was unable to make it to the meeting, had a different perspective. He stated his opinion clearly in a scathing letter to C. R. Singleton, secretary/treasurer of the Genetics Society:

> I doubt very much whether, under the hampering conditions imposed at the last meeting . . . the committee can any longer accomplish anything of value in this critical and fragile situation. . . . If . . . the members of the Genetics Society itself cannot rally wholeheartedly and honestly to the support of their own subject . . . the situation is not at all hopeful that they will in time be aroused to their duty as scientists, as human beings, and as responsible social agents, interested in forestalling the collapse of intellectual and moral progress.
>
> It is a matter of great regret to me that circumstances made it impossible for me to attend the business meeting . . . at which these matters were discussed. (Muller to Singleton, February 17, 1950)

Sonneborn ended his presidential year with an address to the American Society of Naturalists entitled "Heredity, Environment and Politics," in which he addressed several issues left largely untreated by many of the American discussants of the Lysenko issue. First he attempted to explain the attitude of "most professional geneticists" who viewed the inheritance of acquired characteristics as "an outmoded superstition." "Whether right or wrong," Sonneborn reasoned, their attitude was at least understandable in view of the record:

> Of the many previous attempts to demonstrate experimentally the inheritance of acquired characteristics, all have failed. In most cases, the attempts yielded negative results. When positive results were claimed, the work later proved to be fraudulent, indecisive, or incompletely performed; repetition with unobjectionable methods always failed to establish the claims. No wonder most geneticists consider the matter closed. (Sonneborn, 1950a, p. 529)

Now that Lysenko had rooted out and removed the opposition by political force, the subject of the inheritance of acquired characteristics was placed in a new position. It no longer could be considered merely as a biological controversy. Nevertheless, for the purpose of analysis, the biological and political aspects of the matter had to be considered separately. This was so, Sonneborn pointed out, since "the political support given to a biological theory and its agreement with a particular philosophy may be irrelevant with respect to its scientific validity" (Sonneborn, 1950a, p. 529). He then attempted to present a summary of what evidence had been presented by the Lysenkoists and the responses of Lysenkoists to their critics.

Treating the issue at this level gave Sonneborn the opportunity to carefully defend and distinguish his work and its theoretical and experimental basis from that of Lysenkoism. First, he attempted to dispel the belief that the demonstration of cytoplasmic determinants or "plasmagenes" supported Lysenkoist views and show that they were in fact at variance with Lysenkoism. Lysenkoism, he pointed out, denied the existence of any special substances of heredity such as "plasmagenes" and nuclear genes, whereas the demonstrated cases involving cytoplasmic inheritance were, at least in most cases, ascribable to cytoplasmic particles. As for the second apparent compatibility—that is, the inheritance of acquired characteristics—Sonneborn agreed that the work on plasmagenes showed that acquired characters could be inherited. However, he made it clear that this did not undermine "neo-Mendelian" genetics, which dealt with an entirely separate category of phenomena. The inheritance of antigens in *Paramecium,* which did not involve visible particles, in Sonneborn's opinion also represented a case of the inheritance of acquired characters:

> The main feature of the antigen work is that specifiable environmental conditions can force upon the cells specifically adapted and directed responses which are thereafter inherited through the cytoplasm (Sonneborn, 1950a, p. 536)

Nonetheless, he claimed that these cases of "the inheritance of acquired characters" did no support Lysenkoism either, since in both cases the decisive genetic factors were localized in the cytoplasm and never transmitted by the nucleus. Such localization, Sonneborn repeatedly argued, was contrary to Lysenkoism, which, he claimed, maintained that each part of the cell was involved in the inheritance of all traits. To Sonneborn, then, Lysenkoist views confronted genetic views in much the same way as the Hippocratic humoral theory of disease confronted localistic pathology.

As others had done before him, Sonneborn concluded his essay by challenging Lysenkoists to provide detailed descriptions of procedures so that the validity of these results could be independently tested by others, and to repeat their experiments taking into account the critiques of Mendelian geneticists. Finally, he concluded his attack with a discussion of the larger issues, which involved the freedom of science from political dictation. Summing up the situation, Sonneborn (1950a, p. 536) quoted from Julian Huxley:

> The issue could not be stated more clearly: Do we want science to continue as the free pursuit of knowledge of and control over nature, or do we want it to become subordinated to political theory and the slave of national governments? It is a crucial question, on which the general public as well as the professional scientist must make up its mind.

The drive of the Genetics Society of America to counteract Lysenkoist propaganda culminated in 1951 with the text *Genetics in the Twentieth Century*. In structure the text was comparable to the Lysenkoist text *The Situation in Biological Science,* which had appeared two years earlier and contained papers written by fifty-eight Soviet scientists and two by Lysenko himself. *Genetics in the Twentieth Century* was composed of twenty-six papers written by members of the society. The papers had been presented at the program on the "Golden Jubilee of Genetics" which was sponsored by the American Institute of Biological Sciences. The Rockefeller Foundation also made a grant of $15,000 to the Genetics Society of America to protect "the freedom of science." The money was to be used for travel expenses of speakers to the meeting, largely for foreign speakers, and for publication of the conference papers. The objectives of the celebration were summarized by one of the members of the committee in charge of the organization of the meeting as follows:

> It is the considered opinion of the program committee that the best answer to the anti-genetics propaganda, and anti-science propaganda, in general, is to present a program at this Golden Jubilee which will put principal emphasis on positive achievements of genetics, both theoretical and applied. This should not be done in a spirit of either boastfulness or complacency, but nevertheless should stress accomplishments among scientists under which the progress was made. No single science has a better opportunity to speak for the cause of the freedom of science in general than has genetics during this Golden Jubilee year (Irwin to Sonneborn, March 24, 1950).

On the practical side, the relationships between Mendelism and practical breeding was emphasized. Hybrid corn, which resulted in a billion-dollar industry in the United States, especially its scientific development and its practical success, was held up in defense of the practical reality of Mendelism and in support of "pure" science. On the theoretical side, of course, advances in Mendelian genetics were highlighted with historical treatments of the development of research in various new specialities. For Sonneborn, the publication of the text would provide a forum for him to relate the work on cytoplasmic inheritance to nuclear genes. About the title, Sonneborn (April 22, 1950) wrote Irwin:

> I'll suggest now "the hidden role of the genes in cytoplasmic inheritance." Please let me know if this is the sort of thing you want and if you would like the title "softened" or modified in any way. Perhaps the word "hidden" should be deleted.

Ultra-conservative Anti-genetics in France

> As I look back on our conversations, I still feel that I presented to you as honestly as possible my opinion concerning the risks of leftist Lysenkoism, but I am not quite sure that I emphasized sufficiently the dangers to French genetics coming from the representatives of the still traditional French Lamarckism, more frequently associated with political ultra-conservatism. A few days ago, we learned the results of the first round of elections of the new C.N.R.S. committees. Although it is impossible from these to predict the exact composition of the future "Directoire," it is not altogether unlikely that it will be weighted with some very conservative elements. This might recreate some of the difficulties which retarded the development of genetics in France in pre-war days. (Boris Ephrussi to Warren Weaver, March 16, 1950)

In France, where many leading biologists were members of the Communist Party, the situation was especially complex. Before Lysenko obtained hegemonic power in the Soviet Union, there already had been attempts to relate Marxist theory to problems of genetic regulation and the primary role of the cytoplasm in heredity and evolution. This, of course, could be done without recourse to a belief in the inheritance of acquired characteristics. Indeed, the embryological idea that evolution involved macromutations of the structure of the egg's cytoplasm was fully compatible with dialectical materialism. In his text *Biology and Marxism* (1943), for example, the Marxist biologist Marcel Prenant, professor of zoology at the Sorbonne, reviewed the embryological arguments for the claim that the material structure of the cytoplasm played a primary role in heredity and evolution. Prenant (1943, p. 124) claimed, like many other embryologists, that the traits that followed Mendelian laws were

> hereditary characters of relatively small importance, such as the differences between individuals of the same species. But the characters which concern the general forms of the body and which form the basis for the separation of the great zoological and botanical groups depend on the material structure of the cytoplasm.

He concluded that macromutations involving changes in the structure of the egg cytoplasm brought about by extracytoplasmic influences, such as by an accumulation of many gene mutations, conformed very closely to Marxist principles:

> This is an hypothesis only, and one which will be hard to test for a long time to come. Changes of this kind must certainly be far less frequent than simple mutation. However, it has two merits. It is of the dialectical type which applies to the whole realm of known human experience; and it makes the sudden developments which it implies not miracles but revolutions following long preparations through crisis. (Prenant, 1943, p. 150)

Prenant's work was one of the targets of the committee for "scientific freedom" and was attacked by Dobzhansky (1949) in a review entitled "Marxist Biology, French Style."

When Lysenkoism emerged on the French scene, many French biologists embraced it, Marxists and non-Marxists alike. On the one hand, the traditional neo-

Lamarckian biologists of France were quite excited about Lysenkoism inasmuch as it was allied with their belief in the inheritance of acquired characteristics. On the other hand, Communist biologists such as Prenant and Georges Teissier, who had been director of the C.N.R.S., attempted to make a compromise between Lysenkoism and Mendelism. Like many other Western Marxists, they initially claimed that, although classical genetics should not be dismissed altogether, Lysenko had demonstrated the inheritance of acquired characteristics and its practical importance to the great benefit of the Soviet people.

One could in principle go far in allying the physiological conception of the "cell as a whole," the genetic work on cytoplasmic inheritance, and genetic regulation with Marxist principles. Ephrussi himself was not a Marxist and did not associate his work with Marxist views. At the same time, he had no major conflicts with his leftist colleagues in France. His primary conflict was with the authoritarian control of the antigenetic neo-Lamarckians in the universities, who were more frequently allied with the extreme political right.

As discussed in the last chapter, Ephrussi and his collaborators carried out their genetic work at the Rothschild Institute for Physico-Chemical Biology. However, his laboratory at the Rothschild Institute was cramped. In Ephrussi's view, successful competition in the rapidly developing postwar genetics required an expansion of laboratory facilities. It required large, new, and costly equipment. Adequate space was needed for collaboration with other French scientists and with foreign visiting researchers, which in turn was necessary for rapid exchange of fresh ideas and approaches. In spite of the lack of space in his laboratory, Ephrussi insisted that L'Héritier and his collaborators install themselves in his laboratory.

The C.N.R.S. promised to overcome some of these difficulties and to compensate for the situation at the universities. In 1946 an ordinance had been issued by the director of the C.N.R.S, Teissier, for the construction of new buildings for the Institute of Genetics to be situated about 20 kilometers outside of Paris, at Gif-sur-Yvette. However, the plans for the new institute matured slowly and obtaining adequate support for it was a difficult task.

The problems to be encountered became immediately apparent when on July 22, 1949, Ephrussi applied to the Rockefeller Foundation for a grant of $54,000 for research equipment to be installed at the new institute, then under construction at Gif.

On February 15, 1950, after studying Ephrussi's request for financial assistance, Warren Weaver wrote back to him describing the reluctance of the officials to fund the research project under the C.N.R.S., whose director they believed to be the Communist Teissier:

> The record of The Rockefeller Foundation makes it hardly necessary to preface my remarks by pointing out that we exclude any question of race, religion, or politics in judging whether we shall make a requested appropriation. This necessarily implies, however, that such considerations will also be excluded from any scientific research that we support. Otherwise, we should in fact—and against our policy—be indirectly involved in supporting racial, religious, or political interests.
>
> Until recently this position has been one of principle and of general interest, and assurances have not been required in specific cases. Indeed, for over three centuries

activities in science have been specially free from the influence of such considerations. No one, however, can sensibly disregard the fact that genetics and politics have recently been inter-related, in some quarters, in a most confusing, a most disappointing, and indeed a most fantastic way.

There is no necessity to go into detail, for the whole matter has been widely discussed by those who have special competence. The essence is well stated by Huxley in the opening paragraph of his NATURE article Soviet Genetics: The Real Issue (June 18, 1949, page 935), "There is now a party line in genetics, which means that the basic scientific principle of the appeal to fact has been overridden by ideological considerations. A great scientific nation has repudiated certain basic elements of scientific method, and in so doing, has repudiated the universal and supranational character of science."

The Rockefeller Foundation is not prepared to aid research in genetics anywhere in the world unless it has assurance that this research can and will be carried out in the true spirit of universal science. There must be, of course, a complete dedication to the unbiased discovery of facts—all the facts and not merely certain misleading or partial facts which conform to a predetermined code. It goes without saying that we could only be interested in aiding geneticists who are in a position to affirm their devotion to the concept of properly controlled experiments, fully described and fully open to critical judgement of the scientific world, carefully interpreted by modern quantitative standards as to reliability and significance.

You will not, I am sure, think that I am to the slightest degree suggesting that any special set of scientific ideas are so sacred that they must not be questioned. On the contrary, we all know that vigilant skepticism and a steady willingness to shift ideas—or even wholly to abandon old ideas—is one of the proud characteristics of science. But true science shifts or abandons ideas on the basis of valid evidence and logical reasoning, not on the basis of confused and obscure polemics.

The new Institute of Genetics at Gif is attached to the C.N.R.S., under the control and management of its officers, and is not a corporate part of the University of Paris with its centuries of tradition and its large but well recognized system of authority. Furthermore, it is planned that the Institute will have three main laboratories, each with its own director and with a considerable staff. This is thus a project which may well reflect rather more sensitively than is usual the scientific philosophy of its leaders and the opinion of the men in its own higher levels of administration.

Thus before we proceed further in our consideration of your request, we would appreciate knowing whether and in what form we could be furnished with assurance from the authorities of the C.N.R.S. that the men in this Institute will be free to carry on their work in the true spirit of modern universal science; and assurance from the geneticists involved that their scientific work will be uninfluenced by political considerations or party loyalties.

Weaver's letter certainly came as a shock to Ephrussi. As discussed in the previous chapter, the Rockefeller Foundation had invested considerably in the development of biology in France, in the C.N.R.S., and in Ephrussi. Under a Rockefeller grant of $250,000 to the C.N.R.S. in 1946, $18,000 had been allocated to the Institute of Genetics for equipment. Ephrussi himself had an International Board Fellowship in 1926–1927, and was a foundation fellow when he visited Morgan's laboratory in 1933 and 1935. He shared an appropriation of $50,000 at the Institute of Physico-Chemical Biology in 1936. As a refugee scholar at the New School for Social Research in New York, he received a part of his salary between 1941 and 1943 under a foundation appropriation.

The situation concerning Marxism and its relations to genetics in France was of a much more difficult, important, confused, and subtle character than Weaver and the other Rockefeller Foundation officials had imagined. Weaver's intention to interfere in what many, at first at least, viewed as a scientific controversy and the controls he demanded were potentially dangerous to French science. Ephrussi showed the letter only to Professor Pérès, vice-director and present acting director of the C.N.R.S., and to Professor Terroine, in charge of the 4th Bureau (Foreign Relations) of the C.N.R.S. All three of them agreed that "great harm might result if too many people knew the content of this letter" (Pomerat, diary, February 28–March 3, 1950). Ephrussi obtained permission from the C.N.R.S. Directorate to pay for the expenses of an emergency visit to the United States.

Ephrussi flew to New York and discussed the various issues raised in Weaver's letter during four days of exhaustive meetings with Weaver and Gerard R. Pomerat, an assistant director of the National Science Division, who was primarily responsible for Rockefeller Foundation activities in Europe. As the conversations progressed it became clear to the Rockefeller officials that they had entered this difficulty with four major concerns: the political structure of the C.N.R.S. and the political convictions of L'Héritier, Ephrussi, and Teissier. Ephrussi reviewed the legal status of the C.N.R.S., the methods which were used to elect the members of its directorate, the relation of the Institute of Genetics to the C.N.R.S., and his own responsibilities as director of that institute. He persuaded the Rockefeller officials that the C.N.R.S. could be fully compared with other French educational agencies, including the universities, at least in terms of "freedom of election," "freedom of appointment," and "freedom of decision." In fact, Weaver himself came to the opinion that academic freedom at the C.N.R.S. surpassed that of the French universities:

> We became completely convinced, . . . that the structure of the C.N.R.S. and the way in which they select personnel are such as to assure that French academic traditions will in fact be followed, an that intellectual freedom will actually be protected, within the C.N.R.S., by a more broadly based academic control than applied, in point of fact, to any one French university. (Weaver, diary, February 28, 1950)

Within the first hour of the interviews with Ephrussi, the Rockefeller officials recognized that L'Héritier, instead of being a Communist, was its exact antithesis, a rabid Royalist. However, there was no doubt that Ephrussi considered himself at least partially suspect of having Marxist sympathies, and he went to great lengths to provide Weaver and Pomerat with legitimate reasons why he himself had not until then made, and would not in the immediate future make, a public statement on his attitude towards the Lysenko situation. He pointed out that his French colleagues knew perfectly well the private stand he took in this matter, and he explained how he motivated in the first instance Huxley's investigation of the Lysenko situation, which resulted in two articles he wrote in *Nature* and the fuller account which eventually appeared in his book. Ephrussi stressed the need to translate the issues of Marxist biology and inheritance of acquired characteristics into French terms which allowed for French attitudes toward Resistance, toward Lamarckism, and so on. First he pointed out that as the country of Lamarck, France remained a land whose scientists still accepted to a very large degree a belief in the inheritance of acquired characteristics, and Ephrussi claimed that many of them "emotionally wish to accept Lysenkoism" because it tended to support Lamarckism. He said that this was true of Caullery, of Cuénot, and of other "grand old men of French science." Second, while "worship" of the men of the Resistance was now decreasing in France, there nevertheless remained very deep loyalties for the "few who sacrificed so much at a time when so many Frenchmen were willing to lie down under the pressures of Occupation." Ephrussi stressed the role that Teissier played in the Resistance from its earliest days, and he pointed out that Prenant never failed in his duties to the Resistance, nor betrayed any of his compatriots there when he was tortured by the Germans (Pomerat, diary, February 28–March 3, 1950).

Weaver noted a few other issues concerning Ephrussi's relations to Communism (Weaver, diary, February 28, 1950). First, in the teacher's union to which Ephrussi belonged there was a strong movement for a strike against the Marshall Plan. Ephrussi not only refused to have anything to do with that strike, but tried to force the hand of the authorities of the union by demanding that they expel him from the union for his refusal to strike. Second, when it was planned to arrange deputations to visit the minister of education to protest the dismissal of Teissier as director of the C.N.R.S., Ephrussi agreed to head the first such deputation but only provided that there not be a Communist in the group. Third, Ephrussi had given a talk, just a few weeks previously, to some 600 high school teachers in France. Many of them obviously came ready to ask him difficult and searching questions; and in a long question period after his lecture, he told them exactly where he stood on the whole issue of Lysenkoism.

This left only the fourth issue, namely, the case of Teissier. Ephrussi explained that until then Teissier had continued to publish scientific papers that reported research in modern Western aspects of genetics along Mendelian-Morganist lines. He also emphasized that he could not guarantee in any way that in the future Teissier would not produce scientific studies having a Lysenkoist bias or that he had not made and would not make public statements or write popular articles with a Lysenkoist bias. However, Ephrussi explained that Teissier had been ousted as director of the C.N.R.S. (on false pretexts) and that he could make a much more effective contribution towards proselytizing for Marxist views in the lectures he gave as professor at the Sorbonne than as a member of the staff of the Institute of Genetics at Gif.

At the same time, Ephrussi refused to lay down a policy that scientific papers emanating from the Institute of Genetics would have to be approved by him. He claimed that the Rockefeller Foundation officials could only make their decisions in terms of their estimation of the "moral qualifications" of the men who were concerned with the operation of the project. In Weaver and Pomerat's view the whole problem of Rockefeller Foundation aid to the institute was less difficult since Teissier was no longer director of the C.N.R.S. As Pomerat (diary March 2, 1950) wrote:

> It seems fairly clear that Teissier would be most reluctant to give written assurance that his own relation to the Institute of Genetics would not carry Communist bias; that compelling him to make such a statement might upset an already unstable situation and perhaps thereby force him to go completely over toward Communist genetics if for no other reason than that he has already been ousted from the C.N.R.S. on a pretext which was not completely honest; and that Ephrussi is not at the moment and probably will not be prepared or willing to make such a request of Teissier on our behalf (GRP now feels that we must not force this issue because it isn't really vital, it is potentially dangerous, and it will not accomplish anything very significant).

In the end, the officials had, as usual, to place their confidence in the person who would direct the project. Ephrussi made no compromises. As a result of the meetings they were prepared to lay Ephrussi's request for an appropriation of $54,000 in "the lap of the gods." Weaver summarized his judgment about the meetings and Ephrussi as follows:

> It should be said, for of all, that these conversations were extraordinarily satisfactory, particularly when one takes into account the extremely difficult, important, confused and subtle character of the situation. E. was a remarkably well-qualified person for the French to send over on this particular mission. Being a Frenchman by choice, rather than by the accident of birth, he shares with his illustrious comrade, Louis Rapkine, an enthusiasm for France and for French culture which exceeds that of many native-born Frenchmen, balanced by a kind of external objectivity and understanding which would be almost impossible for any Frenchman to achieve. E. also combines the emotional appreciation of the Latin mind with a sort of Russian stubbornness and sense of reality. In addition to all this, he is deeply devoted to the Rockefeller Foundation, to which he feels a great debt of gratitude.
>
> There was absolutely no sense of bargaining in any of our conversations, and it was also possible to conduct them on a plane of directness and frankness which would have been completely impossible had they sent over a native-born Frenchman. Throughout the conversation, moreover, E. grew steadily in both intellectual and moral stature. WW thinks that a very large measure of trust and confidence in E. is justified. (Weaver, diary, February 28, 1950)

Although the Rockefeller Foundation officials were willing to support the development of genetics in France in the face of the threatening leftist Lysenkoism, the plans for the new Institute of Genetics continued to mature slowly, impeded largely by problems originating from the political right. In his meetings with the Rockefeller Foundation officials in New York, Ephrussi had outlined the risks involved in supporting the Institute of Genetics in the clearest of terms. There were the dual problems of the Communist Lysenkoism on the one hand and the authoritarian control of the antigenetic neo-Lamarckism on the other. Although Teissier was no longer director of the C.N.R.S., the development of genetics remained threatened by a new, conservative, and what Ephrussi considered to be an antigenetics directorate elected in 1950. This second problem was stressed by Ephrussi (March 16, 1950) in a letter to Weaver which followed the meeting in New York.

Throughout the first half of the 1950s, the Rockefeller Foundation officials would find themselves engaged in another series of problems which centered around a struggle for control over genetic research in France, which in turn revolved around a conflict between Ephrussi and the new ultraconservative director of the C.N.R.S., G. Dupouy. In brief, the situation was as follows. By 1951, six large laboratory buildings for the Institute of Genetics, designed to house at least eighty researchers, had been completed at Gif. However, by 1953 the laboratories at Gif were working at about five percent of their capacity. L'Héritier occupied one of the buildings with four or five research workers and a few technicians. According to Ephrussi, he was doing good work but complained bitterly about isolation, about the lack of a decent reference library, and about the difficulty of getting people to go to work at Gif, about 20 kilometers from Paris. Teissier had two or three people working in two other buildings, whereas he himself was much more interested in the Marine Biological Station at Roscoff, and spent only one afternoon in every two weeks at Gif. The fourth building, which was to be a combined administration, library, and conference center, was not being used for anything but administration. The remaining two buildings which had been planned for Ephrussi's laboratories were unoccupied and unused.

Ephrussi and his co-workers continued to work at the makeshift laboratories loaned to him by friends at the Rothschild Institute in Paris, with little communication between him, Teissier, and L'Héritier, and having to turn down many applications from scientists wanting to work in his laboratories. In 1953 his research group consisted of fourteen research people (including foreign guests) plus six or seven technicians. Out of about twenty-one, some fifteen received all of their salary from the C.N.R.S. Ephrussi's going to Gif hinged on his being assured of adequate lodging for L'Héritier, perhaps Teissier, and himself and seven of his collaborators. Commuting 20 kilometers every day from Paris to Gif was difficult.

According to Ephrussi, Dupouy had assured him in 1951 that housing would be ready for occupancy during the following year. By 1952, Dupouy claimed that no adequate lodging at Gif could be found. In Ephrussi's opinion the original statements about providing housing were simply "a lie." In the meantime, Dupouy was accused of mismanagement and fostering his personal interests as director of the C.N.R.S. Ephrussi viewed Dupouy as an inflexible personality who had succeeded in antagonizing just about everyone with whom he came in contact; a "Napoleonic dictator" who would brook no interference from anybody. In 1952, when describing the problem to Pomerat of the Rockefeller Foundation, Ephrussi wrote:

> I doubt very much that it is Mr. Dupouy's intention to offer me and my workers lodgings. . . . I am told that Teissier is now fully aware of the difficulties of the commuting system, and that he and L'Héritier are rather eager to obtain lodgings for their workers. . . .
>
> . . . The above description may appear somewhat unfair to the efficiency of the C.N.R.S. I must admit that some of the C.N.R.S. undertakings are proceeding at a much faster rate. Just as an example: the C.N.R.S. purchased for [for 20,000,000 francs] last spring a lovely three-story "hotel particulier" in the Rue Pierre Curie, in Paris, in order to install there the Photographic Department of the "Service de Documentation." Work was rapidly started for adapting the interior and adding two new stories occupied by one or two apartments. This work is now almost completed.

According to public rumor, the apartment will be occupied by Mr. Dupouy. (Ephrussi to Pomerat, September 28, 1952)

There is a good deal of evidence to support Ephrussi's claims. By 1953, when Dupouy claimed that the C.N.R.S. had no money for a building to lodge researchers at Gif, he had, in fact, moved into the apartment, bringing in furniture which he obtained from the National Museum. "In the meantime" as Weaver recorded in his diary, "he heats a chateau out at Gif all week, since he spends his weekends there" (Weaver, diary, January 7, 1953).

Dupouy's notorious activities reached the scientific public. In the newspaper *Combat,* for example, published by the association of scientific workers and headed by Prenant, one reader raised the question of what was to be done about "Le Scandale à Gif," where, although French scientists were crying desperately for laboratories in which to work, there were huge laboratories built for an Institute of Genetics which were almost unoccupied and a chateau which served as a weekend pleasure place for the director of the C.N.R.S. This letter was taken up for much more extensive and detailed discussion in the May 1953 number of the journal. There was also an abundance of evidence of Dupouy's mismanagement of research funds (see Pomerat, diary, June 5, 1953).

It should be stressed, however, that the very idea of transplanting three University of Paris professors to Gif was a controversial one in itself. According to Ephrussi and Pomerat, several members of the new C.N.R.S. directorate, including Dupouy, were opposed to the Gif scheme from the very beginning. This, of course, was in virtual conflict with Ephrussi's arguments for the necessity of the relative autonomy of genetic research from the university administration. To Pomerat, however, it seemed wrong to have three University of Paris professors doing research so isolated from "the young students who ought to be inspired every day by the sight of their 'Masters' at work in constructive research." As Pomerat saw the situation:

Ephrussi says that he turns down about two applications per month from men who want to work in his labs, but he is age 52, he is happily married to a woman he describes as a superb scientist in her own right, he has a team of good youngsters who are already able to direct groups of their own, he isn't too badly installed here for the work he loves. To GRP it seems clear and reasonable that Ephrussi has much more than many other French scientists and that if he were a more reasonable man he could be content to stay here and work effectively until the pressure of circumstances forces the University of Paris to build adequate labs for its professors. But the die is cast and GRP has tried to play up to it. (Pomerat, diary, June 5, 1953)

Despite Pomerat's statement, geneticists faced both bureaucratic and direct opposition in the French universities. Although Ephrussi had received the first chair of genetics in 1945, the teaching of genetics in the French universities developed extremely slowly and entailed fierce administrative battles. The difficulties to be encountered became immediately apparent with attempts to reform the *Licence d'Enseignement en Science Naturelle* so as to include a prominent place for genetics. Such powerful neo-Lamarckian biologists as P.-P. Grassé at the Sorbonne were extremely active in suppressing the expansion of genetics in the university curriculum. During the negotiations of the mid-1950s, Grassé coldly proposed a decrease in the number of genetic *leçons* for the *License*. In response to his proposal, Ephrussi (December 22, 1956) wrote to him a scathing letter:

> It's a repressive change that cannot be sanctioned by the titular of genetics at the moment when it appears to all the world that this science forms the basis of all progressive biology and where its place should in consequence be enlarged and not diminished. . . . If this backward evolution must take place, I am happy to leave the responsibility to you. . . .
>
> I await therefore the day, which I hope is soon, where, preoccupied as you are in the interest of French biology, you will recognize that it is your duty to ask of your geneticist colleagues their cooperation for establishing a collective program for teaching general biology. (my translation)

It should be stressed, however, that the struggle between geneticists and naturalists was not unique to France. Indeed, throughout the development of genetics naturalists had voiced opposition to the threat of genetics to dominate biology departments. This hostility to genetics can be traced to various struggles over the way science was to be done: individualism versus team research, theoretical disputes, and institutional power. The Sorbonne was not unique in its resistance to genetics. As discussed in Chapter 3, biologists in German universities also resisted change and in several of the elitist and traditional universities outside France and Germany there was also resistance to geneticits. The situation at Harvard was similar. When Guido Pontecorvo considered leaving Glasgow for Harvard in the late 1950s, Ephrussi described the attitudes toward genetics as follows:

> In so far as Science in general is concerned, Harvard is an excellent place. Biology is the weak and difficult spot: Bundy, who *is* a good and reliable man discussed with me in detail the situation. I am sure that when he told you that he wishes to improve the Biology Department, he was sincere. What he may not have told you is that the Department is extremely conservative and extremely divided, and does not want to be improved: this is the reason why, for years, they had no reasonable policy with respect to Genetics. As you probably know, the appointment of a second geneticist which they are now trying to make is against the wishes of the majority of Biology professors and is forced upon them from outside. It is clear therefore, that, although there will be, as Bundy told you, several retirements in the near future replacement will be a most difficult problem.
>
> As it is, the Department of Biology contains few good *and* active people, especially in fields of interest to a geneticist, and contacts within the Department are extremely poor. (Ephrussi to Pontecorvo, January 9, 1958)

Paul Doty also provides testimony to the difficulties at Harvard as late as 1957 after J. D. Watson was appointed:

> Jim's first months here have gone very well. I don't mean that he has made any discoveries in the lab, but rather that he has dealt with the various grand-dads in the Biology Dept. with unexpected diplomacy and tact, has done a good job in lecturing as well as in setting up and running a laboratory course and has shown just the maximum tolerable reform zeal and no more. Even Carpenter remarked to me that he would never have believed it possible. So by doing well I mean he has solved the problems of living amicably in the Biology Dept., getting his lab set up

> and initiating a number of things that need initiating, e.g. reorganization of courses, modernization of the labs etc. (Doty to Ephrussi, January 4, 1957)

At Cambridge, England, the situation seemed to be even worse. When Pontecorvo was offered J.B.S. Haldane's Chair in 1958, he refused the position. The laboratory conditions were quite unsuitable and according to Pontecorvo, he was told he would not be permitted to build a new one with money already promised from outside the university. He wrote to Ephrussi:

> Just to let you know that, contrary to your predictions, I am staying on in Glasgow. I was elected to the Chair at Cambridge last May but the conditions were quite inadequate. I now realise that, contrary to what everybody believes, it is not Fisher's fault that conditions at Cambridge are so bad but the combined results of the hate for geneticists that both zoologists and the botanists have, and the incredible ways in which the administration of that University works. . . .
> . . . . At Cambridge the impression was that I was being forced upon them by the electors. (Pontecorvo to Ephrussi, September 5, 1958)

However, as discussed in Chapter 5, the government administration in France constituted a major additional constraint for the development of genetics in France because it controlled the creation or extension of laboratories, teaching staff, and research equipment. Innovation was difficult and Ephrussi found himself in constant struggle against the central government control. In an unpublished report entitled "Défense de la Science Française" (1949) he wrote:

> A new chair can be created only by a law, a law is equally necessary for creating a *post d'assistant* or a *garçon de laboratoire*. A discipline as important, both in theoretical and practical terms, as genetics has been taught in the United States largely for a quarter of a century. Here, the first chair of genetics was created at the Sorbonne in 1945 and still doesn't possess a laboratory in the university building (my translation).

During the first half of the 1950s the number of students who took genetics at the Sorbonne was around twenty-five per year. The major difficulty, according to Ephrussi, was the lack of premises, particularly of a room for laboratory work. Each candidate for the *certificat* in genetics, Ephrussi maintained, required his or her own table and individual instruments and must be able to work outside of the seminars. Even ten years after its establishment, the chair of genetics at the Sorbonne possessed neither laboratories, workrooms, nor library. Ephrussi complained to the dean of the Faculty of Sciences, Jean Pérès (April 1, 1955), that the Service de Génétique "was without contest the worst served of all the *services* of Biology at the Sorbonne." Genetics was not the only modern discipline that was poorly represented at the University of Paris. As late as 1957, there was no chair of experimental embryology at the Faculty of Sciences, nor was there a chair of microbiology; although there was a chair of biochemistry, the laboratory of biochemistry was located at the *Institut Pasteur*.

It was not until 1956 when conditions improved that Ephrussi began to move part of his group to Gif, and within another two years the remainder of the team was able to join them. From then on it was possible for him to expand his staff and accept large numbers of foreign guest investigators who applied to work in his laboratories. With the laboratories of Ephrussi at Gif and the very strong groups led by Jacob, Monod, and Lwoff at *Institut Pasteur,* France was playing a dominant role in microbial genetics in Europe. While its total effort in these fields fell well below the sum of what was being done in the United States, the activities of the French groups attracted large numbers of foreign workers who at Paris or at Gif found many skilled theoreticians and a wealth of fresh ideas.

In 1965 Monod, Jacob, and Lwoff became the first French scientists to be awarded a Nobel Prize in 35 years (see Judson 1979, pp. 590–591). When they returned from Stockholm they gave a joint interview in which they called for a complete reorganization and decentralization of the control of education and research in French universities. The centralized career of control of Sorbonne professors was soon largely diminished when the French universities were transformed, triggered by the student protests of 1968.

CHAPTER 7

Problems with "Master Molecules"

> One point at least already seems to be quite clear: namely that biochemical differentiation (reversible or not) of cells carrying an identical genome, does not constitute a "paradox," as it appeared to do for many years to both embryologists and geneticists. (F. Jacob and J. Monod, 1961, p. 397)

> This statement tells us nothing about the nature of the primary causes responsible for the orderly, divergent biochemical differentiation of different cell lineages derived from a single egg (whether its mechanism be based on self-maintaining regulatory states or, for that matter, on any mechanism of differential gene activation or amplification). *The real problem is that of the origin (seat) of the asymmetrical causes which bring about these asymmetrical effects.* (Boris Ephrussi, 1972, p. 113, referring to the above quotation)

A superficial inspection of the cases of cytoplasmic inheritance reported by 1955 leaves one with the impression of a collection of genetic oddities. Aside from their non-Mendelian transmission and manifestations as stable differences between cell lines, cytoplasmically inherited characteristics seemed to follow no common rule. There seemed to be almost as many patterns of manifestations, variations, and transmissions as there were individual cases. But as long as genes were thought to govern the cell largely influenced by extranuclear activities, and development was seen as a nucleo-cytoplasmic dilemma, the location of the various non-Mendelian phenomena in the cytoplasm bestowed upon them an apparent unity of biological purpose. By responding differentially to changing environmental circumstances, cytoplasmically inherited characteristics offered an intelligible explanation of somatic cell differentiation. To account for permanent cellular differentiation, geneticists had postulated the existence of various sorts of plasmagenes (as independent genetic elements or as gene products) sorting out at cell division, multiplying at various rates, responding differentially in different environments, and interacting with each other in various modes of competition and cooperation.

During the 1940s and 1950s direct nuclear control of the structures of some of the most important cytoplasmic components—the proteins—and of their specificity had not yet been firmly established. The "gene," which had been attributed with the power to direct the formation of enzymes, remained the indivisible, formal, abstract Mendelian hereditary unit, whose physical nature had no relevance to the interpretation of the experimental results. The gene theory of classical genetics had nothing to say about how the gene actually directs the formation of the enzymes said to be under its domination. It remained possible that protein specificity was under cytoplasmic control, a possibility which found support in its compatibility with the embryological arguments that cellular differentiation was largely cytoplasmic. These theoretical considerations, together with the genetic evidence indicating cytoplasmic control of such vital physiological functions as respiration and photosynthesis, lent support to the notion that nuclear genes and their mutations were concerned primarily with trivial characteristics of the organism.

By not yielding to the generalizations of the chromosome theory, cytoplasmic inheritance as a genetic phenomenon required rationalization and subordination to the major genetic synthesis. Some of the most common rationalizations of the irregular transmission patterns during the first half of the century were that they were due to parasites, complex symbiotic organisms, "delayed nuclear effects," or other chromosomal aberrations. Yet one of the main reasons for postulating a cytoplasmic basis for the biochemical differentiation of cells, that is, the complete autonomy, randomness, and rarity of gene mutations, did not seem to allow any explanation of the orderly and directed process of ontogeny. The gap between genetics and embryology remained perhaps wider than that between any two fields of research in all biology. Embryologists and the few geneticists who worked on the problem of differentiation repeatedly stated the unsolved problem of how cells with identical genomes could become differentiated.

With the advent of molecular genetics and the transformation of the chromosome theory to the nucleic acid theory, the boundaries between cytoplasm and nucleus began to dissolve. First, with the structure of DNA reported by Watson and Crick in 1953, the basic mechanism for gene replication immediately became apparent and was soon followed by a molecular explanation of the process of protein synthesis. It was becoming recognized by the end of the 1950s that sequences of four kinds of bases in DNA spelled out messages specifying protein. The machinery for protein specificity was located in the cytoplasm but the control of protein specificity was in the DNA, thought to be located solely in the nucleus.

The elucidation of the physical nature of the gene and its role as an "information" element was the work of molecular genetics. During the 1940s the evidence for DNA stemmed from its location in the nucleus, the site of Mendelian genes. When molecular biologists rose to an authoritative position in the field, the physicochemical structure of DNA bestowed upon it its hereditary qualities, without reference to breeding experiments. The argument had come full circle and genetic properties were accorded to the nucleus because it contained DNA. Thus, the basis of heredity switched from one based on cellular *location* to one based on *information* encoded in the structure of macromolecules. The molecular notion of the gene was conceptually different from that of classical geneics. Genes could no longer be seen as "beads on a string," nor simply in terms of units of segregation and recombination. Heredity was understood in terms of a message, a language encoded in the structure of macromolecules. The clear recognition that the genetic material was nucleic acid rather than protein led to a refinement in the relations between the genotype and the phenotype. It permitted a clear conceptual distinction between a change in structural information and a change in the expression of genetic potentialities. A concept of genic regulation which was largely excluded from classical genetics came to be a central constituent in the soup of molecular genetic concepts.

Indeed, the concept of genes as reservoirs of information represented only one half of the new molecular meaning of heredity. By the end of the 1950s, it was becoming clear that in higher organisms as in microorganisms, the information transfer from genes to proteins could be turned "off" or "on." The genome not only contained a series of blueprints but was capable of systematic and programmed regulation. The most thoroughly analyzed and most influential study which led to this view was based on the genetic control of the enzyme galactosidase in the bacterium *Escherichia coli*. This work culminated in the early 1960s, led by Jacques Monod and François Jacob and their many collaborators at the *Institut Pasteur* (Judson, 1979; Grmek and Fantini, 1982).

As discussed in Chapter 5, in 1946 when Monod began to systematically investigate bacterial enzyme synthesis, *E. coli* was known to possess β-galactosidase activity when growing in medium containing lactose as a carbon source, and to lack this enzyme when growing in media in which a natural sugar other than lactose was provided. β-Galactosidase was classified as an "adaptive enzyme." It was formed only in the presence of its substrate in the medium. Monod and Cohn (1952, p. 68) abandoned the use of the expression "enzyme adaptation" since "adaptation" was commonly used to describe the selection of spontaneous mutations in a microbial population. To prevent any confusion with selection, they renamed "enzyme adaptation," "enzyme induction." "Inducers" were defined as those compounds (e.g., lactose) to whose presence a cell responds with the formation of an enzyme.

Spiegelman, it will be recalled, had attributed the non-Mendelian inheritance of enzyme adaptation in yeast to plasmagenic action and Ephrussi and Slonimski presented evidence indicating enzyme activity in *petites* was due to a loss of a plasmagene—a mutation affecting cytoplasmic proteins. Throughout the 1950s Monod and his collaborators investigated the environmental and genetic control of enzyme formation with various techniques involving kinetic and nutritional studies and trace incorporation experiments. By 1956 Monod concluded that the induced enzyme formation involved the complete *de novo* synthesis of the enzyme protein molecule from amino acids. It was not until a half a decade later still that Jacob and Monod began to formulate their theory of the "operon."

Synthesizing their work in 1961, they postulated the existence of different kinds of genes and chromosomal elements. "Structural genes" had the classical function of specifying enzymes and were thought to be under the influence of an "operator." Together the structural genes and the operator, which was at one extremity of a linkage group, made up an operon. The operator was in turn under the control of a "regulatory gene." The functional relations of these sorts of genes in the hierarchy were thought to be as follows: Regulatory genes produce an unidentified cytoplasmic product, the "repressor" (possibly protein), that can repress the action of the operator. The operator, when so repressed, cannot stimulate the structural genes to produce messenger RNA, without which there can be no synthesis of corresponding proteins. This chain of inhibition was thought to be broken by certain substances in the cytoplasm, "inducers" (e.g., lactose), which block the repressors produced by regulatory genes. Thus, the operator can exist in either of two states, "opened" or "closed." It is open when it is free of repressor, and it is closed as soon as it has combined with the repressor. (The repressor, for its part, can combine with the operator only if it has not interacted with the inducer.)

In the end, the molecular meaning of heredity represented two concepts, messages and feedback regulation. Feedback regulation allows the nuclear genetic system to adjust its activity, not only in terms of what it has to do, but also in terms of what it is doing. It operates by introducing into the genome the results of its past activity and keeps the cell informed of the results of its own operation. It was only after messages and "gene regulation" came to the center of genetic concepts and when heredity lost its major reference to transmission and exchange that geneticists could recognize that genes were *differential factors* or conservative elements of heredity rather than *central controlling* elements. These concepts brought genes back to the attack on the problem of cellular differentiation. Cellular differentiation ultimately depended on specific cytoplasmic substances which activate or repress the genes that make the differentiating proteins, for it soon appeared that many genes remained inactive unless specifically derepressed. The reorientation was really quite simple, involving a change in the basic assumption of the role played by genes. Formerly, it was assumed that the whole set of genes was active in every cell. Hence, cells that had the same set of genes could not become diverse by reason of direct genic action.

When genes became endowed with the dual functions of regulating and specifying proteins, the biochemical differentiation of cells possessing an identical genome no longer represented a "paradox" as it appeared to do for so many years to both embryologists and geneticists. The secret to the puzzle of directed and persistent modifications during ontogeny in the presence of genomic equivalence could no longer lie hidden simply in the distinction between cytoplasm and nucleus. The control over cellular differentiation could not be accounted for simply by the kinds of genes present in the nucleus or occurring in the cytoplasm and conditioned by a variety of intracellular and environmental circumstances. If the cytoplasm was to play a key role in development, distinct from that of the nucleus, as had long been expected, then a novel cytoplasmic mechanism of heredity would have to be discerned. In effect, the transformation of the chromosome theory to the nucleic acid theory, and notions of "gene regulation," led to a new problematic for investigations of cytoplasmic inheritance. That is, it brought about a decisive change in the conceptual framework which conditioned the manner and defined the limits by which elements of the investigator's discourse were to be understood.

Opposing "Dictatorial Elements": Democratic Steady States

> It appears unlikely that the role of genes in development is to be understood so long as the genes are considered as dictatorial elements in the cellular economy. It is not enough to know what a gene does when it manifests itself. One must also know the mechanisms determining which of the many gene controlled potentialities will be realized. (D. L. Nanney, 1957, p. 140)

The turning point for the plasmagene theory and for the research of Ephrussi and Sonneborn came in 1957, at a conference on "Extra-Chromosomal Heredity" called by Ephrussi at Gif-sur-Yvette and moderated by Jacques Monod. It was there that a former student of Sonneborn, David Nanney (1958a), provided a convincing argument against the usual classification on what he called "the geographical basis" and argued for the "developmental importance" of "epigenetic control systems" as "a part of the physical basis of inheritance."

Since the early 1950s, Nanney's genetic investigations had been primarily concerned with mating type determination in the ciliates (see Nanney, 1953, 1954, 1958b). The determination of mating types in *Paramecium* remained a poorly understood phenomenon despite extensive work by Sonneborn and his students. Nonetheless, the non-Mendelian inheritance of mating type charateristics in *Paramecium* was considered to be an excellent case of cytoplasmic inheritance due to plasmagenic action in the 1940s. However, by 1951 Nanney had reached the conclusion that the non-Mendelian inheritance of mating type specificity through both vegetative and sexual reproduction was not due to the transmission of a "plasmagene." Instead, he embraced the concept of self-perpetuating metabolic patterns, or the "steady-state" concept which had been formulated in general terms by Wright (1941) and applied to the case of antigenic determination in *Paramecium* by Delbrück (1948).

According to this model the cytoplasm would be an active partner in cell heredity. It would be an important factor determining the manner in which the genes were expressed, but the cytoplasmic conditions were in their turn determined by the kind of nucleus that previously occupied the cell. "By the term 'steady state,'" Nanney (1957, p. 136) wrote,

> we envision a dynamic self-perpetuating organization of a variety of molecular species which owes its specific properties not to the characteristics of any one kind of molecule, but to the functional interrelationships of these molecular species.

Throughout the 1950s, Nanney protested against the conception of the gene as a dictatorial "Master Molecule" directing the activities of the cell (whether they were in the nucleus or in the cytoplasm) as an adequate explanation of cellular differentiation. In contrast to the particulate model of the "Master Molecule," which he likened to a "totalitarian government," Nanney continued to lend his support to a "democratic organization" in the cell, "composed of cellular fractions operating in self-perpetuating patterns" (Nanney, 1957, p. 136).

The use of this political metaphor served not simply as persuasive rhetoric in a genetic milieu colored by a Cold War inside and outside of biology. It reflected well the struggle between nuclear and cytoplasmic geneticists represented by Muller versus Sonneborn, and it also reflected Nanney's own perceptions of his personal relationship with his mentor. After leaving Indiana in 1951, Nanney abandoned *Paramecium* and turned to another ciliate, *Tetrahymena*. As discussed in Chapter 4, *Paramecium,* which is a bacteria-feeder in nature, was having difficulties in becoming incorporated into the molecular-microbial mainstream occupied by bacterial and fungal genetics. The biochemical markers and methodologies used with *Neurospora* and *E. coli* required a defined and preferably synthetic culture medium. Like *Paramecium, Tetrahymena* generally eats bacteria. However, it is much easier to cultivate on artificial media, and its nutrient requirements had been explored and a defined medium was developed by protozoologists in the 1940s, making it susceptible to biochemical manipulations (see Nanney, 1980).

These attributes made *Tetrahymena* a logical alternative to *Paramecium* and offered a plausible and convincing rationale for developing *Tetrahymena* genetics and receiving funds for research. However, Nanney's reasons for leaving *Paramecium* were not informed by such logical arguments. In the early 1950s the explosive success of molecular biology in the prokaryotic area was not yet entirely certain, and the inherent difficulties of *Paramecium* research had not yet been generally perceived. Nanney's relations with Sonneborn have to be taken into consideration when understanding his decision to abandon *Paramecium*. As he wrote in an appendix to an application for a National Institutes of Health grant in 1980, when describing the origins of *Tetrahymena* genetics:

> Sonneborn was a powerful and demanding personality, a perfectionist, an idealist, a disciplinarian. He was my chief introduction to serious science after a desultory undergraduate education of mixed liberal arts. He was a father figure, an idol, a domineering force that could scarcely be faced directly, and yet one that had to be retained in my scientific and personal life. Perhaps my motive in leaving *Parmecium* but staying with ciliates was to deflect the force of our interaction, to obtain space to develop my own style and value system, without however disengaging entirely from an essential source of energy and stability. Perhaps I didn't need distance from *Paramecium,* but from Sonneborn.

As the authoritative leader of the school, Sonneborn's control over *Paramecium* genetics and its interpretations paralleled that of Morgan and his followers, and Muller and his. The inner nature of this relationship may be illustrated by a heated dispute between Sonneborn and Geoffrey Beale. It will be recalled that Beale and Sonneborn had worked together on the inheritance of antigenic specificity in *Paramecium* during the late 1940s. Beale subsequently was expanding *Paramecium* genetics and establishing a laboratory at the Institute of Animal Genetics in Edinburgh. However, in direct conflict with Sonneborn, who interpreted the non-Mendelian transmission of antigenic specificity to involve cytoplasmic genetic properties perhaps located in the cortex of the cell, Beale came to interpret the results without recourse to cytoplasmic genetic elements and adopted the "steady state" concept. In 1951 Beale attempted to express his views in a book-length manuscript, *The Genetics of Paramecium Aurelia,* which was eventually published in 1954. When he considered leaving the field and investigating mosses because of what he considered to be a threat to his relationship with Sonneborn, Sonneborn wrote to him:

> Of course, I was upset . . . about your little book, but it was mainly because I felt that you had not presented the material in a way that I thought was clearly exposing the situation. I would be the last one to feel or think that you did not have the right to your own views and to express them freely and in any way you wanted. . . . The fact that your views are more nearly of the classical type than mine is, of course, your own business and not mine; and I think you are completely entitled to them and to express them at any time and in any way you wish. This is what is necessary for proper evaluation by others. The revised copy of your book . . . is, in my opinion a very much better product. . . . Naturally, I don't agree with everything and there are parts of it I would have written differently, as you undoubtedly know; but that is certainly no cause for any bad feelings. On the contrary, I think my feelings would be much more hurt if I felt that I were in any way responsible for you leaving the field. (Sonneborn to Beale, September 20, 1954)

Although Nanney, like Beale, recognized that genetic systems probably existed in cytoplasmic organelles such as blepharoplasts, kinetosomes, plastids, and mitochondria, and controlled "*essential*" cell functions such as respiration and photosynthesis, he did not find these very useful in explaining cellular differentiation or very different in principle from DNA-based systems in the nucleus. He was persuaded that cellular differentiation was a problem of nuclear gene regulation, and that nuclear regulatory systems must be maintained by something more than DNA replication.

The concept of self-perpetuating metabolic patterns did not represent a refutation of the plasmagene theory of cellular differentiation. As Nanney recognized, most cases of non-Mendelian inheritance could be formally explained by some behavior of plasmagenes. But the plasmagene theory of cellular differentiation confronted the major genetic synthesis and was highly criticized as a mechanism of cellular differentiation. In Nanney's view, the most significant attribute of the concept of self-perpetuating metabolic patterns was that it brought somatic and germinal variations and transmissions into convergence with a "unifying" perspective.

To accommodate both mechanisms for maintaining "cell specificity," the notion of heredity, which was bound to problems of evolution and breeding, became a principal stake. In his attempt to formulate a conceptual basis for a synthesis between these two competing conceptions, Nanney struggled to free the term "heredity" from its restricted reference to particulate mechanisms implied by breeding experiments and the analysis of recombination. An "older and broader interpretation" was called for, in order to bring together "phenomena of fundamental similarity without restriction to analytic devices." The term "heredity," he proposed,

> may be used to describe the more general capacity of living material to maintain its individuality (specificity) during proliferation. . . . "Heredity" in this sense is a type of homeostasis, similar to physiological homeostasis but implying more, since it includes regulation during protoplasmic increase. (Nanney, 1957, p. 134)

In an attempt to support the importance and generality of a nonparticulate physical basis of heredity, Nanney suggested a reinterpretation of some cases of cytoplasmic heredity claimed to be due to "plasmagenes," but for which, he claimed, a particulate basis had not been established. These included the "barrage" phenomenon in *Podospora,* serotypes in *Paramecium,* poky in *Neurospora,* and *Plasmon* characters described by Michaelis in *Epilobium* (Nanney, 1957, p. 141).

Since critical experimental evidence against the alternative explanations of cell differentiation via plasmagenes was not available, Nanney attempted to undermine the logical necessity for maintaining a cytoplasmic genetic basis for cellular differentiation. One of the chief theoretical reasons for the belief that somatic variations and heredity were primarily under the control of cytoplasmic genetic elements rested on the assumption that nuclear differentiation did not occur. However, during the 1950s some studies were reported which contradicted this classical belief. Nanney quickly grouped together the limited and scattered evidence which suggested a nuclear basis for cell differentiation. The experiments of the embryologists T. J. King and R. Briggs at the Institute for Cancer Research in Philadelphia provided instrumental evidence. It will be recalled that in 1914, Spemann was able to show that a single nucleus from a salamander embryo in the sixteen- to thirty-two-cell stage was capable of developing into a complete embryo. These experiments had helped to convince most biologists of the functional equivalence of nuclei during early development. However, the actual evidence was restricted to the first few cleavages, and it remained possible that nuclear differentiation occurred. However, technical problems prevented testing the developmental potentialities of older nuclei.

In the early 1950s, King and Briggs returned to the classical studies of the early embryologists, and using new methods of investigation, tested the hypothesis of nuclear equivalence at later stages of development—at times when irreversible cellular changes were known to occur. This was accomplished by removing nuclei from embryonic cells at various times during develoment, injecting these nuclei into enucleated eggs, and following the course of differentiation. Their results with nuclei from early cleavage stages corroborated the earlier experimental results. However, nuclei from progressively later stages of development were progressively less capable of maintaining normal development. Moreover, the differences detected with nuclei from differentiated cells were reproduced upon successive serial transfers, i.e., they were hereditary.

These results were supported by other reports by geneticists indicating chromosomally localized control systems in various organisms. In 1956, Barbara McClintock spoke of "controlling elements" in the genome of maize (Fox Keller, 1983). Long-lasting but impermanent variations in *Salmonella* serotypes were shown by the Lederbergs (1956) to be associated with bacterial "chromosomes." Nanney pointed out that all of these cases violated to some extent the classical notion concerning the behavior of genetic systems. He claimed that these control systems had features similar to the physiological "hereditary" systems in the ciliates. "It might appear, therefore," he wrote in 1957,

> that the dichotomy between germinal and somatic inheritance, between cytoplasmic and nuclear bases was after all a mistake, and that investigations may now converge with a unified perspective. (Nanney, p. 143)

At Gif in the fall of 1957 Nanney formulated his argument against the significance of the "geographical question" as an explanation of development by juxtaposing the two types of "regulatory systems": a "genetic system" based on DNA replication by a template, and a second which he called a "paragenetic" or "epigenetic" system operating by "self-regulating metabolic patterns."

His claim that the usual classification of hereditary mechanisms on "the geographical basis" was misleading and his argument for the importance of epigenetic regulatory systems in cell heredity caused a great deal of alarm for cytoplasmic genetic investigators. He not only denied a logical theoretical necessity for a nucleo-cytoplasmic dichotomy in cellular differentiation, but further challenged the very experimental basis for establishing the possible existence of cytoplasmic genes. The existence of nuclear genes had been based on recombination analysis and mutations which demonstrated their particulate nature. However, the cytoplasm was transmitted primarily uniparentally in nature, and it was not yet possible to artificially alter this sexual pattern of transmission so as to demonstrate recombination of cytoplasmic genetic particles. And when few cases of "cytoplasmic mutations" existed in any one organism, there were grave difficulties in distinguishing the "genetic" from the "epigenetic," and cytoplasmic variations due to changes in structural information from nuclear variations.

As long as "heredity" maintained its reference to evolution, a conceptual distinction between the genetic and the epigenetic was essential. Specific directed adaptive hereditary changes could be induced by new environments, as had long been supposed and observed in tissue cultures. However, these inherited changes did not result from a change in the information in DNA, recognized to be responsible at least for specifying proteins, but from a differential expression of genetic potentialities, i.e., they were persistent phenotypes. It was necessary, therefore, to distinguish epigenetic mechanisms that regulated the *expression of genetic potentialities* from the "truly genetic mechanisms" that regulated the *maintenance of the structural information*. As Nanney argued:

> A recognition of the existence of the two types of systems, and even the difficulties in distinguishing between them, may be useful in avoiding confusion in discussing cytoplasmic inheritance, developmental alterations, inheritance of acquired characters, mutations, and genetic recombination. (Nanney, 1958a, p. 717)

Sonneborn, with whom Nanney had discussed these matters in detail and who commented extensively on his manuscript, remained aloof from the controversy and was unwilling to easily trivialize the place of the cytoplasm in heredity and development. On the other hand, Nanney's argument received strong public support from Ephrussi. By the late 1950s, when nuclear control over protein specificity was an inescapable genetic reality and when few instances of non-Mendelian inheritance had been reported, Ephrussi was prepared to agree with the epigenetic argument. In 1958 at Gatlinburg, Tennessee, where he was invited to speak on the role of the cytoplasm in development, Ephrussi reiterated Nanney's argument and gave his approval to the importance of the distinction between genetic and epigenetic systems. It was necessary to recognize that "not everything that is inherited is genetic," and to avoid making a distinction of mechanisms based on nuclear and cytoplasmic localization. "In my opinion," he wrote,

> this has been a major source of confusion in the past, and it is not going to be easy to avoid it in the future because we have all been trained to regard the problem of differentiation as a nucleus/cytoplasmic dilemma. (Ephrussi, 1958, p. 49)

Although the role of the cytoplasm in heredity was displaced somewhat with the recognition of gene regulation, as stressed above, cytoplasmic genetic systems were not excluded from hereditary phenomena in general or from cell heredity in particular. Assuredly, it could be claimed that the cytoplasmic differences were due in the first instance to the action of nuclear genes setting up a cycle of events. Ultimate control of development would then lie with the kinds of genes present in the nucleus. However, cytoplasmic geneticists were not willing to concede that the nucleus monopolized this control.

During the 1960s cytoplasmic genetic investigations took two primary routes, based on two often competing concepts, both of which continued to challenge the classical "evolutionary synthesis." One was based on investigations of cytoplasmic nucleic-acid-based genetic systems, not subject to modification by nuclear genes, but acting either directly or in conjugation with nuclear gene products in determining cell characteristics. Within this scheme both nucleus and cytoplasm would contain two kinds of hereditary systems: genetic systems proper, considered as the complex of mechanisms responsible for maintaining the library of genetic specification, and epigenetic systems responsible for regulating the expression of particulate specification. During the 1960s and 1970s, with the development of new methods for detecting cytoplasmic genes, investigations of cytoplasmic genetic systems began to take the form of a systematic program with a coherent conceptual framework led by the work of Ruth Sager, Nicholas Gillham, and others.

A second research front was based on investigations of extragenic sources of information. The notion that all structures in the cell could be built up *de novo* by gene action and that the entire pattern of every organism encoded by DNA implied an ideal situation which many biologists considered to be too much to expect of DNA.

The demonstration of regulatory systems and the possibility that the genome differentiates during development represented a solution to only one of the problems of development encountered by cytologists and embryologists since the late nineteenth century. Cytologists, for example, spoke of organelles arising only from preexisting organelles of the same kind and many biologists were not convinced that all three-dimensional structures were encoded in nucleotide sequences.

Embryologists and physiologists had often claimed that submicroscopic structures or "fields," that is, the arrangement of molecules, had some hereditary content. Since the early 1950s, Sonneborn and Ephrussi had protested against the biochemical view of the cell as a "bag of enzymes," and maintained that the organization of the cell was a part of the genetic system. Throughout the 1960s and 1970s, they continued to claim that genetic regulation based on nuclear mechanisms of gene activation or self-maintaining regulatory states said nothing about the nature of the primary causes responsible for the orderly, divergent biochemical differentiation of different cell lineages derived from a single egg (see Ephrussi, 1972, pp. 113–116). They insisted that it was necessary to superimpose on the genotype a spatial principle, usually called polarity, which was placed in the rigid ectoplasm or cell cortex.

During the 1960s these views were grouped into a model which proposed that in addition to DNA-based inheritance, cells contained complex supramolecular "templates": two- or three-dimensional structures composed of different macromolecules, which were regulated by a copying process, by a mechanism which remained unknown. The study of the inheritance of cell organization, based on the ordering or arrangement of new structures under the guidance of the old, began to emerge into an active and coherent research program, led by Sonneborn and several of his former students and associates including Janine Beisson, Ruth Dippell, D. L. Nanney, Joseph Frankel, and E. D. Hanson.

Meanwhile, Nanney's epigenetic interpretation for many cases of non-Mendelian inheritance was greeted with hostility by several geneticists who would later become prominent investigators of cytoplasmic genetic systems. During the 1960s and 1970s, Ruth Sager in the United States and J. L. Jinks in England challenged this epigenetic interpretation of many cases of "cytoplasmic heredity" reported prior to 1958 in organisms other than ciliates. The physical basis of many cases of non-Mendelian inheritance, especially *Plasmon* characteristics in higher plants, those concerning male sterility, and other physiological properties, remains obscure to this day.

Nineteen fifty-eight was a cold year for cytoplasmic geneticists, and especially for Ephrussi. Beadle and Tatum shared a Nobel Prize with Lederberg; Ephrussi was excluded, and he was hit hard by the decision. As he wrote to Sonneborn (October 6, 1958), "I suddenly felt my life wasted." Many biologists in France and Belgium were also shaken by the decision. Ephrussi had led Beadle to develop biochemical genetics, and they had showed the general relationship of genes to enzymatic control of reactions in their early work on *Drosophila*. But no procedures were as fruitful and as widely applicable to both biochemical genetics and biochemistry as the *Neurospora* methodology developed by Beadle and Tatum and their followers. The decline of the plasmagene theory of cellular differentiation only added to the "failure" measured on the basis of winning a Nobel Prize. Sonneborn, who had a glimpse of a Nobel Prize before biochemical genetics emerged into prominence, had come to realize that "top recognition" could never be his (see Sonneborn to Ephrussi, November 10, 1958).

Ephrussi left yeast genetics after 1958, and turned the work on *petite* mutations, which he believed would ultimately be demonstrated to have a particulate basis, over to Piotr Slonimski. He subsequently made important contributions to developing an alternative, more direct approach to analyzing mechanisms of cellular differentiation based on a new technique involving vertebrate somatic cell hybridization. The controversial and precarious state of cytoplasmic genetics in the early 1960s is best represented by the following quotations from leading investigators of cytoplasmic inheritance:

> We must admit that we know almost nothing about the action or mechanism of differences inherited in the cytoplasm. (Jinks, 1963, p. 352)

> Cytoplasmic inheritance is a little bit like politics and religion from several aspects. First of all, you have to have faith in it. Second, one is called upon occasionally to give his opinion of cytoplasmic inheritance and to tell how he feels about the subject. (Preer, 1963, p. 374)

"The Cell as an Empire": Cytoplasmic DNA

> What I was looking for were genes that appeared to be cytoplasmic because of their genetics—that they didn't show Mendelian ratios—but otherwise were stable and permanent and that had all the properties that one would expect that a gene would have, regardless of its location. So that was the hypothesis: that there were in the cytoplasm and probably in the organelles . . . small genetic systems like the nuclear system for some reason located in the organelles (Ruth Sager, interview, August 15, 1981).

The analysis of cytoplasmic inheritance made slow progress, often under severe attack, since the first genetic report of non-Mendelian inheritance in 1909. The scarcity of evidence for a genetic system of cytoplasmic factors controlling a variety of traits in any one organism contributed to the widespread skepticism about the possible importance of cytoplasmic inheritance. As discussed in the previous chapters, to geneticists who supported the existence and importance of cytoplasmic genetic systems, the rarity of the reported cases of non-Mendelian inheritance was due in part to technical obstacles. First, the difficulty in obtaining mutations of "nonchromosomal genes," either spontaneous or induced, severely limited the kinds of experiments that could be done and consequently the elucidation of such a genetic system. It will be recalled that the attempt to explore natural variability of non-Mendelian hereditary factors by *Plasmon* investigators by intervarietal and interspecies crosses had led, for the most part, to hopelessly complex results involving multifactorial interactions.

Second, the typical pattern of maternal inheritance, used in the initial discrimination between Mendelian and non-Mendelian inheritance, posed a great obstacle to the analysis of segregation and recombination. The genetic investigations of cytoplasmic inheritance in higher plants had provided what some considered to be ample evidence of non-Mendelian genes influencing chloroplast development, pollen sterility, and a host of other morphogenetic properties. However, these studies were based on the analysis of the interactions of the cytoplasm and the nucleus as a whole. With the lack of a particulate concept, the demonstration of a genetically autonomous hereditary system in the cytoplasm remained a difficult task. A particulate, genic approach to cytoplasmic inheritance could only be constructed by a study of sexual exchange of genetic material: recombination and linkage. In no instance was any evidence adduced of linkage or linked recombination, and consequently no further genetic analysis was achieved beyond the recognition of many phenotypes under cytoplasmic influence.

During the 1960s both of these "obstacles" were overcome by the work of Ruth Sager (b. 1918), who played a prominent role in the development of a coherent conceptual framework and a suitable experimental method for the investigation of "cytoplasmic genes" in microorganisms. Her classical investigations on cytoplasmic inheritance culminated in 1972 with the appearance of her text *Cytoplasmic Genes and Organelles*. Sager was a student of M. M. Rhoades in the late 1940s at Columbia University. She was first introduced to cytoplasmic inheritance through Rhoades's investigations of pollen sterility (see Chapter 3) and other cytoplasmic traits in the corn plant *Zea mays*. Like other geneticists who investigated cytoplasmic inheritance in microorganisms, she strove to construct a molecular understanding of cytoplasmic inheritance. In fact, in 1961, together with Francis Ryan, a well-known bacterial geneticist at Columbia University, Sager wrote what might be considered to be the first molecular genetics textbook, *Cell Heredity*, subtitled *An Analysis of the Mechanics of Heredity at the Cellular Level*. In spite of the radical change in outlook that accompanied the arrival of biochemical and molecular genetics, most textbooks continued to follow a conventional and chronological pattern. That of Sager and Ryan, however, was the first to attempt to provide a new synthesis of the domain of molecular genetics which had occurred since World War II.

After completing her Ph.D. in 1949, Sager worked for six years at the Rockefeller Institute for Medical Research in New York. Her first two years were spent investigating chloroplast biogenesis as Merk Fellow of the National Research Council. During her final period at the Rockefeller Institute she was invited to work as an assistant biochemist to Sam Granick, a "senior researcher" who worked on biochemical pathways in chlorophyll production. Granick wanted to use mutants for study, and Sager's appointment allowed her to turn to the study of the most widely known class of cytoplasmic mutations, those affecting the chloroplasts.

The history of this research begins with the selection of *Chlamydomonas* as an object of genetic investigation. The choice of *Chlamydomonas* "for analysis of heritable alterations affecting the chloroplast and other cytoplasmic bodies" was based on a number of criteria (Sager, 1955). A single-celled microorganism containing one or more discrete chloroplasts was desired. *Chlamydomonas* contains one chloroplast and several mitochondria per cell. It was able to grow in the dark on a reduced carbon source; it exhibited a readily controlled complete sexual cycle, and was able to carry out all stages of its life cycle on a chemically defined medium so that biochemically distinct mutant phenotypes could be detected.

During the 1950s *Chlamydomonas* offered a new technology for various biological investigations. The early 1950s witnessed a burst of genetic and biochemical studies on *Chlamydomonas* in the United States. At Stanford, one of the leading algologists, G. M. Smith, who supplied Sager with her first mating type strains of *Chlamydomonas,* had turned to the study of sexuality in *Chlamydomonas reinhardi*. R. A. Lewin at Yale, C. S. Gowans at the University of Missouri, and others had also turned to biochemical genetic investigations of *Chlamydomonas* by the early 1950s. However, the use of *Chlamydomonas* as a subject of biochemical genetic studies was not an American invention; its popularity in the United States after World War II was due largely to the work of a German biologist, Franz Moewus.

As mentioned in Chapter 3, Moewus worked in Max Hartmann's division of the *Kaiser-Wilhelm-Institut für Biologie* in the late 1920s. Like many biologists in Germany between the two World Wars, Moewus carried out his genetic studies largely in isolation from the mainstream of classical geneticists in Britain and the United States. In the late 1930s, Moewus moved to Heidelberg where he worked on the genetics of sexuality in algae. He subsequently developed methods for the Mendelian analysis of *Chlamydomonas,* making it into a potentially fruitful tool for genetic investigations. At Heidelberg, where he worked in close collaboration with Richard Kuhn, Nobel Prize winner in biochemistry, Moewus developed some basic concepts and methods for the biochemical genetic analysis of microorganisms.

Moewus's first report along these lines was published three years before Beadle and Tatum's first paper on biochemical genetics in *Neurospora*. In 1940, a year before their first paper appeared, Moewus published a series of remarkable papers dealing with the general theory of biochemical genetics with its applications to a theory of sexuality in *Chlamydomonas*. To many, it seemed that he had swung wide open the doors to the revolutionary biochemical genetic analysis of microorganisms. However, his work soon became the center of a great deal of controversy, and during the decade following World War II some geneticists claimed that he had set the pattern for the work for which Beadle and Tatum were receiving great acclaim in the United States, and for which they ultimately were awarded a Nobel Prize in 1958. On the other hand, Moewus's published results were relentlessly criticized and defamed by many other geneticists who claimed that his data were faulty and unreliable, and that he was unwilling to send his stocks to others so that his experiments could be repeated. By the mid-1950s, attempts to reproduce many essential aspects of his published results, carried out by Moewus himself (under the supervision of Ryan) and by many others, were unsuccessful. Moewus was ultimately regarded as one of the most ambitious cases of fraud in the history of science. His published results on *Chlamydomonas* were seen to be wholly unreliable and evidently made up to fit his theories (see Sapp, 1986).

Chlamydomonas had to be genetically reconstructed, restoring its biochemical and genetical integrity. Sager had obtained *Chlamydomonas reinhardi* from G. M. Smith, who isolated this particular species and worked out its life cycle under laboratory conditions. During the 1950s and 1960s, she repeatedly and carefully described the life cycle of the organism. She demonstrated that it inherited characteristics in the expected Mendelian pattern of 2:2 segregation among the four products (tetrads) of a zygote, just as if Moewus's work did not exist.

During the 1950s, Sager maintained close contact with Ephrussi and Sonneborn, whose genetic results and methodologies had a significant influence on her work. *Kappa* in *Paramecium,* for example, which turned out to be a symbiont, was highly useful to investigations in *Chlamydomonas,* as Sager later recalled:

> Before we knew what it [*Kappa*] was, the methods that Sonneborn had worked out were absolutely magnificent methods for studying cytoplasmic genes. And he wrote some extremely important papers in which the general thinking was right, even though the specific case was wrong. So he had an extremely important influence on me. (interview, August 15, 1981)

The genetic system in yeast also appeared to have much in common with that of *Chlamydomonas*. As early as 1954, Sager reported a case of uniparental nonchromosomal inheritance of resistance to the antibiotic streptomycin which resembled the *petite colonie* in yeast and poky in *Neurospora*.

The identification of chloroplast mutations in plants with mutations involving mitochondria was an alliance that had both sociointellectual and financial benefits. Between 1956 and 1966 Sager worked as a research associate in zoology at Columbia University. She could not obtain a permanent teaching position before her appointment as professor of biological sciences at Hunter College. She not only had the stigma of working on cytoplasmic inheritance, which had a long, checkered history, with many geneticists discrediting it. She, like McClintock (Fox Keller, 1983), had the stigma of being a woman, and there was much gossip about "Ruth's defense of the egg." Although Sager's work was handicapped by her having only two assistants during her ten-year period as research associate, she did enjoy ample funding from the National Institutes of Health, which provided her with both financial support for her research and a salary.

This was at the time when the General Medical Section of the National Institutes of Health was founded. Unlike the other institutes, which were tied to particular diseases, the General Medical Section could fund research that was not obviously and directly related to a particular medical problem. Chloroplasts in themselves were not of great concern to the NIH, but mitochondria, found in all eukaryotic cells, were. Chloroplasts and mitochondria did appear to be very similar organelles with respect to both their function in the cell, which has to do with energy production, and the cytochromes that they contained. The basic mechanisms could then be assumed to be the same, and Sager's argument in obtaining funds was that chloroplasts were simply much easier to study than mitochondria (interview, August 15, 1981).

By the early 1960s, Sager's analysis of "cytoplasmic genes" in *Chlamydomonas* began to take the form of a systematic study with the invention of two general procedures. First, she developed a technique for readily acquiring various nonchromosomal mutations by studying the conditions under which they occurred with streptomycin. The mutagenic method simply involved growth of cells on agar containing sublethal concentrations of streptomycin. Under these conditions, nonchromosomal mutations were found in almost every colony. As a mutagen, streptomycin produced an extensive series of nonchromosomal variations, which indicated the presence of an extensive nonchromosomal genetic system and provided new material with which to reinvestigate the role and origin of nonchromosomal genes. By the mid-1960s, Sager had reported some 40 different nonchromosomal mutations which exhibited the same pattern of maternal transmission. The phenotypes of the mutations concerned such diverse biochemical capabilities as the loss of ability to grow photosynthetically, poor growth on all media, temperature sensitivity (i.e., ability to grow at 25° C but not at 35° C), and resistance to a number of different antibiotics.

The second step came with the genetic demonstration of segregation and recombination with a number of nonchromosomal mutations. The particulate conception of Mendelian inheritance had been based on (1) the transmission of characteristics unchanged through hybrid generations, (2) their segregation as pure parental types out of the hybrids, and (3) their independent assortment in crosses involving several different factors simultaneously. The application of the Mendelian method to the nonchromosomal mutations in *Chlamydomonas* required the acquisition of hybrids carrying presumed nonchromosomal genes derived from both parents. However, in *Chlamydomonas,* as in other organisms, most cases of cytoplasmic inheritance were detected by their strict maternal inheritance, which precluded Mendelian analysis.

In higher plants, it will be recalled, maternal inheritance had been simply attributed to the unequal contribution to the fertilized egg of cytoplasm from the female and male (pollen) parents. In *Chlamydomonas reinhardi,* however, both parents contributed equal amounts of cytoplasm to the zygote by complete fusion of the two gametes. The "exclusion hypothesis" did not hold up as an explanation of maternal inheritance in *Chlamydomonas,* nor for the other rarely detected instances of biparental, non-Mendelian inheritance in higher plants. One could therefore attempt to control and inhibit the mechanism of maternal inheritance in *Chlamydomonas* in an attempt to overcome the "drastic impediment" which it represented to the construction of a formal genetic analysis. In 1953 Sager found "exceptional zygotes" which exhibited biparental inheritance but with non-Mendelian ratios. Using these exceptional zygotes, she was able to study the segregation and recombination of nonchromosomal genes, thereby demonstrating their particulate nature. By 1967 Sager had established a procedure for converting maternal inheritance to a form of biparental inheritance by a variety of means, most dramatically by exposure of the maternal parent to ultraviolet irradiation just before mating.

The formal genetic analysis of nonchromosomal genes, then, involved a two-step process: first, by using streptomycin as a mutagen and observing the pattern of maternal inheritance, cytoplasmic genes could be distinguished. Then, by converting the pattern of inheritance from uniparental to biparental, Mendelian analysis could be carried out. Using this procedure, Sager and her co-workers argued that nonchromosomal genes could exhibit independent segregation and recombination and therefore could "obey Mendel's laws" in a qualitative sense; they behaved "like particulate units of heredity" (Sager, 1966, p. 292). By 1972 she and her co-workers constructed a single circular cytoplasmic genetic linkage group in *Chlamydomonas* (analogous to the circular "chromosome" of some phages and bacteria) (Sager, 1972, pp. 90–96).

Although Sager was not given the opportunity to develop her own school, during the 1960s Nicholas Gillham and his collaborator, John Boynton, did train many students in *Chlamydomonas* genetics and confirmed many of her results. The school developed by Gillham at Duke University extends investigations of "organelle heredity" to the present day (see Gillham, 1978). Gillham's primary strategy in attacking the "*Chlamydomonas* problem" was to apply the better-established principles of bacterial genetics. In fact, in his earliest investigations of *Chlamydomonas,* Gillham applied bacterial genetic methods to test some of the striking phenomena reported by Sager which had clear Lamarckian implications.

In the early 1960s, Gillham was attempting to establish whether certain predictions concerning the effects of various mutagens (e.g., nitrous acid, bromouracil, ultraviolet light, and alkylating agents) in prokaryotes and bacteriophage could be verified in a unicellular eukaryote such as *Chlamydomonas*. Since the classical work of Luria and Delbrück in 1943, all permanent hereditary manifestations in prokaryotes were claimed to have a spontaneous origin. However, Sager's claim that streptomycin could induce nonchromosomal streptomycin-resistant mutations was in direct conflict with the dogma that selection and mutation were independent processes. As Gillham and R. P. Levine (1962, p. 1463) remarked:

> We are confronted, therefore, with the possibility that a selective agent induces characteristic change in the genotype of a cell which allows the cell and its progeny to become "adapted" to existence in the presence of the agent.

Gillham and Levine (1962) confirmed Sager's claim that mutations which exhibit Mendelian inheritance in *Chlamydomonas* arose spontaneously, while those which are non-Mendelian in transmission arose, for the most part, only in the presence of streptomycin. However, they resisted the hypothesis that the chloroplast mutations were due to streptomycin induction of chloroplast genes. Instead, they speculated that the adaptive character of the non-Mendelian mutations might be due to selection of a cytoplasmic particle at the intracellular level. The origin of mutations in cytoplasmic organelles remains largely unknown and continues as an active research front to this day.

Mitochondrial genetics developed slowly after the classical investigations on respiratory-deficient *petite* mutations by Ephrussi, Slonimski, and their collaborators. It was not until the late 1960s, after the classical work of Sager, that "mitochondrial genetics" emerged into a systematic research front. The new era of mitochondrial genetics imitated the developments in chloroplast genetics, and emerged with a new class of non-Mendelian mutations conferring resistance to various antibiotics. Several hundred mitochondrially inherited mutants conferring resistance to various antibiotics or drugs have been discerned and analyzed. (see review by Dujon, *et al.*, 1977) A second breakthrough in the genetic study of mitochondrial inheritance occurred in the mid-1970s when respiratory mutants deficient for one function were isolated by a new type of mutagenesis. Since that time, hundreds of these mutants have been isolated and analyzed. By the late

1970s investigators of cytoplasmic genes could state that organelle heredity is "one of the fastest-flying fields of genetics," (Birky, 1979, p. 761).

Formal genetic analyis (that is, the analysis of offspring in a cross in terms of segregation and recombination patterns) was not enough to satisfy those who were skeptical of the existence of cytoplasmic genes. It was through cytological investigations that tangibility was given to Mendelian concepts. However, with the rise of the authority of molecular biology, an argument for the existence of cytoplasmic genes required evidence of information-bearing molecules: nucleic aids. Thus, A. H. Sturtevant (1965, pp. 124–125) wrote:

> Quite recently it has been found that there is DNA in plastids and in at least some mitochondria. It may therefore be supposed that these bodies carry genes of the same nature as those in the chromosomes.

However, the problem of identifying cytoplasmic DNAs proved to be a formidable task, and generated well over seventy articles during the 1960s. Many invetigators claimed to have detected DNA in various organelles and in an extensive array of organisms. Cytochemical, electron microscopic, and biochemical lines of evidence were offered in support of the presence of DNA in chloroplasts, mitochondria, and kinetosomes. Some investigators even claimed that the amount of cytoplasmic DNA in amphibian eggs was many times that of nuclear DNA. The problem of distinguishing cytoplasmic DNA from nuclear DNA led to the development of a standard procedure whereby cytoplasmic DNAs were not only detected but classified as distinct "molecular species."

Despite Sturtevant's assertion, theoretically speaking, the presence of DNA in cytoplasmic organelles *per se* did not in itself demonstrate its information content. As discussed in Chapter 6, the informational attribute of DNA was bestowed upon it by genetic methods—by its presence in chromosomes in the nucleus, the seat of Mendelian genes. The evidence that DNA was the physicochemical basis of nonchromosomal genes was indirect and based mainly on the occurrence of recombination, the existence of nonchromosomal DNA, and the mutagenic effects of streptomycin. Theoretically, cytoplasmic DNA could have some other function. A genetically autonomous RNA could take its place as the basis of cytoplasmic genes, especially in view of the existence of RNA alone as the genetic material in some viruses. Indeed, throughout the 1960s an experimental correlation was required between the biochemical and the genetic data in order to demonstrate the proposed primary genetic information carried by organelle DNA.

Nonetheless, in the context of a celebrated molecular biology, the recognition of cytoplasmic DNAs in chloroplasts and mitochondria played a crucial role in the acceptance of cytoplasmic genes. As Sager (1972, p. 2) remarked:

> The pendulum of opinion had swung from one extreme—cytoplasmic genes do not exist because we do not see cytoplasmic chromosomes—to the other extreme—cytoplasmic DNA's exist, and therefore there must be cytoplasmic genes.

Though the DNA contained in the cytoplasmic organelles is now recognized to be relatively little in quantity compared to that of the nucleus, the information it contained, and the manner in which it was organized, continued to indicate a special, distinct, and important role for cytoplasmic genes in the economy of the cell and for evolution. The cytoplasmic control over important characteristics involving energy availability—respiration and photosynthesis—was enough to indicate to several geneticists in the 1950s that cytoplasmic genes controlled "fundamental" characteristics of the organism. During the 1960s, this argument was maintained and reinforced.

The behavior of nonchromosomal genes seemed to suggest to Sager (1966), as it did to Ephrussi and Sonneborn, that nonchromosomal genetic systems had been "designed" for the maximum conservation of identity among organisms. In Sager's view these were the very properties which had, in retrospect, made the system so impenetrable to genetic investigations in the first place: the pattern of maternal inheritance and the stability towards conventional mutagens. Moreover, all eukaryotic cells have multiple copies of their chloroplast and/or mitochondrial linkage groups; the process by which one could ever obtain nonchromosomal mutations remains a mystery. The fact that most cytoplasmic gene mutations were obviously defective also lent support to the unique role of the cytoplasmic genetic system. "The existence of a nonchromosomal genetic system designed to minimize variability," Sager (1966, p. 296) remarked, "leads one to wonder whether NC genes control particular traits of crucial survival value to the organism." Sager was quite reluctant to speculate about evolutionary processes. She was more concerned with relating her work to biomedical problems, including the etiology of cancer and the molecular basis of aging (Sager, 1972, p. 372). She is presently professor of cellular genetics at Harvard Medical School and is also head of the Division of Cancer Genetics at the Sydney Farber Cancer Institute, Boston. Nonetheless, as she recognized, her work did pose unorthodox questions regarding evolutionary processes.

At the level of subcellular organization as well as at the level of the organism as a whole, a vast number of cytoplasmic traits show striking similarity, if not identity, from one organism to another. In this sense, it is thought that cytoplasmic gene mutations which in the laboratory result in a defective or lethal phenotype are not highly represented by variations encountered in natural populations. As J. L. Jinks (1964, p. 162) phrased it, they are like "major gene mutations," and in Sager's view remain outside of the synthetic theory of evolution which has little to say about invariant traits (Sager, 1966, p. 296). The cytoplasmic control of "fundamental" or "crucial" characteristics continues to suggest a process of macroevolution brought about by unknown mechanisms. However, all investigators of cytoplasmic genes stress the integration between nonchromosomal genes and chromosomal genes, an integration that had to be borne in mind in any consideration of the semiautonomy of organelles. Balancing out the problem, Jinks (1964, p. 167) suggested that the most important influence of cytoplasmic inheritance from an evolutionary point of view was its ability to produce more or less effective crossing barriers.

During the late 1960s the old question of the role of cytoplasmic genetic elements in the evolution of higher taxonomic groups and the possibility of environmentally induced adaptive mutations became overshadowed by competing theories concerning the evolutionary origin of organelles and the eukaryotic cell.

Recently the symbiont hypothesis has captured the interest of biologists. The traditional view of cell evolution, "direct filiation," holds that cell organelles such as mitochondria and plastids evolved by compartmentalization inside cells. On the other hand, it will be recalled that the symbiotic origin of organelles had been sporadically proposed by several researchers early in the century (see Chapter 3). With the recognition that chloroplasts and mitochondria have all the essential equipment for "life": DNA, transcription enzymes for making RNA, and a full protein synthesis apparatus, the endosymbiotic theory reemerged, led by the writings of Lynn Margulis. The particular theory set forth by Margulis holds that

> mitochondria developed efficient oxygen-respiring capabilities when they were still free-living bacteria and that plastids derived from independent photosynthetic bacteria. Hence, the functions now performed by cell organelles are thought to have evolved long before the eukaryotic cell itself existed. (Margulis, 1981, p. 3)

The theory of endosymbiosis is recognized today by leading ecologists, such as Evelyn Hutchinson, as representing a "quiet revolution" in biological thought. Hereditary endosymbiosis is presented as a "modern synthesis" of the mechanisms and processes of cell evolution. Its systematic investigation is emerging as a major biological research front in relation to evolutionary and cell biology, microbiology, geology, and environmental science.

Although investigators of cytoplasmic genes continue to create what they consider to be new problems of evolutionary divergence and speciation, cytoplasmic genetics has not escaped from its controversial past. Some older evolutionists who played leading roles in the classical evolutionary synthesis continue to dismiss it as a minor curiosity. Ernst Mayr (1982, pp. 786–790), for example, excludes the non-Mendelian inheritance of respiratory deficiency, investigated by Ephrussi and Slonimski, as being due to an infection. While mitochondrial genetics is one of the most rapidly growing specialities of modern genetics, Mayr can only state that mitochondria "may" have their own DNA. Cytoplasmic geneticists claim that genes in mitochondria, chloroplasts, and other cytoplasmic organelles (such as the centriole, which directs the migration of sister chromosomes to opposite poles of the cell during mitosis) control "crucial" and "fundamental" cellular functions. However, nucleo-centric molecular biologists take a different perspective. In his text *An Introduction to Molecular Genetics,* Gunther Stent (1971, p. 622) expressed his view of the power relations in the cell clearly when he wrote:

> Thus a eukaryotic cell may be thought of as an empire directed by a republic of sovereign chromosomes in the nucleus. The chromosomes preside over the outlying cytoplasm in which formerly independent but now subject and degenerate prokaryotes carry out a variety of specialized service functions.

While molecular geneticists such as Stent were busy building empires with bacteria and viruses and rapidly expanding and reproducing molecular biology laboratories, Sonneborn and his collaborators were investigating morphogenesis. At the same time, they were quietly challenging "the unwritten dogma" of molecular genetics, according to which biological evolution is solely the evolution of nucleotide sequencies.

Structural Guidance: "A Virus is Far from a Cell"

> Perhaps, as many people think, polarity represents something that was invented only once and has evolved since on its own.
>
> . . . Perhaps, contrary to what many people think, polarity resides exclusively in the gene-determined structures of polar molecules, and only genes evolve. If so, the only cell theory worthy of the name is wrong. (A. D. Hershey, 1970, p. 700)

By the 1960s, the story of DNA as "the basis of life" had been told at every level of scientific discourse, from research papers, through review articles, to textbooks and the latest issue of news magazines. The story is well known. The DNA molecule is a code which contains all the information required to specify the heritable characteristics of the organism. The information is translated into protein structures by a process in which DNA dictates the specificity of protein synthesis. Once the information has been translated, all of the chemical reactions of the cell—which are wholly determined by the structures of enzyme proteins—have also been specified. The genes, consisting of DNA, regulate the inherited characteristics of the species, and the self-duplication of DNA and its regulation is the basis of heredity.

However, not all biologists agreed that genes were the only hereditary agents and that biological evolution was solely the evolution of nucleotide sequences. The work of Monod and Jacob indicated how genes could directly participate in cellular differentiation. However, genetic regulation and stable cell states did not touch on the major and fundamental problem of morphogenesis, which lay in how differentiated cells or parts of an organism come to be arranged in space in different ways. As we have seen, many embryologists throughout the century protested against the neo-Weismannian reduction of the organism to the expression of genes. Instead they claimed that the pattern of morphogenesis was the result of epigenetic interactions between differentiated embryonic cells, and that morphogenesis was ultimately conditioned or constrained by a specific structural organization in the egg cytoplasm.

Embryologists frequently claimed that there was a specific character or "organizing principle" in the "ground substance" of the cytoplasm. Some similar sort of structural organization had been considered by *Plasmon* investigators. In the 1950s Sonneborn and Ephrussi, it will be recalled, also claimed that the organization or structure of the cell was a hereditary property which could not be directly or explicitly programmed in the genome. In all cases, the "ground plan" roughly engraved in the egg cell and expressed by its polarity was held to be responsible for the pattern of cellular change in time and space during the course of development.

Development was an orderly process, and even within the cell, genes had never been shown to control the development of structures at the supramolecular and microscopically visible levels. Sonneborn (1960, pp. 160–161) phrased the problem of translating chemistry into biological structures and cell organization as follows:

> "How are the chemical substances which the genes make, and the products of their interactions, translated into organized structures?" For the cytoplasm is more than a bag of chemicals. It is highly structured, even on the purely chemical level. The enzymes resulting from the action of the various genes often form systems that operate within millisecond speeds in ordered sequences; and this calls for their precise organization in ordered spatial sequences. On the grosser levels of visibility in the light microscope, the distinctive structures of diverse cell types are of course obvious and well known, and in recent years the fantastically powerful electron microscope has yielded much more insight into the structural organization of the cytoplasm. How is this organization, especially the difference in organization between different cells of the same organism, determined?

Genetic investigations of cell organization during the 1960s and 1970s emerged in direct conflict with the "doctrine of self-assembly" which was gaining ground among molecular biologists. As we have seen, during the first half of this century the field of heredity had been characterized in part by a struggle between those who believed biological organization could be reduced to principles of physics and chemistry and nothing more and those who believed new unknown theoretical principles were necessary. However, the attempts of molecular biologists to understand organization in terms of constituent parts were, in fact, based on new theoretical principles. Prior to the emergence of molecular biology those who had believed genes alone could not direct the building up of a cell or organism worked within the theoretical confines of colloid chemistry. As discussed in Chapter 1, colloids were substances that did not manifest specific characteristic structures but changed their shapes, forming structurally undefined gels. Within the confines of colloid chemistry, Jacques Loeb and others had insisted that the orderliness of chemical reactions in the cell had to be due to cell structure, and for the phenomena of life to persist that structure had to be preserved. Order could not be made anew in each cell generation. To many, the orderliness of chemical reactions in the cell therefore had to be due to the nature of the ill-defined colloidal mesh: protoplasm.

The principles of molecular biology are much more specific and, in fact, partly contradict the beliefs of the physicochemical school of Loeb and others at the beginning of the century. What was new in molecular biology was the claim that the essential properties of living beings could be interpreted in terms of the structures of their molecules. The principles of molecular biology and the doctrine of self-assembly did not simply emerge out of an internal academic development and convergence of genetics, physics, and chemistry. As Francois Jacob has emphasized molecular biology was part of a much larger theoretical synthesis which began to emerge out of World War II:

> With the development of electronics and the appearance of cybernetics, organization as such became an object for study by physics and technology. The requirements of war and industry led to the construction of automatic machines in which complexity increased through successive integrations. In television sets, an antiaircraft rocket or a computer, units are integrated which already result themselves from integration at a lower level. Each of these objects is a system of systems. In each of them, the interaction of the constituent parts underlies organization of the

> whole. . . . Until then, the coordination of components was considered as a property that existed only in certain systems. Thereafter, organization and integration of the components were inseparable. (Jacob, 1976, p. 247)

The chief molecular biological assumption concerning the control over cell structure was based on a belief in a transformation of disorder into order due to three factors: the physicochemical properties of reactants, their random collisions, and the ionic and molecular constitution of the cell "soup." In sum, the self-assembly hypothesis ultimately traces the building of all cell configurations, organelles, and other subcellular structures to molecular contributions from the milieu and genes and to random collisions of previously unarranged reactants (see Sonneborn, 1963b; Monod, 1972; Jacob, 1976, pp. 279–286). The strength of this argument was fortified by the startling demonstration that a linear genetic code could be translated into three-dimensional structure, as in the assembly of virus organization (Sonneborn, 1964).

Sonneborn became convinced that the doctrine of self-assembly overlooked a most important element, namely, the existing supramolecular architecture of the cell, which he believed played a decisive part in determining where the products of gene action became located in the cell and what they formed. The nucleic acid control of virus organization indicated that genic action somehow directed an amazing degree of precise nonrandom structural patterning. But Sonneborn (1964, p. 924) protested against the extension of such a mechanism to account for all cellular organization and denied the cellular nature of a virus:

> Yet a virus is far from a cell. . . . A virus does not grow and divide like a cell. Its nucleic acid replicates and its other structures are separately formed, the parts later coming together in the final organization. On the contrary the integrity of non random cell structure persists throughout growth and division which immediately suggests that the pre-existing structure plays a decisive role that may not be explicable by mere random self-assembly of genic products.

By resisting the generalizations of molecular biology, the research on cellular organization found itself competing with the mainstream of microbial genetic research programs. The forefront of the widely acclaimed progressive biology during the late 1950s and 1960s was those fast-paced research domains in which the biological problem had been reduced to chemical and physical terms. Biochemical and biophysical investigations of this period were characterized by rapid developments in an understanding of various biological problems, such as metabolism, photosynthesis, the biosynthesis of macromolecules, an the structure of genes and viruses. When biological problems could be stated in molecular terms, the enormous theories and technologies of modern chemistry, physics, and engineering could be brought to bear upon them.

On the other hand, the fast-paced and rapidly expanding molecular biology laboratories, densely packed with expensive electromehanical apparatus, students, and postdoctoral fellows, which represented the glamorous cutting edge of biological research, threatened to overshadow research on biological problems which continued to resist biophysical and biochemical approaches, such as analysis of cellular development. There were published statements in the scientific and popular literature on the 1960s which attempted to discourage research on nonmolecular problems. Summarizing the state of biologial sciences in a book written for "the intelligent man," Isaac Asimov (1960, p. 1) wrote that "modern science has all but wiped out the borderline between life and non-life" and that life began "through chance combinations, of a nucleic aid molecule, capable of inducing replication" (p. 542). Summing up the text, one reviewer wrote about Asimov's position in *Science:*

> For him . . . biology is a system that proceeds from biochemistry to the associated subjects of neuro-physiology and genetics. All else, as they used to say of the non-physical sciences, is stamp collecting. . . . I happen to agree firmly with Asimov about what is central in science and what is not and will defend him to the death against traditionalists who may deplore his not starting with 'heat, light and sound' or his giving short shrift to natural history. (de Solla Price, 1960, p. 1830)

Molecular geneticists working on viruses and bacteria often took for granted that all "important" genetic phenomena observable in higher organisms would be found in microbes where they were "stripped to essentials" and most amenable to analysis at the molecular level. This was their main reason for considering investigations of complicated higher organisms (except when they were carried out at the molecular level) to be less rewarding and even unnecessarily wasteful of effort. Sonneborn and other genetic investigators of cell structure who worked on ciliated protozoa recognized the value of the new revolutionary molecular genetics, and they appreciated that the more fundamental aspects of genetics were best studied in microbes. But they also recognized that higher organisms, while retaining the fundamental features observable in simpler microbes, had evolved new genetic mechanisms that do not occur at all in the simpler microbes, and which extend beyond the current principles of molecular biology (see Sonneborn, 1965).

However, the school of ciliate genetics developed by Sonneborn was not able to sustain its growth. During the late 1950s and 1960s, young geneticists rush to exploit the biochemical and molecular genetic technologies offered by bacteria and its many phages. During the 1960s a trickle of students continued to be trained in the genetics of the ciliated protozoa *Paramecium* and *Tetrahymena*. However, many of them failed to find positions in major research institutions in the United States, and according to Nanney (1983), ciliate geneticists in the United States today struggle against what they see as the threat of fading into "total obscurity." *Paramecium* was not a suitable organism for the techniques of biochemical and molecular genetics that could be applied to viruses and bacteria, but it was highly suitable for the genetic dissection and observation of cell structure. Not surprisingly, genetic work on ciliated protozoa prospered more on the international scene than in the United States. According to Nanney (1983), Sonneborn's laboratory at Indiana University continued to be a mecca for *Paramecium* genetics during the 1960s and 1970s. Investigators from all over the world came and returned to their native countries where they continued to work relatively free from the stigma of studying the cytoplasm of a eukaryotic organism.

Doubts as to the efficiency of so simple a hypothesis as self-assembly had long been proposed by investigators of the ciliated protozoa. Although ciliates may not be more highly organized in any fundamental sense than many other cells, their complex patterns of organization were more readily observable with the optical microscope. The conspicuous, constant, normal organizational features of the ciliates made them excellent tools for the experimental analysis of the function of preexisting structure in genic action.

The most impressive structural organization in the ciliates concerns the complex structures that make up the cell surface, i.e., the skeleton, ectoplasm, or cortex. The cortex is composed of linear arrays of a large number of fundamentally similar "ciliary units" arranged in a precise repeating pattern. A ciliary unit is a sophisticated structure that includes a kinetosome (ciliary basal body), cilium, a variety of subcortical fibers, and specialized membranes. At each level of organization observable within the limits of resolution of the optical microscope, the cortical pattern is remarkably constant and reproduces faithfully through a regular sequence of events during growth and fissions. It was difficult to imagine how this organization could arise *de novo* by genic action in any noncellular milieu or unorganized cell "soup."

All of the protozoological work of the 1950s and 1960s pointed to the same conclusion, that the guiding mechanism for the elaboration of formed parts in the cytoplasm was to be sought neither in the nucleus nor in a flowing endoplasm (ground substance), but in the most solid portion of the cell, namely, the ectoplasm. The role of the cortex and its parts in determining the production and ordering of diverse cytoplasmic structures had been investigated by Fauré-Fremiet in various ciliated protozoa in the 1940s and 1950s. André Lwoff, it will be recalled (see Chapter 5), had written a succinct account, in 1950, of the role of kinetosomes in morphogenesis. Based on his protozoological observations of the life cycle of ciliates, he concluded that cortical elements were pluripotent: they were able to form different structures at different times and places within the cell. Vance Tartar (1961) had created cortical differences experimentally by grafting techniques and studied them in some of the larger, complex, unicellular, ciliated protozoa.

Sonneborn himself traced the origin of his belief in a self-perpetuating supramolecular pattern to his earliest research as a student of H. S. Jennings. It will be recalled (see Chapter 4) that in his first work on the flatworm *Stenostomum* and the ciliated protozoan *Colpidium,* Sonneborn had investigated the inheritance of abnormal "doublet organisms" which contained duplicate sets of part or all of the animal's structures. Both studies seemed to show that the number and arrangement of the structures were nongenically inherited during asexual reproduction. The early studies of Jennings on the shell of the protozoan *Difflugia* were also held up in the 1960s as providing indicative evidence that biological information could be stored and transmitted by supramolecular mechanisms (Nanney, 1968).

Although various observations and experiments in ciliates suggested that more than self-assembly was involved in the formation of cellular structure, Sonneborn believed a crucial step was lacking. Genetic analysis was required to exclude the possible role of genes or genic action. During the 1960s the translation of chemistry into morphology was scarcely touched experimentally in higher organisms. On the other hand, in many ciliates, where structural organization had been experimentally investigated, there was no reliable means of controlling mating and carrying out cross-breeding analysis.

Sonneborn returned to the problem of the genetics of structural abnormalities in the early 1960s, by which time he had developed sophisticated procedures for genetic analysis of *Paramecium aurelia*. Doublet animals were obtained which differed from "singlets" in size and in the structures in the cortex of the cell. Through a series of sophisticated and highly acclaimed manipulations, he showed that the character bred true to type through sexual and asexual reproduction, free from both nuclear intervention and the control of the fluid part of the cytoplasm (Sonneborn, 1963b). The genetic basis for the structures therefore seemed to be contained in the cortex of the cell, which bears the intricately arranged pattern of cilia-bearing structures. This hypothesis was then tested by grafting experiments. A piece of the cortex, when torn off from one cell and implanted into the surface of another cell, led to the development of an entire "supranumary oral segment" which perpetuated itself during fission of the abnormal animals.

In further grafting experiments, Janine Beisson (a former student of Georges Rizet at the *Centre de Génétique Moléculaire* at Gif-sur-Yvette) and Sonneborn (Beisson and Sonneborn, 1965) inverted a small patch of unit territories. Subsequently, the inverted patch grew during cell division until its rows extended full-length along the body surface. Thereafter, the progeny inherited the inverted row or rows. The only cells containing the inverted patch were those derived by fission from preexisting cells with the inverted patch, and again it was shown genetically that neither the nucleus nor the free-flowing cytoplasm had any influence on the transmission of this trait.

The theoretically important conclusion was that structural information could be maintained in, and transmitted by, supramolecular structures. The ciliate cortex of *Paramecium* seemed to carry information for its gross organization and transmitted the organization to progeny independently of the genes. In the words of Beisson and Sonneborn (1965, p. 282):

> Our observations on the role of existing structural patterns in the determination of new ones in the cortex of *P. aurelia* should at least focus attention on the information potential of existing structures and stimulate explorations, at every level, of the developmental and genetic roles of cytoplasmic organization.

These results have been, and continue to be, reproduced and extended by several of Sonneborn's former students and associates working on *Paramecium* and *Tetrahymena* (see Nanney, 1980). Based on these studies, ciliate geneticists generally concluded that the location of the cortical parts is not random, but, at the time of their development, some forces exterior to the organelle itself dictate their location, orientation, and number. The interpretation is supported by cytological studies of the growth of individual surface organelles, especially the ciliary basal bodies or centrioles (which are considered to be identical structures). Detailed electron microscope studies by Ruth Dippell (1968), working in Sonneborn's laboratory, showed that the basal bodies in *Paramecium* arise not by division, as might have been expected to account for the physical continuity of organelles. The precise position and orientation of a new basal body during development is defined in relation to the existing cortical structure, which seems to act as a supramolecular template of some sort. Basal bodies or centrioles are not restricted to protozoa but exist in the cytoplasm of almost all animals and lower plants and are implicated in the formation of flagella, cilia, and certain sensory structures, and in the organization of the mitotic apparatus.

From the genetic studies of cell organization in ciliated protozoa it became evident that no one-to-one relationship existed between molecules and cellular form. Large pieces of the cell cortex are equipotent with respect to regeneration of cortical patterns—patterns that are, moreover, subject to metastable variations. The structural elements of the ciliate cortex provide the "scaffolding" for the insertion of new organelles. Cells may reproduce in several stable configurations differing not at all in their molecular composition but only in their pattern of organization. Conversely, cells with essentially the same hereditary structural pattern may have entirely different genes.

Genetically speaking, ciliate geneticists divided the cell into two parts: genes and gene products, and a cell cortex or skeleton that manifests supramolecular properties. Certainly, this did not mean that genes played no important role in the determination of cellular structures. Reproduction of cortical structures is recognized to be typically very indirect, involving a complex series of events. The kinds and quantities of molecules directly and indirectly resulting from genic action are considered to be absolutely essential factors, but they are held to be insufficient. Preexisting cortical structures would play a role in determining where some gene products go in the cell, how these combine and orient, and what they do.

The relation of genes and supramolecular structures in the formation of new structures was expressed by Nanney (1968, p. 497) in the following figurative terms:

> In an extreme polar interpretation, one might postulate that nucleic acids specify only proteins, which must be appropriate for cellular design, but not decisive. In this case the cellular architects (that is, preexisting structures) might be required to determine whether the eventual edifice constructed of the building blocks would be a railroad or a cathedral. I doubt the value of this extreme analogy, but some intermediate position may be more consonant with the larger biological realities than either extreme.

The ordering and arranging of cell structure under the influence of the old has been called "cytotaxis" by Sonneborn, "structural guidance" by Frankel, and more recently, "structural inertia" by Nanney. The differences in terminology are not gratuitous, but relate to variations in the proposed mechanism(s) underlying the formation and inheritance of structural patterns. Although the preliminary investigations of cell patterns led to the view that the microscopically visible, preexisting structures played a *direct role* in the perpetuation of structural information, this conclusion began to fall into question in the 1970s (see Frankel, 1983).

It soon became apparent that all structural patterns of the cortex do not depend on visible structures. Some patterns were reported to persist under circumstances in which major cortical elements had been disassembled. These results led to the possibility that the visualized structures may not play a directive role in the perpetuation of cell organization but represent only its manifestations. An invisible, unknown guiding force, which lies beyond the reach of current molecular principles, may be responsible for the perpetuation of morphogenetic patterns. The following attempt of the celebrated phage geneticist A. D. Hershey (1970, p. 700) to conceive the problem of cell organization highlights current difficulties in providing a plausible model for the emergence and perpetuation of cell structure:

> If cells draw on an extragenic source of information, a second abstraction must be invoked, another vital principle superimposed on the genotype. A likely candidate already exists in what is usually called cell polarity, which tradition places in a rigid ectoplasm for good reason—it's a spatial principle and as such requires mystical language. Seemingly independent of the visible structures that respond to it, polarity pervades the cell much as a magnetic field pervades space without help from the iron filings that bring it to light. Biological fields are species specific, as seen in the various patterns and symmetries of growing things.

Despite differences in terminology, and the nuances they imply as to divergent interpretations of the unknown mechanism(s) underlying the phenomena, all investigators of ciliated protozoa agree that cell organization does not reside exclusively in the gene-determined structures of polar molecules. This generalization was summarized by Sonneborn (1963b, p. 202):

> Without cytotaxis an isolated nucleus could not make a cell even if it had all the precursors, tools, and machinery for making DNA and RNA and the cytoplasmic machinery for making polypeptides. Self-assembly of genic products can go only so far; to go the whole way, cytotaxis must be added on. Strong evidence now confirms the old dictum that only a cell can make a cell.

The inheritance of cell organization clearly represented a challenge to the claim that biological evolution is solely the evolution of nucleotide sequences. Sonneborn (1965) remarked in effect that just as biological evolution is separate from cultural evolution, so is DNA evolution separate from the evolution of cell organization. A similar view was maintained by the influential viral geneticist Salvador Luria (1966). Although parallel, independent, and selectively correlated evolution of genome and cortex is a favored possibility, no direct evidence is available to support it.

Throughout the 1960s and 1970s, the evolutionary significance of cortical inheritance was largely ignored by neo-Darwinian evolutionists. Indeed, for the most part it still is. However, in recent years with renewed interest in macroevolutionary events, developmental constraints of evolutionary changes, and "hopeful monsters," led by the writings of Stephen Gould and Niles Eldredge, the role of cytotaxis in macroevolution has begun to attract some attention and concern. Ciliate geneticists and embryologists who have continued to resist reductionistic neo-Darwinian conceptions of heredity and evolution continue to find cytotaxis to be particularly important for understanding the nature and role of intracellular spatial organization in metazoan egg cells, macroevolutionary changes, and the basis of their inheritance (Frankel, 1983; Horder, 1983; Nanney, 1984). Even some leading neo-Darwinian evolutionists have admitted their concern about the significance of "the phenomenon of cortical inheritance in ciliates." At a symposium on Development and Evolution held at the University of Sussex in 1982, the centenary year of Darwin's death, John Maynard Smith (1983, p. 39) stated, "Neo-Darwinists should not be allowed to forget these cases, because they constitute the only significant experimental threat to our views."

CHAPTER 8

Patterns of Power

> The history which bears and determines us has the form of war rather than that of a language: relations of power, not relations of meaning. History has no "meaning," though this is not to say that it is absurd or incoherent. On the contrary, it is intelligible and should be susceptible of analysis down to the smallest detail—but this in accordance with the intelligibility of struggles, of strategies and tactics. (Michel Foucault, 1980, p. 114)

It is usually stated in the canonical accounts of the rise of genetics that Mendelism was generally accepted by 1915 with the appearance of *The Mechanism of Mendelian Heredity* by Morgan *et al.* The present account introduces an element of ambiguity into this assertion. As we have seen, the belief that Mendelian genes were the sole agents of heredity was challenged throughout the formal genetics period by various groups of biologists in various countries. The idea that the "fundamental" differences that distinguished higher taxonomic units were due to cytoplasmic properties and that Mendelian genetics applied only to characteristics that distinguished individuals or perhaps species, was maintained by many biologists well into the 1950s. Even outside this extreme view, the history of the research and theories of cytoplasmic inheritance requires a clarification of the meaning of the Mendelian-chromosome theory.

It is not enough to know whether biologists generally agreed with the principles of Mendelian segregation or accepted the existence of particulate genetic factors situated in chromosomes. To understand the significance of Mendelism one has to consider the doctrines and theories with which it was associated. The belief maintained by many American geneticists that chromosomal genes were the sole basis of evolution and directed the synthetic processes of the cell, never fully gained support throughout the formal genetics period. As discussed in Chapter 7, it was not until the rise of molecular biology and the transformation of the chromosome theory into the nucleic acid theory during the 1950s and 1960s that the major threat of cytoplasmic inheritance to the general genetic synthesis was laid to rest. Nonetheless, as discussed in the last chapters, in the years that followed, extranuclear genetic systems continued to be investigated and results were reported, and continue to be, which contradict the views of geneticists and classical neo-Darwinian evolutionists who upheld the "nuclear monopoly."

A second major historiographical issue stemming from the present work concerns the belief in the inheritance of acquired characteristics and its relations to Mendelian genetics during the formal genetic period. The purging of this belief is generally recognized to be the result of two major steps: the distinction between the germ plasm and the somatoplasm as theorized by Weismann and others (see Chapter 1) and the genotype-phenotype dichotomy as articulated by Johannsen in 1909 and 1911 (see Chapter 2). However, as we have seen in Chapter 6, when discussing cytoplasmic inheritance and its relations to Lysenkoism, the belief in the inheritance of acquired characteristics found experimental support in several examples of environmentally directed adaptive changes inherited through the cytoplasm, reported during the 1940s and 1950s. The significance of inherited environmental influences through the cytoplasm was unaffected by the two major dichotomies mentioned above. It was also unaffected by the Luria-Delbrück fluctuation test of 1943, which also could be listed among the obstacles in the way of the belief in the inheritance of acquired characteristics.

As discussed in Chapter 7, the molecular biological dichotomy between a change in structural information and a change in the expression of genetic information was crucial in reevaluating the evolutionary significance of the inheritance of acquired characteristics and cytoplasmic inheritance. Indeed, it was difficult to account for the orderly and environmentally directed changes during ontogeny within the confines of the classical conception of the gene, since gene action was held to be largely uninfluenced by extranuclear events. It must be emphasized that the principles of molecular biology partly contradicted those of formal genetics, not only in the recognition of extranuclear inheritance based on cytoplasmic DNAs, but also in the recognition of nuclear differentiation. Molecular biology introduced a new conception of the gene and the genome. In general, the recognition of problems of gene regulation, cytoplasmic inheritance, and the principles of somatic cell differentiation may provide a broader theoretical framework for appreciating the views of those biologists who maintained a belief in the inheritance of acquired characteristics throughout the classical genetics period. This is to say nothing about the existence of a submicroscopic "ground plan" or spatial principle which many embryologists and protozoologists claimed controlled where the products of gene action became located in space and what they formed.

The phenomena and concepts associated with cytoplasmic inheritance throughout the twentieth century cannot be considered simply as obstacles to the progressive development of Mendelian genetics. Biologists who opposed the "nuclear monopoly" of the cell cannot be dismissed as being "wrong" or "irrelevant." Their significance in history does not lie in revealing a scientific methodological moral or in explaining periods of nonproductive science. On the other hand, the present account does not represent a simple addition to the traditional historiography, which is largely concerned with the steps leading from the Mendelian-chromosome theory to the evolutionary synthesis and to the nucleic acid doctrine.

The history of the research and theories of cytoplasmic inheritance illustrates various important patterns in scientific activity. These patterns have to be reconstructed to make the history of genetics clear and meaningful.

All along in this book I have argued that scientists are engaged in a struggle for scientific authority. What is at stake in this struggle is the power to impose a definition of the field: what questions are important, what phenomena are interesting, what techniques are suitable, and what theories are acceptable. Science is a social activity whose outcome is constrained by, but not determined by, the inner logic of its subject matter. In other words, although the view of scientific activity I have employed assumes the existence of a "material reality," this "reality," which is independent of what we think about it, never acts alone in the production of scientific knowledge. It provides only the conditions for the possibility of various scientific interpretations and opinions. Two other restrictions actively form the system of relations out of which scientific knowledge is constructed. One results from a competitive struggle for power among individuals and disciplines within science (the internal politics of science). Another results from a struggle between institutionalized science and other organized bodies in a larger culture (the external politics of science). All three constraints are always involved in the production and acceptance of scientific ideas. Nonetheless, throughout the history of genetics discussed in this account (and perhaps in all of modern institutionalized science) power relations within and among scientific disciplines have usually played the predominant role.

The struggle for scientific authority is manifested in a social hierarchy within science. However, the significance of this hierarchy is not simply that scientists *receive* disproportionate recognition or unequal financial and institutional support for their research. It is rather that the higher up one moves in the hierarchy the more power one has in *bestowing* power, that is, deciding what sort of scientific work deserves recognition and credit. In other words, power in science does not result from a relation between a scientist and "truth" or "nature." It arises socially from negotiations between scientists. A scientist does not act in response to "nature"; he or she acts in relation to peers who grant recognition. The more recognition (symbolic profit) one receives the more credit one can bestow upon (invest in) the work of others.

The effect of this system, as Bourdieu (1975) originally argued, is that a scientist enhances his or her own position by directly changing the field of problems, techniques, and theories to suit his or her interests. The power to define the field extends along a continuum from graduate student to Nobel laureate. There are no neutral opinions in scientific controversy, since all socially recognized judges, by definition, have an "interest" (in both senses of the word) in the outcome. These power relations chiefly determine the nature of scientific results, their interpretations, and the way in which scientific knowledge claims come to be certified and accepted as "true." This means that social interests are always involved in the content of scientific knowledge, and as Albury (1983) has argued (when suggesting the value of the above approach to an understanding of sociobiology), it is in the internal politics of scientific truth that the sociology of scientific knowledge finds its strongest foundation.

It is important to emphasize here that the struggle for authority does not *depend* on opinions about the motivations of individual scientists. Whether an individual's action is shaped by a "search for truth" or by personal power is not at issue. Nor is it implied that a scientist clearly recognizes or acknowledges what is in his or her individual or collective interest. We do not need to fall into the functionalist trap of replacing a "disinterested search for truth" with a "disinterested search for power and influence" and claim that scientists always act in their best interest. What are of concern are the effects of their actions in a given context and the structure of the field which shapes those actions. What is implied is that in a highly developed scientific field (one that has acquired a high degree of autonomy), an individual who does not act according to the rules of the game sketched above will fail to influence the structure of the field and the nature and validity of scientific knowledge.

In the complex structure of the scientific field, in the midst of continuous conflict and competition, scientists have recourse to a variety of means through which they can enhance their position by shaping the structure of the field. This is done through a series of activities which have often been perceived as altruistic or disinterested. They include teaching undergraduate and graduate students and building a "school," writing review papers and "historical" accounts of one's field, refereeing grant applications and papers submitted to scientific journals, editing journals, examining doctoral dissertations, and writing letters of reference in support of candidates for employment or promotion (job control). These kinds of activities represent the "mundane" politics of truth; we have seen many of them at play in the present account.

In Chapter 2, for example, we have seen—when discussing Sumner's genetic work refused from the American journal *Genetics*—the role of the referee system at play as a mechanism of control to maintain orthodoxy. That this form of control was strictly practiced by Morganist geneticists and that judgments of competency in American genetics were often indistinguishable from judgments of orthodoxy is suggested by the fact that few papers on cytoplasmic inheritance were ever published in the journal *Genetics*. Instead, the relatively few American geneticists who voiced opposition to the "nuclear monopoly" throughout most of the present century made heavy use of other vehicles: the few nonrefereed American biological journals, those of related specialities, monographs, and foreign genetic journals. We have seen this form of control on the other side of the controversy in Chapter 7, when discussing Nanney's and Beale's dissent from the plasmagene theory of cellular differentiation and Beale's conflict with Sonneborn over interpretations Beale made in his book.

Job control also represents a major constraint limiting the freedom of thought and expression of scientists who rely on established figures for refereeing grants and writing letters of reference on behalf of scientists applying for positions. We have seen this at work in the laboratory situation in Chapter 4, when discussing Sonneborn's relations with Jennings and Jennings's dispute with Raffel over interpretations of results. We have seen this kind of control in operation at a broader institutional level when discussing neo-Lamarckism in France and Lysenkoism in the Soviet Union. And, as discussed in Chapter 6, when the Genetics Society of

America set up a committee for "scientific freedom" there was an attempt to suppress the teaching of Lysenkoist ideas in American colleges and universities by direct political action. Various issues, including "race," nationality, and unorthodox evolutionary beliefs, converged to confront Jollos in America, who, as we have seen in Chapter 3, was abruptly removed from the field.

These forms of social control through which individual scientists persistently shape the field of production and the nature of scientific knowledge are commonplace. However, there is still a larger collective institutional form of struggle; that is the formation of the discipline itself. Discipline formation represents a central strategy through which groups of scientists attempt to establish those objectives and explanatory standards which can be embraced by their technical procedures and theory. The success of this strategy results from the ability of organized groups to develop rapidly those lines of inquiry and to define the field in a way that gives them the competitive advantage.

As discussed in the present account, the nuclear monopoly of the cell resulted largely from the ability of Mendelian geneticists to form their own discipline with their own objectives, technical methods, explanatory standards, doctrines, journals, and societies, and to restrict their field of inquiry to problems which could be dealt with effectively by "the Mendelian method." This disciplinary structure entailed reductionistic, mechanistic thought and was allied with neo-Darwinism. Reductionism, neo-Darwinism, and their corollary, narrow disciplinary structure, were all challenged by many upholders of the cytoplasm. As a result, the controversy over the roles of the cytoplasm and the nucleus in heredity continued as a dispute understood by the participants in terms of scientific values which included the motivations of researchers and the problem of establishing priorities in scientific research.

We have seen also that the institutional context of scientific research can play an instrumental role in inhibiting or facilitating the emergence of a new discipline. The role played by institutional structure can be detected by international comparisons. This was discussed in Chapter 3 when introducing the work on the *Plasmon* and the concept of "the cell as a whole" in Germany between the two World Wars. It has been seen in its most striking form in the gross differences between the university system of France and that of the United States as described in Chapters 5 and 6. The institutional context of the American university system typified by departments and competition between universities and the intimate socioeconomic relations between genetics and agricultural research programs were highly favorable to the early and rapid development and institutionalization of Mendelian genetics. In France, on the other hand, the highly bureaucratic central structure of the university system facilitated the maintenance of traditional neo-Lamarckian authority. The institutionalization of genetics was retarded and the development of genetics research impeded.

This is not to suggest that competition between researchers is due to competition between universities and that the way in which genetics developed in France, Germany, and the United States can be reduced to this factor. The development of the research on cytoplasmic inheritance is exemplary in demonstrating how different "traditions" may emerge in different countries. By the turn of the century, the study of heredity was recognized by biologists to be highly valuable in view of its importance to the already established problem of evolution. Mendelian analysis was subsequently rapidly deployed and Mendelian genetics was able to grow rapidly in England and especially in the United States. As Mendelian genetics rapidly acquired followers it became increasingly difficult for an individual to make a major discovery in that research domain and thereby acquire prestige and recognition. In this light, one can easily understand the departure of a fraction of the researchers toward other objects and problems such as the study of the role of the cytoplasm in heredity, where competition was less intense. The development of the research on cytoplasmic inheritance can be understood as a strategy (objectively at least) directed toward the maximization of scientific recognition and prestige. As we have seen in Chapter 5, when discussing the emergence of genetic research at the Institute of Genetics in France after World War II, the choice of centering studies on cytoplasmic inheritance emerged as an explicit strategy to be competitive with the highly developed Mendelian genetic research programs of the United States and England.

In their struggle to impose a definition of the field, scientists also have recourse to an elaborate body of rhetorical tactics. Indeed, "method talk" itself has to be considered as rhetoric. Many historians, sociologists, and philosophers of science realize that there is no single efficacious scientific method (Schuster and Yeo, 1986). However, although method does not exist in the widely believed sense, formal method doctrines (inductivism, hypothetico-deductivism, falsificationism, etc.) do play a discursive role in the social constitution of science. Formal method doctrines are deployed in diverse and conflicting ways. These discursive doctrines are deployed in diverse and conflicting ways. These discursive resources are used as polemical tools to legitimate divergent objectives and techniques. They are summoned to help on both sides of scientific controversy, that is, in warding off and making room for alternative theoretical possibilities. They are also used in attempts to establish property rights in disciplinary disputes and in individual priority disputes. But formal methodological doctrines are not the only resources available for these purposes. Scientists also use nonformal discursive resources to accomplish these functions, including accounting for knowledge claims. Indeed, one can widen the category of "methodological" accounting recourses to include "social accounting," as discussed by Mulkay and Gilbert (1982), as well as a new category of resource involving the ascriptions of "technique-ladenness of observation."

All these types of resources function in similar ways in the social construction of science, and hence all can be treated within a broadened category of "scientific method" once formal methods themselves are understood as rhetorical devices. We have seen these polemical tools in use throughout the controversy over the relative importance of the nucleus and cytoplasm in heredity. For example, those who upheld the "nuclear monopoly" accounted for the "false" views of their opponents in terms of their lack of adherence to formal methodological doctrines (such as pragmaticism, empiricism, falsificationism, etc.), their "refusal" to accept the dominant role of the nucleus, and their training in terms of "false theories" (e.g., Lamarckism and/or the belief that the nucleus controlled only relatively trivial traits while the cytoplasm controlled the fundamental characteristics of the organism).

Those who supported the importance of the cytoplasm in heredity accounted for the "false" beliefs of Morganists in terms of "defensive attitudes," "failure to put enough effort," the ascription of technique-ladenness of observations, and the sociopolitical nature of scientific activity. Their accounts included claims of both quantitative and qualitative misrepresentation of reality resulting from specific experimental procedures and materials. In the view of cytoplasmic geneticists, cross-breeding analysis was responsible for perpetuating the idea that genes were self-autonomous agents acting independently from the rest of the cell. The facility of Mendelian procedures was largely responsible for the attention focused on the nucleus. The kinds of materials, i.e., organism, employed conditioned the range of results that could be obtained. Social accounting, the technique-ladenness of observations, and formal methodological doctrines all play similar roles in scientific controversy. None of these accounting resources is any less "scientific" in any *a priori* sense, that is, any less likely to be deployed in the constitution of knowledge-making and knowledge-breaking claims.

Social accounting feeds back into the scientific enterprise and allows competitors to adjust their activity, not only in terms of what they are doing, but also in terms of what they have to do. In the midst of continuous conflict and competition within the scientific field where scientists attempt to make their opinions pass for "scientific facts," the manner in which a person propounds his or her views may be decisive, at least in terms of persuading competitors to adopt one of various alternative theories. In this context, when critically evaluating contributions, scientists necessarily have to consider the history of the development of an idea, the specific indoctrination and point of view of the contributor, the technical nature of the experimental procedure, and the data presented. These issues are especially salient when attempts are made substantially to modify a dominant theory in order to reconcile it with contradictory observations and theoretical problems which are persistently ignored by members of dominant groups.

The struggle for authority has been the operative social mechanism underlying the present account; it has furnished the driving force for scientific development. Such struggle has led to the development of new techniques in the field of heredity constructed as polemical tools designed with maximum efficiency for equipping the agent, in his or her institutional posture, for fighting his or her competitors in order to win their recognition. This pattern is repeated over and over again throughout the history of genetics. The attempt of competitors to undermine each other has led to the "domestication" of various organisms—from flies and higher plants to protozoa, fungi, unicellular algae, bacteria, and viruses—which have transformed the scientific enterprise as essential technologies underlying much of the recent understanding of biological mechanisms.

To be sure, the technical norms of science are compatible with a wide range of political or ideological orientations. The larger political predispositions of scientists may play a considerable role in influencing their willingness to develop or accept certain ideas. We have seen this relationship functioning in Chapter 6, when discussing cytoplasmic inheritance and Lysenkoism and when describing neo-Lamarckism in France. Perceptions of social relations are also reflected in the metaphors biologists used to construct concepts of the cell, e.g., "nuclear monopoly," "master molecules," "democratic organization," "republic of chromosomes," "the cell as an empire," etc. We have also seen that producers inside the field may gain support for their work through its relations to various socioeconomic problems, i.e., agricultural and biomedical technological programs. However, it would be a gross error to account for the success of genetics (nuclear or cytoplasmic) or the dominance of a particular conception of heredity, the cell or the gene, in terms of the interests of the ruling class in a given society. Although the theoretical and disciplinary development of genetics is constrained by larger class interests, it is not determined by them. The present account relies on the consideration of the combined effects of the internal and external politics of science with the internal politics predominating. This view is expressed beautifully in Bourdieu's concept of *double determination:*

> Ideologies owe their structure and their most specific functions to the social conditions of their production and circulation, i.e., to the functions they fulfill, first for the specialists competing for the monopoly of competence in question . . ., and secondarily and incidentally for the non-specialist. When we insist that ideologies are always *doubly determined,* that they owe their most specific characteristics not only to the interest of the classes or class fractions which they express . . . but also to the specific interests of those who produce them and to the specific logic of the field of production . . ., we obtain the means of escaping crude reduction of ideological products to the interests of the classes they serve (a "short-circuit" effect common in "Marxist" critiques), without falling into the idealist illusion of treating ideological productions as self-sufficient and self-generating totalities amenable to pure, purely internal analysis (semiology). (Bourdieu, 1977, p. 4)

Geneticists did not receive recognition for their work simply on the basis of its socioeconomic value. Indeed, throughout the twentieth century, the field of heredity as a locus of struggle has maintained its specificity from the fact that individual producers have also tended to receive recognition for their product from their competitors, who were the least inclined to grant recognition without discussion and scrutiny. This tendency gives the field its specificity, intellectual order, and technical coherence: "the specific logic of the field of production." Within the scientific field, with its diverse problems, theories, techniques, and research strategies, where several competing groups exist, there are invariably differences in their success as perceived by the competitors themselves. At any given moment there is a hierarchy of objects, techniques, and theories which strongly orients practice.

The history of genetics is discontinuous. It is marked by periods of rapid development, sudden take offs, and changes which do not correspond to the calm continuist conception of scientific development that once had been imagined. The important point here is not that these transformations can be extensive or rapid. Their scope and speed are only an indication of something else, that is, a change in the manner in which scientific knowledge claims are constructed, accessed, and accepted as scientifically "true." Thus, these changes are not understood in terms of a "refutation of errors" or "discovery of truths," nor are they gross modifications in theoretical form such as "incommensurable" "Gestalt shifts" (cf. Kuhn, 1970). The present account reveals four phases in the study of heredity resulting from changes in the order of the power distribution among competitors. Each phase is characterized by a range of possibilities defined by the current theories and beliefs about heredity, the nature of the organisms accessible to investigation and the way of observing and discussing them. In isolating these phases we can see clearly that scientific controversies do not rely on a perfect competition of ideas whereby the intrinsic strength of the "true" idea decides the outcome. The acceptance of scientists of an idea is dependent both on its *a priori* attractiveness and on its scientific "evidence." But this is not a matter of simply weighing prejudice against a fixed scientific "proof," for what constitutes scientific proof—what counts as evidence—is determined by a social process of competitive struggle.

A first phase is characterized by the rise of Mendelian genetics and the emergence of the Mendelian-chromosome theory. It is marked by attempts of embryologists to formulate a compromise between the roles of the cytoplasm and the nucleus in heredity. Embryologists suggested that the egg, embryonic, and general phyletic characteristics of any stage of the developing organism were determined in the egg cytoplasm, whereas genetic elements in the chromosomes made their appearance known only through specific or individual adult differences. This view was supported by theoretical arguments concerning the nature of embryonic development, cytological observations of the behavior of chromosomes and the organization of the cytoplasm during cell differentiation, principles of colloid chemistry, and results of various embryological experiments, primarily on echinoderm and amphibian eggs.

A second phase, from about 1920 to 1940, is marked by the dominance of classical Mendelian genetics and the construction of a new concept of heredity which could only be authoritatively investigated and certified by a group of socially recognized experts. By restricting the definition of heredity to problems that could be rapidly investigated by existing techniques of cross-breeding, Mendelian geneticists designated themselves as experts in such a way as to exclude all competing groups. They understood heredity—which previously embraced growth and differentiation—in terms of the sexual transmission of chromosomal genes from one generation to the next. Cross-breeding analysis on higher organisms, especially *Drosophila* and maize, combined with cytological observations of chromosome movements, formed the basis upon which the physical location of genes in the chromosomes was established.

Mendelian geneticists in the United States, claimed nuclear genes as the "governing elements" of the cell, largely immune from extranuclear influence. Changes in the genes, situated in the chromosomes, were held to fuel the process of evolution by natural selection. Problems of embryonic development and cellular differentiation which could not be investigated by Mendelian methods were ignored by Mendelian genetic research programs. Nonetheless, Mendelian geneticists claimed jurisidiction over the entire field of heredity irrespective of their capacity to deal with it effectively. The need for advocating cytoplasmic inheritance was dogmatically denied by leading American Mendelian geneticists such as T. H. Morgan and H. J. Muller.

This framework defined the limits by which investigations of heredity could maneuver during the period. When knowledge was lacking concerning what it was that genes actually do in the cell, cytoplasmic heredity represented a threat and a challenge to the doctrines of Mendelian genetics and to the importance of Mendelian genetic research programs. Genetic investigations of cytoplasmic inheritance were carried out primarily in Europe and in direct conflict with Mendelian genetic research programs. Their primary significance was in accounting for the regulatory qualities of epigenetic development, the organization of the cell and the spatial pattern of the organism as a whole.

The genetic challenge to the "nuclear monopoly" of the cell failed during this period, but the contest was not decided by the intrinsic strength of a true idea. The power of the chromosomal genes as dictatorial elements in the cell was bestowed upon them by the technical capacity of Mendelian analysis and the institutional power of Mendelian geneticists. Mendelian geneticists had an effective technique and attracted many new recruits who worked on problems that were both easily accessible to analysis and that had been judged to be important by producers endowed with a high degree of legitimacy, such as the *Drosophila* group headed by the Nobel-Prize-winning Morgan.

Cytoplasmic inheritance was also judged to be important by the celebrated biologists Carl Correns, Jacques Loeb, and many others. However, with the rise of Mendelian genetics to an authoritative position in the field, biologists who postulated the existence of cytoplasmic hereditary elements had to subject their views to the scrutiny of their chief competitors in the United States, who were the least inclined to grant recognition without intense scrutiny. As discussed in Chapter 3, the existence of cytoplasmic hereditary properties during this period required demonstration through cross-breeding methods which had been especially devised for Mendelian analysis: chromosomal mapping, segregation ratios, chromosomal recombination, etc. It also required a correlative demonstration of the existence of cytologically visible self-reproducing cytoplasmic bodies similar to chromosomes.

Plasmon theorists in Germany developed a breeding strategy for investigating cytoplasmic heredity based on the study of maternal effects resulting from crosses between various organisms—*Epilobium,* mosses, and *Oenothera*—selected for their maximum efficiency for outcrossing and satisfying the genetic criteria for a hereditary role of the cytoplasm. However, genetic methods for investigating cytoplasmic inheritance in higher plants were tedious and difficult. One could not produce results comparable to those for analyzing chromosomal inheritance and successfully compete with Mendelian genetic research programs. It was difficult to distinguish the "hereditary element" of the cytoplasm as due to particulate or nonparticulate forces with the existing techniques. In the context of the dominant Mendelian-chromosome theory, it was difficult, if not impossible, to defend maternal inheritance from being formally interpreted in terms of the physiological action of genes. The evidence for cytoplasmic heredity was not only largely based on "vague principles." It also became allied with the belief in the inheritance of acquired characteristics and the view that nuclear genes controlled only trivial characteristics.

When reviewing the evidence for cytoplasmic heredity in the 1930s, the American geneticist E. M. East claimed that the idea that the cytoplasm controlled the fundamental properties of the organism was not a very satisfying scientific hypothesis, since it could not be effectively tested by existing techniques. Mendelian geneticists did have an effective technique for investigating chromosomal inheritance, but as East fully recognized, the possibility existed that Mendelian genes themselves might be concerned only with trivial characteristics and have little to do with evolution. Neo-Lamarckian biologists in France, it will be recalled, also maintained this view and simply chose the option of not investigating chromosomal inheritance. Cytoplasmic inheritance represented a threat to the importance of Mendelian research and the evolutionary synthesis and was therefore criticized and denied by Mendelian geneticists in the United States and by evolutionists who upheld the predominant, if not exclusive, role of genes and natural selection in heredity and evolution.

A third phase from about 1941 to 1958 is marked by the rise of biochemical genetics to an authoritative position in the field. During this phase, microbial genetic technology successfully competed with genetic investigations based on *Drosophila* and higher plants and the main thrust of genetic research was turned to investigate the nature of the gene and the means by which it affected biochemical processes. With the entrance of microorganisms into genetic study, the notion of heredity was extended by cytoplasmic geneticists from the sexual transmission of hereditary properties of the organism as a whole to account for asexual hereditary transmission and perpetuation of specificity at the cellular level. However, the genetic acknowledgment of "cell heredity" brought with it the apparent paradox of identical genomic reproduction in the fact of phenotypic change, as it had to embryologists since the end of the nineteenth century. With the invention of microbial genetic technology, genetic investigations of cytoplasmic inheritance arose anew, and to a prominent position in the United States and France during the 1940s and 1950s. They found their chief significance in attempts to understand the principles of somatic cell differentiation and embryonic development and continued to be carried out in conflict with the dominant genetic research based on Mendelian genes.

The research programs based on plasmagenes competed with the work on Mendelian genes for control over the synthetic processes of the cell and for control over what were considered to be the most important cytoplasmic constituents—the proteins and their specificity. During this period, the gene remained the formal, abstract Mendelian hereditary unit. There was no correlation between its structure and its function. It was possible that the control of the basic structure of cytoplasmic proteins was a function of cytoplasmic elements, a view which continued to find support in the embryological claims that development was largely a cytoplasmic phenomenon. The gene concept itself remained largely immutable to the program of biochemical genetics and was simply extended into the cytoplasm, witnessed by such terms as plasma*gene,* cyto*gene,* and *gen*oid. To account for cellular differentiation in the face of nuclear equivalence, geneticists had postulated the existence of various sorts of plasmagenes (as independent genetic elements or as gene products) sorting out at cell division, multiplying at various rates, and responding to each other in various modes of competition and cooperation.

The struggle to gain recognition for the importance of cytoplasmic inheritance was modified in accordance with a new change in the power relations of the field. Geneticists who advocated the existence of cytoplasmic genetic elements required recognition for their work by their chief competitors. Morgan's school of genetics was succeeded by that headed by George Beadle, which maintained the intellectual conformity to Morganist doctrines, which included the exclusive nuclear control over the synthetic processes of the cell. *Neurospora* became the capital biochemical genetic organism. The members of the school led by Beadle at the California Institute of Technology became chief authorities. During this period the existence of cytoplasmic genetic elements became more convincing when their efforts were demonstrated by biochemical genetic methods in microorganisms, particularly in *Neurospora* and at the California Institute of Technology. The dialogue analyzed in Chapter 5 between Ephrussi at the Institute of Genetics in France, Sonneborn at Indiana University, and Horowitz at Cal Tech over the significance of *petites* in yeast and poky in *Neurospora* is illustrative of the social negotiations that take place in the process of scientific discovery. Each participant attempted to impose the greatest significance to the scientific results best suited to his special interests (that is the significance most likely to enable him to occupy the dominant position by attributing the highest value to the results which he personally or institutionally possessed).

Throughout this period the dogma of the "nuclear monopoly" persisted. Investigators of cytoplasmic inheritance lacked the scientific techniques required to effect a major change in the "political economy of the cell." It will be recalled that *Paramecium* was domesticated for genetic analysis by Sonneborn, before the rise of biochemical genetics, to study the relations of the nucleus, cytoplasm, and environment. It was developed as a technology to be competitive with research programs based on higher organisms. *Paramecium* could not be properly domesticated for biochemical genetic analysis. Yeast, on the other hand, was chosen by Ephrussi for its known biochemical properties to be competitive with *Neurospora* genetics in the study of the means by which genes affect biochemical processes. It was not initially domesticated for its efficiency in analyzing the role of the cytoplasm in heredity.

During this time, only relatively few cases of non-Mendelian inheritance were reported. Yet, the question of cytoplasmic heredity and whether or not cytoplasmic elements controlled the fundamental characteristics of the organism, whereas Mendelian genes controlled only relatively minor characteristics, remained a theoretical possibility. Within this context, the evidence for the cytoplasmic genetic elements had to be ignored or rationalized and was vigorously and persistently attacked by leading Mendelian geneticists who excluded it as due to viruses or parasites, or reinterpreted it formally as due to physiological effects of nuclear genes.

It was at this time that we saw the controversy over the technical capacity of Mendelian genetics and theoretical scope of cytoplasmic inheritance, with its intimate relations with the environment, become entangled in the Lysenko affair. Although the nucleo-cytoplasmic dispute cannot be considered in any way to have been value-free, it must be distinguished from the Lysenko affair. Sonneborn, Ephrussi, and their followers did not argue on grounds that most Western geneticists would exclude as "philosophical" or "political"; they did not cite authority figures outside the field such as Marx or Engels; they did not dismiss the Mendelian-chromosome theory entirely. They attempted to make an "orderly revolution." Lysenkoists, on the other hand, ignored the social and technical norms of the discipline and attempted to establish new technical norms and initiate a scientific revolution in a rapid and wholesale way.

A fourth phase, from about 1960 to the present, is marked by the rise of the authority of molecular biology in the field of heredity and the transformation of the chromosome theory to the nucleic acid theory, based primarily on investigations of the bacterium *Escherichia coli* and its viruses. During this period, as discussed in Chapter 7, heredity was released from its strict reference to transmission and segregation of genetic material. It is now described in terms of information, message, and code. What is transmitted from generation to generation is understood in terms of instructions specifying molecular structures, the building blocks of the future organism. When heredity was reduced to molecular mechanisms and molecular structure, the gross distinction between hereditary principles no longer rested on the location of genetic elements in the cell. As reservoirs of information, genes could be "turned on and off" and cytoplasmic heredity lost its strategic theoretical niche as a primary basis for understanding cellular differentiation. "Plasmagenes" were supplanted by genomic regulation and by self-perpetuating regulatory systems ultimately traceable to the effects of nuclear genes. When Mendelian genes could fully participate in developmental phenomena, cytoplasmic heredity no longer posed a major threat to nucleocentric genetics.

As the hierarchy of authority in the field changed with the emergence of molecular biology to a prominent position, the required demonstration for the existence of cytoplasmic genetic systems was altered accordingly. As discussed in Chapter 7, the existence of cytoplasmic genetic elements as evidenced by cross-breeding methods and the correlative demonstration of cytologically visible self-perpetuating cytoplasmic bodies was no longer satisfactory. The general recognition of cytoplasmic genes depended on a satisfactory demonstration of cytoplasmic DNAs. During this period sophisticated techniques were established for the genetic dissection of cytoplasmic organelles based primarily on investigations of the unicellular alga *Chlamydomonas* and fungi. Research on organelle heredity in *Chlamydomonas* and yeast successfully competed with nucleocentric genetics and was carried out in full accordance with the principles of molecular biology. Although cytoplasmic genes differ little from nuclear genes, the information they contain and their special mode of transmission and variation are held by some to be unaccounted for by the present synthetic theory of evolution.

Systematic genetic investigations of the supramolecular properties of the cell also emerged during this period, led by Sonneborn and his colleagues using genetic technology based on ciliated protozoa, primarily *Paramecium* and *Tetrahymena*. During the rise of biochemical genetics, the question of cellular organization as a hereditary property was discussed, but not investigated by genetic procedures. Investigators of cytoplasmic inheritance concentrated their efforts on the study of the cytoplasmic control of biochemical properties which had been at the center of genetic research. Microbial technology based on *Neurospora,* bacteria, and viruses had outcompeted genetic research based on ciliates. By the 1960s the basic underlying questions of the nature of the gene and genic action were thought to be nearing solution. Research on *Chlamydomonas* was making a bid as the primary technology for dissecting the genetic basis of organelle heredity. It was at this point that Sonneborn turned his research strategies to investigate the next higher level of biological complexity, the nature, development, and inheritance of structures at the supramolecular and microscopically visible levels of the cell.

Paramecium was technically highly suitable for the genetic dissection and study of observable cell structure. Nonetheless, the school of ciliate genetics headed by Sonneborn was unable to sustain its growth during the 1960s and 1970s. Genetic research on cell organization in ciliates developed slowly and was carried out in direct confrontation with the highly competitive, commercially valuable research programs of molecular biology and the "doctrine of self-assembly," which found experimental support from viral technology. Whether cell structure emerged only once and has evolved since on its own, or whether it is ultimately controlled by gene-determined polar molecules and only genes evolve, remains a subject of contention.

It has been through this adversary procedure that we have seen the weaving back and forth across the boundary between what is, and what is not, "known" or "knowable." The direction of genetic research, the kinds of questions asked, and the knowledge of heredity produced, though conditioned by technology, are intimately informed by the social struggles, dichotomies, and hierarchies in the field. And as we have seen, these social relations, which constitute the field of production (the internal politics of science), ultimately find their expression and reflection in the conception of the cell itself.

The struggle for authority is above all a struggle for reproduction. In the scientific field, researchers are constantly and automatically tested for their ability to produce followers who can exist in certain sociopolitical and technical conditions and perpetuate and develop their ideas and research interests. The technical advantage one research program may have over its rivals in producing results useful outside and/or inside the field may be enough to tip the scales in its favor. But technical advantages themselves do not exclusively determine success or failure: disciplinary, institutional, national, and political interests all have to be taken into consideration. Patterns of power are complex and involve strategies and tactics in all aspects of scientific endeavor.

Bibliography

LETTERS AND UNPUBLISHED MANUSCRIPTS

Beale, G. H., to author, July 17, 1981.

Cleland, R., to M. R. Irwin, November 6, 1948; Sonneborn papers, Manuscripts Department, Lilly Library, Indiana University, Bloomington, Indiana.

Cole, L. C., to members of the executive committee, American Society of Naturalists, October 25, 1929; Herman J. Muller file, Manuscripts Department, Lilly Library, Indiana University, Bloomington, Indiana.

Cook, R. C., to M. R. Irwin, December 4, 1949; Sonneborn papers.

Darlington, C. D., to T. M. Sonneborn, October 7, 1948; Sonneborn papers.

Darlington, C. D., to T. M. Sonneborn, October 16, 1948; Sonneborn papers.

Dobzhansky, T., to T. M. Sonneborn, December 29, 1948; Sonneborn papers.

Doty, P., to B. Ephrussi, January 4, 1957; Ephrussi papers, Centre de Génétique Moléculaire du C.N.R.S., Gif-sur-Yvette, France.

Ephrussi, B., "Notice sur les titres et travaux—Curriculum vitae de Monsieur Boris Ephrussi." Undated manuscript written in the mid-1970s, p. 4. Ephrussi papers.

Ephrussi, B. (1949) "Défense de la Science Française." Unpublished, Ephrussi papers.

Ephrussi, B. (1949) "Rapport sur l'activité de l'Institut de Génétique; Lu à la séance du Conseil de direction de l'Institut de Génétique." Unpublished 10 pages, Ephrussi papers.

Ephrussi, B., to P.-P. Grassé, December 22, 1956; Ephrussi papers.

Ephrussi, B., to N. Horowitz, June 26, 1953; Ephrussi papers.

Ephrussi, B., to J. A. Jenkins, December 4, 1956; Ephrussi papers.

Ephrussi, B., to H. K. Mitchell, March 29, 1952; Ephrussi papers.

Ephrussi, B., to J. Pérès, April 1, 1955; Ephrussi papers.

Ephrussi, B., to G. R. Pomerat, September 28, 1952; Record Group 1.2, Series 500D, The Rockefeller Archive Center, Tarrytown, New York.

Ephrussi, B., to G. Pontecorvo, August 25, 1952; Ephrussi papers.

Ephrussi, B., to G. Pontecorvo, January 9, 1958; Ephrussi papers.

Ephrussi, B., to T. M. Sonneborn, November 1943; Sonneborn papers.

Ephrussi, B., to T. M. Sonneborn, May 24, 1952; Sonneborn papers.

Ephrussi, B., to T. M. Sonneborn, October 6, 1958; Sonneborn papers.

Ephrussi, B., to W. Weaver, March 16, 1950; Record Group 1.2, Series 500D, The Rockefeller Archive Center.

Gilbert, S. F. (1985) "Cellular Politics: Goldschmidt, Just, Waddington, and the Attempt to Reconcile Embryology and Genetics, 1938–1940." Unpublished.

Glass, B., to T. M. Sonneborn, December 17, 1949; Sonneborn papers.

Hansen, F. B., diary, July 1, 1939; Record Group 1.1, Series 200D, Rockefeller Archive Center.

Hansen, F. B., diary, July 13, 1939; Record Group 1.1, Series 200D, Rockefeller Archive Center.

Hansen, F. B., diary, January 19, 1940; Record Group 1.1, Series 200D, Rockefeller Archive Center.

Hansen, F. B., diary, April 1–18, 1940; Record Group 1.1, Series 200D, Rockefeller Archive Center.

Hansen, F. B., diary, July 13–August 31, 1940; Record Group 1.1, Series 200D, Rockefeller Archive Center.

Hansen, F. B., diary, January 9, 1945; Record Group 1.1, Series 200D, Rockefeller Archive Center.

Hansen, F. B., diary, January 28, 1946; Record Group 1.1, Series 200D, Rockefeller Archive Center.

Hansen, F. B. (1945, April 4) "Resolved." Record Group 1.1, Series 200D, Rockefeller Archive Center.

Hansen, F. B., to H. J. Muller, February 23, 1945; Record Group 1.1, Series 200D, Rockefeller Archive Center.

Hansen, F. B., to H. J. Muller, March 5, 1945; Record Group 1.1, Series 200D, Rockefeller Archive Center.

Horowitz, N., to B. Ephrussi, June 10, 1953; Ephrussi papers.

Irwin, M. R., to T. M. Sonneborn, March 24, 1950; Sonneborn papers.

Jollos, V., to L. Noland, February 25, 1940; Sonneborn papers.

Lillie, F. R., to V. Hamburger, May 2, 1944; Lillie letters, Marine Biological Laboratory, Woods Hole, Massachusetts.

Lillie, F. R., to J. Huxley, March 19, 1928; Lillie letters.

Lillie, F. R., to H. Spemann, December 28, 1931; Lillie letters.

Loomis, W. F., diary, February 13, 1951; Record Group 1.1, Series 200D, Rockefeller Archive Center.

Lynch, R., to T. M. Sonneborn, September 6, 1946; Sonneborn papers.

McClintock, B., to B. Ephrussi, December 29, 1945; Ephrussi papers.

Muller, H. J., to C. R. Singleton, February 17, 1950; Sonneborn papers.

Nanney, D. L. (1980, August) "*Tetrahymena* Genetics: Toward the Fourth Decade." Appendix to National Institutes of Health grant application.

Nanney, D. L. (1982) "Personal Interactions: Footnotes to a Biography of T. M. Sonneborn." Lilly Library Archives, Indiana University, Bloomington, Indiana.

Noland, L., to V. Jollos, February 1, 1940; Sonneborn papers.

Noland, L., to V. Jollos, February 23, 1940; Sonneborn papers.

Pomerat, G. R., diary, February 28–March 3, 1950; Record Group 1.2, Series 500D, Rockefeller Archive Center.

Pomerat, G. R., diary, June 5, 1953; Record Group 1.2, Series 500D, Rockefeller Archive Center.

Pontecorvo, G., to B. Ephrussi, September 5, 1958; Ephrussi papers.

Sapp, J. (1986) "Who was Franz Moewus and Why Was Everybody Saying Such Terrible Things About Him?" *History and Philosophy of the Life Sciences,* In preparation.

Sonneborn, T. M. (1965, September) "The Evolutionary Integration of the Genetic Material into Genetic Systems." Unpublished talk presented at the Mendel Centennial Celebration of the Genetics Society of America, Fort Collins, Colorado, 29 pages, p. 1. Sonneborn papers.

Sonneborn, T. M. Undated Autobiographical essay, Sonneborn papers.

Sonneborn, T. M. (1978)"My Intellectual History in Relation to My Contributions to Science." Unpublished autobiography, Sonneborn papers.

Sonneborn, T. M., to G. H. Beale, September 20, 1954; Sonneborn papers.

Sonneborn, T. M., to R. C. Cook, December 10, 1949; Sonneborn papers.

Sonneborn, T. M., to C. D. Darlington, October 8, 1948; Sonneborn papers.

Sonneborn, T. M., to C. D. Darlington, October 13, 1948; Sonneborn papers.

Sonneborn, T. M., to T. Dobzhansky, October 18, 1948; Sonneborn papers.

Sonneborn, T. M., to T. Dobzhansky, January 11, 1949; Sonneborn papers.

Sonneborn, T. M., to L. C. Dunn, November 15, 1943; Sonneborn papers.

Sonneborn, T. M., to B. Ephrussi, November 17, 1952; Sonneborn papers.

Sonneborn, T. M., to B. Ephrussi, November 10, 1958; Sonneborn papers.

Sonneborn, T. M., to B. Glass, November 23, 1948; Sonneborn papers.

Sonneborn, T. M., to M. R. Irwin, April 22, 1950; Sonneborn papers.
Sonneborn, T. M., to H. S. Jennings, October 17, 1944; Sonneborn papers.
Sonneborn, T. M., to V. Jollos, February 12, 1940; Sonneborn papers.
Sonneborn, T. M., to D. F. Jones, July 5, 1950; Sonneborn papers.
Sonneborn, T. M., to R. Lynch, June 12, 1948; Sonneborn papers.
Sonneborn, T. M., to W. MacDougall, December 17, 1931; Sonneborn papers.
Sonneborn, T. M., to Captain Weir, December 6, 1948; Sonneborn papers.
Spiegelman, S., to B. Ephrussi, November 16, 1946; Ephrussi papers.
Spiegelman, S., to T. M. Sonneborn, September 22, 1947; Sonneborn papers.
Sturtevant, A. H., to T. M. Sonneborn, January 7, 1946; Sonneborn papers.
Weaver, W., diary, February 28, 1950; Record Group 1.2, Series 500D, Rockefeller Archive Center.
Weaver, W., diary, November 17, 1950; Record Group 1.1, Series 200D, Rockefeller Archive Center.
Weaver, W., to B. Ephrussi, February 15, 1950; Record Group 1.2, Series 500D, Rockefeller Archive Center.
Wettstein, D. von, to author, February 3, 1981.

INTERVIEWS

Adoutte, André, Centre de Génétique Moléculaire du C.N.R.S., Gif-sur-Yvette, France, January 14, 1982.
Brachet, Jean, Naples, Italy, December 10, 1981.
Lwoff, André, Paris, France, January 15, 1982.
Melchers, Georg, Max-Planck-Institut für Biologie, Tübingen, Germany, December 12 and 14, 1981.
Nanney, D. L., University of Illinois, Champaign-Urbana, Illinois, October 16, 1981.
Preer, J. R., Indiana University, Bloomington, Indiana, October 15, 1981.
Rhoades, Marcus, Indiana University, Bloomington, Indiana, October 19, 1981.
Sager, Ruth, Marine Biological Laboratory, Woods Hole, Massachusetts, August 15, 1981.
Slonimsky, Piotr, Centre de Génétique Moléculaire du C.N.R.S., Gif-sur-Yvette, France, January 13, 1982.

PUBLISHED SOURCES

Abir-Am, Pnina (1982) "The Discourse of Physical Power and Biological Knowledge in the 1930s: A Reappraisal of the Rockefeller Foundation's 'Policy' in Molecular Biology." *Social Studies of Science* 12:341–82.
Abir-Am, Pnina (1984) "Beyond Deterministic Sociology and Apologetic History. Reassessing the Impact of Research Policy upon New Scientific Disciplines. (Reply to Fuerst, Bartels, Olby and Yoxen)." *Social Studies of Science* 14:252–263.
Adams, M. B. (1977–78) "Biology After Stalin: A Case Study." *Survey: A Journal of East/West Studies* 23:53–80.
Albury, W. R. (1983) "The Politics of Truth: A Social Interpretation of Scientific Knowledge, with an Application to the Case of Sociobiology." In Michael Ruse, ed., *Nature Animated*. Dordrecht, Holland: D. Reidel Publishing Company, pp. 115–129.
Allen, G. E. (1968) "Hugo De Vries and the Reception of the 'Mutation Theory.'" *Journal of the History of Biology* 2:55–87.
Allen, G. E. (1975) *Life Science in the Twentieth Century*. Cambridge: Cambridge University Press.
Allen, G. E. (1978) *Thomas Hunt Morgan: The Man and His Science*, Princeton: Princeton University Press.
Allen, G. E. (1979) "Naturalists and Experimentalists: the Genotype and the Phenotype." In W. Coleman and C. Limoges, eds., *Studies in the History of Biology* 3:179–209.
Altenburg, Edgar (1945) *Genetics*. New York: Henry Holt and Co.

Altenburg, Edgar (1946a) "The 'Viroid' Theory in Relation to Plasmagenes, Viruses, Cancer and Plastids." *American Naturalist* 80:559–567.

Altenburg, E. (1946b) "The Symbiont Theory in Explanation of the Apparent Cytoplasmic Inheritance in *Paramecium*." *American Naturalist* 80:661–662.

Altmann, Richard (1890) *Die Elementarorganismen*. Leipzig: Veit and Company.

Anderson, E. G. (1923) "Maternal Inheritance of Chlorophyll in Maize." *Botancial Gazette* 76:411–418.

Ash, M. G. (1980) "Academic Politics in the History of Science: Experimental Psychology in Germany, 1879–41." *Central European History* 13:255–286.

Asimov, Isaac (1960) *The Intelligent Man's Guide to Modern Science. Vol. 2., The Biological Sciences*. New York: Basic Books.

Avery, O. T., MacLeod, C. M., and McCarty, M. (1944) "Studies on the Chemical Nature of the Substance Inducing Transformation of Pneumococcal Types." *Journal of Experimental Medicine* 79:137–158.

Baitsell, G. A. (1940) "The Cell Theory. II. A Modern Concept of the Cell as a Structural Unit." *American Naturalist* 74:5–24.

Baltzer, Fritz (1967) *Theodor Boveri*. Translated by Dorthea Rudnick, Berkeley: University of California Press.

Bartels, Ditta (1984) "The Rockefeller Foundation's Funding Policy for Molecular Biology: Success or Failure?" *Social Studies of Science* 14:238–247.

Bateson, William (1911) "Address to the Agricultural Sub-Section." *British Association for the Advancement of Science,* pp. 1–10.

Bateson, William (1912) *The Methods and Scope of Genetics. An Inaugural Lecture Delivered 23 October 1908*. Cambridge: Cambridge University Press.

Bateson, William (1913, reissued 1979) *Problems of Genetics*. New Haven: Yale University Press.

Bateson, William (1914) "Address of the President of the British Association for the Advancement of Science." *Science* 40:287–302.

Bateson, William (1926) "Segregation: Being the Joseph Leidy Memorial Lecture of the University of Pennsylvania, 1922." *Journal of Genetics* 16:201–235.

Bateson, William, and Gairdner, A. E. (1921) "Male Sterility in Flax, Subject to Two Types of Segregation." *Journal of Genetics* 11:269–276.

Bather, F. A. (1927) "Biological Classification: Past and Future." *Proceedings of the Geological Society of London* 83:62–104.

Baur, Erwin (1909) "Das Wesen und die Erblichkeitsverhaltnisse der 'Varietates albomarginatae hort' von *Pelargonium zonale*." *Zeitschrift für induktive Abstammungs und Vererbungslehre* 1:330–351.

Baur, Erwin (1911) *Einfuhrung in die experimentelle Vererbungslehre*. Berlin: Borntraeger.

Beadle, G. W. (1948a) "Genes and Biological Enigmas." In G. A. Baitsell, ed., *Science in Progress,* 6th series. New Haven: Yale University Press, pp. 184–248.

Beadle, G. W., and Ephrussi, B. (1936) "The Differentiation of Eye Pigments in *Drosophila* as Studied by Transplantation." *Genetics* 21:225–247.

Beadle, G. W., and Tatum, E. L. (1941) "Genetic Control of Biochemical Reactions in *Neurospora*." *Proceedings of the National Academy of Sciences* 27:499–506.

Beale, G. H. (1948b) "The Process of Transformation of Antigenic Type in *Paramecium aurelia,* variety 4." *Proceedings of the National Academy of Sciences* 34:418–423.

Beale, G. H. (1951) "Nuclear and Cytoplasmic Determinants of Hereditary Characters in *Paramecium aurelia*." *Nature* 167:256–258.

Beale, G. H. (1954) *The Genetics of Paramecium Aurelia*. New York: Cambridge University Press.

Beale, G. H. (1982) "Tracy Morton Sonneborn 1905–1981." *Biographical Memoirs of Fellows of the Royal Society* 28:537–574.

Becker, G. (1932) "Experimentelle Analyse der Genom- und Plasmonwirkung bei Mossen III. Osmotischer Wert heteroploider Pflanzen." *Zeitschrift für induktive Abstammungs und Vererbungslehre* 60:17–38.

Beisson, Janine, and Sonneborn, T. M. (1965) "Cytoplasmic Inheritance of the Organization of the

Cell Cortex in *Paramecium aurelia*." *Proceedings of the National Academy of Sciences* 53:275–282.

Benda, C. (1901) "Die Mitochondriafärbung." *Verhandlungen der Physikalische-Medizinischer Gesellschaft zu Wurzburg* 19:155–174.

Ben-David, Joseph (1968–69) "The Universities and the Growth of Science in Germany and the United States." *Minerva* 7:1–35.

Ben-David, Joseph (1971) *The Scientist's Role in Society: A Comparative Study*. Englewood Cliffs, N.J.: Prentice Hall.

Bergson, Henri (1907) *L'évolution créatrice*. Paris: F. Alcan et Guillaumin.

Bernard, Claude (1885) *Leçons sur les phénomènes de la vie*. Paris.

Billingham, R. E., and Medawar, P. B. (1948) "Pigment Spread and Cell Heredity in Guinea Pigs' Skin." *Heredity* 2:29–48.

Birky, C. W. (1979) "Extrachromosomal Genetics." *Science* 204:761.

Boesiger, Ernest (1980) "Evolutionary Biology in France at the Time of the Evolutionary Synthesis." In Ernst Mayr and W. B. Provine, eds., *The Evolutionary Synthesis*. Cambridge: Harvard University Press, pp. 309–320.

Bonner, D. (1946) "Biochemical Mutations in *Neurospora*." *Cold Spring Harbor Symposia on Quantitative Biology* 11:14–24.

Bourdieu, Pierre (1975) "The Specificity of the Scientific Field and the Social Conditions of the Progress of Reason." *Social Science Information* 1:19–47.

Bourdieu, Pierre (1977) "Symbolic Power," in *Two Bourdieu Texts,* translated by Richard Nice, University of Birmingham Centre for Contemporary Cultural Studies, Birmingham, pp. 1–14.

Boutell, Chip (1950) "Is There Any Scientific Basis for the Lysenko Theory?" *Daily Compass,* April 18, p. 8.

Boveri, Theodor (1889) "Ein geschlechtlich erzeugter Organismus ohne mütterliche Eigenschaften." *Sitzungsberichte der Gesellschaft für Morphologik und Physiologie* 5:73–80.

Boveri, Theodor (1901) "Die Polarität von Ovocyte, EI, und Larve des *Strongylocentrotus Lividus*." *Zoologische Jahrbücher* 14:630–653.

Boveri, Theodor (1903) "Über den Einfluss der Samenzelle auf die Larvencharaktere der Echiniden." *Archiv für Entwicklungsmechanik* 16:340–363.

Boveri, Theodor (1906) *Die Organismen als historische Wesen: Festrede*. Würtzburg: H. Stürtz.

Boveri, Theodor (1910a) "Die Potenzen der Ascaris-Blastomeren bei abgeanderter Furchung. Zugleich ein Beitrag zur Frage qualitative ungleicher Chromosomenteilung." In *Festschrift Z. 60. Geburtstag von R. Hertwig,* Bd. III. Jena.

Boveri, Theodor (1910b) "Über die Teilung Centrifugierter Eier von *Ascaris megalocephala*." *Archiv für Entwicklungsmechanik* 30:101–125.

Boveri, Theodor (1918) "Zwei Fehlerquellen bei Merogonieversuchen und die Entwicklungsfähigkeit merogonischer und partiell-merogonischer Seeigelbastarde." *Archiv für Entwicklungsmechanik* 44:417–471.

Bowler, P. J. (1983) *The Eclipse of Darwinism: Anti-Darwinian Evolution Theories in the Decades around 1900*. Baltimore: The Johns Hopkins University Press.

Boycott, A. E., and Diver, C. (1923) "On the Inheritance of Sinistrality in *Limnaea peregra*." *Proceedings of the Royal Society of London* 95:207–213.

Brabec, F. (1954) "Hans Winkler 1877–1945." *Berichten der Deutschen Botanischen Gesellschaft,* pp. 27–32.

Brachet, Albert (1917) *L'oeuf et les facteurs de l'ontogenèse*. Paris: Gaston Doin.

Brachet, Albert (1935) *Traité d'embryologie des vertébrés,* 2nd ed., author's translation. Brussels: Masson.

Brachet, Jean (1942) "La localisation des acides pentose-nucléiques dans les tissus animaux et les oeufs d'Amphibiens en voie de développement." *Archives de biologie* 53:207–257.

Brachet, Jean (1949) "L'hypothèse des plasmagènes dans le développement et la différenciation." In *Unités biologiques douées de continuité génétique*. Paris: Edition du Centre National de la Recherche Scientifique, pp. 145–162.

Brachet, Jean (1958) "Le rôle du Noyau at du cytoplasme dans la synthèse de proteines." In *Memoriam Methodi Popoff. Papers on Selected Problems of Biology and Medicine*, pp. 125–133.

Brachet, Jean (1946) "Nucleic Acids in the Cell and Embryo." *Symposia of the Society for Experimental Biology*, 1:213–245.

Brachet, Jean (1950) "Le rôle du noyau et du cytoplasme dans les synthèses et la morphogenèse." *Annales de la Société Royale Zoologique de Belgique* 81:185–209.

Braun, W. (1953) *Bacterial Genetics*. Philadelphia: W. B. Saunders.

Buchner, P. (1952) *Endosymbiose*. Basel: Verlag Birkhauser.

Buican, Denis (1984) *Histoire de la génétique et de l'évolutionnisme en France*. Paris: Presses Universitaires de France.

Carlson, E. A. (1981) *Genes, Radiation and Society. The Life and Work of H. J. Muller*. Ithaca and London: Cornell University Press.

Caspari, E. (1948) "Cytoplasmic Inheritance." *Advances in Genetics* 2:1–66.

Caullery, Maurice (1922) "La contribution que les divers pays ont donnée au développement de la biologie. *Scientia* 31:23–36.

Caullery, Maurice (1935) *Les conceptions modernes de l'hérédité*. Paris: Flammarion.

Child, C. M. (1924) *Physiological Foundations of Behavior*. New York: Henry Holt and Company.

Chittenden, R. J. (1927) "Cytoplasmic Inheritance in Flax." *Journal of Heredity* 18:337–343.

Churchill, F. B. (1968) "August Weismann and a Break from Tradition." *Journal of the History of Biology* 1:91–112.

Churchill, F. B. (1969) "From Machine-Theory to Entelechy: Two Studies in Developmental Teleology." *Journal of the History of Biology* 2:165–185.

Churchill, F. B. (1974) "William Johannsen and the Genotype Concept." *Journal of the History of Biology* 7:5–30.

Clark, T. N. (1973) *Prophets and Patrons: The French University and the Emergence of the Social Sciences*. Cambridge: Harvard University Press.

Clausen, J. (1927) "Non-Mendelian Inheritance in Viola." *Hereditas* 9:245–256.

Coleman, William (1971) *Biology in the Nineteenth Century: Problems of Form, Function, and Transformation*. Cambridge: Cambridge University Press.

Coleman, William (1965) "Cell, Nucleus, and Inheritance: An Historical Study." *Proceedings of the American Philosophical Society* 109:124–158.

Coleman, William (1970) "Bateson and Chromosomes: Conservative Thought in Science." *Centaurus* 15:228–314.

Conklin, E. G. (1903) "The Cause of Inverse Symmetry." *Anatomischer Anzeiger Centralblatt* 23:577–588.

Conklin, E. G. (1905a) "Organ-Forming Substances in the Eggs of Ascidians." *Biological Bulletin* 8:205–230.

Conklin, E. G. (1905b) "The Mutation Theory from the Standpoint of Cytology." *Science* 21:525–529.

Conklin, E. G. (1908) "The Mechanism of Heredity." *Science* 27:89–99.

Conklin, E. G. (1912) "Problems of Evolution and Present Methods of Attacking Them." *American Naturalist* 46:121–128.

Conklin, E. G. (1915) *Heredity and Environment in the Development of Men*. Princeton: Princeton University Press.

Conklin, E. G. (1917) "The Share of the Egg and the Sperm in Heredity." *Proceedings of the National Academy of Sciences* 3:101–105.

Conklin, E. G. (1919) "The Mechanism of Evolution in the Light of Heredity and Development." *Scientific Monthly* 7:481–506.

Conklin, E. G. (1920) "The Mechanism of Evolution in the Light of Heredity and Development." *Scientific Monthly* 10:388–403.

Conklin, E. G. (1934) "Fifty Years of the American Society of Naturalists." *American Naturalist* 68:385–401.

Conklin, E. G. (1940) "Cell and Protoplasm Concepts: Historical Account." In G. R. Moulton, ed., *The Cell and Protoplasm*. Washington: The Science Press, pp. 6–19.

Corliss, J. O. (1972) "A Man to Remember, E. Fauré-Fremiet (1883–1971): Three Quarters of a Century of Progress in Protozoology." *Journal of Protozoology,* 19:389–400.

Correns, Carl (1909) "Vererbungsversuche mit blass (gelb) grünen und bluntblättrigen Sippen bei *Mirabilis Jalapa, Urtica pilulifera* und *Lunaria annua.*" *Zeitschrift für induktive Abstammungs und Vererbungslehre* 1:291–329.

Correns, Carl (1928) "Über nichtmendelnde Vererbung." *Zeitschrift für induktive Abstammungs und Vererbungslehre. (Suppl.)* 1:131–168.

Correns, Carl, edited by Wettstein, F. von (1937) "Nicht mendelnde Vererbung." In: *Handbuch der Vererbungswissenschaft.* Berlin: Verlag von Gebruder Borntraeger.

Crabb, A. R. (1947) *The Hybrid-Corn Makers.* New Brunswick: Rutgers University Press.

Cuénot, L. (1941) *Invention et finalité en biologie.* Paris: Flammarion.

Dalcq, Albert (1949) "L'apport de l'embryologie causale au problème de l'évolution." *Portugaliae Acta Biologica, Serie A,* pp. 367–400.

Dalcq, Albert (1951) "Le problème de l'évolution, est-il prêt d'être resolu." *Annales de la Société Royale Zoologique de Belgique* 82:117–138.

Darlington, C. D. (1944) "Heredity, Development and Infection." *Nature* 154:164–169.

Darlington, C. D. (1948) "The Plasmagene Theory of the Origin of Cancer." *British Journal of Cancer* 2:118–126.

Darlington, C. D. (1949a) "Genetic Particles." *Endeavor* 13:51–61.

Darlington, C. D. (1949b) "Les plasmagènes." In *Unités biologigues douées de continuité génétigue.* Paris: Edition du *Centre National de la Recherche Scientifigue,* pp. 123–130.

Darlington, C. D. (1951) "Mendel and the Determinants." In L. C. Dunn, ed., *Genetics in the 20th Century.* New York: Macmillan, pp. 315–332.

Darlington, C. D. (1958) *The Evolution of Genetic Systems,* 2nd ed. New York: Basic Books.

Darlington, C. D., and Mather, K. (1949) *The Elements of Genetics.* London: George Allen and Unwin, Ltd.

Darlington, C. D., and Mather, K. (1950) *The Elements of Genetics.* New York: Macmillan.

Darwin, Charles (1868) *The Variation of Animals and Plants under Domestication.* Vol. 2. London: John Murray.

Delage, Yves (1895) *La structure du protoplasma et les théories sur l'hérédité et les grands problèmes de la biologie générale.* Paris: C. Reinwald et Co.

Delage, Yves (1899) "Etudes sur la Mérogonie." *Archives de Zoologie Experimentale* 3:381–417.

Delage, Yves (1903) *L'hérédité et les grands problèmes de la biologie générale,* 2nd ed. Paris: Schleicher.

Delbrück, M. (1949) Discussion following paper of T. M. Sonneborn and G. H. Beale, "Influence des gènes, des plasmagènes et du milieu dans le déterminisme des caractères antigèniques chez *Paramecium aurelia* (variété 4)." In *Unités biologiques douées de continuité génétique.* Paris: Edition du Centre National de la Recherche Scientifique, pp. 25–36; 35.

Demerec, M. (1927) "A Second Case of Maternal Inheritance of Chlorophyll in Maize." *Botanical Gazette* 84:139–155.

De Solla Price, D. J. (1960) "Review of Asimov's Text." *Science* 132:1830.

De Vries, Hugo (1901) *The Mutation Theory,* translated by J. B. Farmer and A. D. Darbishire, Chicago: Open Court.

De Vries, Hugo (1910) *Intracellular Pangenesis,* translated by C. Stuart Gager. Chicago: Open Court.

Dippell, R. V. (1948) "Mutations of the killer plasmagene, kappa, in variety 4 of *Paramecium aurelia.*" *American Naturalist* 82:43–58.

Dippell, R. V. (1950) "Mutation of the killer cytoplasmic factor in *Paramecium aurelia.*" *Heredity* 4:165–87.

Dippell, R. V. (1968) "The Development of Basal Bodies in *Paramecium.*" *Proceedings of the National Academy of Sciences* 61:461–468.

Dobzhansky, T. (1937) *Genetics and the Origin of Species,* 1st ed. New York: Columbia University Press.

Dobzhansky, T. (1947) *Genetics and the Origin of Species,* revised ed. New York: Columbia University Press.

Dobzhansky, T. (1949) "Marxist Biology, French Style, (a review of *Biologie et Marxisme*, by Marcel Prenant)." *Journal of Heredity* 40:78–79.

Driesch, Hans (1884) *Analytische Theorie der organischen Entwicklung*. Leipzig: Wilhelm Engelmann.

Driesch, Hans (1891) "Entwicklungsmechanische Studien. I. Der Werth der beiden ersten Furchungszellen in der Echinodermentwicklung. Experimentelle Erzeugung von Theil- und Doppelbildungen." *Zeitschrift für Wissenschaftliche Zoologie* 53:160–178.

Driesch, Hans (1908) *The Science and Philosophy of the Organism*. London: Ada and Charles Black.

Driesch, Hans, and Morgan, T. H. (1895–1896) "Zur Analyse der ersten Entwicklungsstadien des Ctenophoreies. I. Von der Entwicklungeinzelner Ctenophorenblastomeren." *Archiv für Entwicklungsmechanik* 2:204–215.

Duesberg, J. (1913) "Plastosomes et 'organ-forming substances' dans l'oeuf des Ascidiens." *Bulletins de l'Académie Royale de Belgique* 5:463–475.

Duesberg, J. (1919) "On the Present Status of the Chondriosome-Problem." *Biological Bulletin* 36:71–81.

Dujon, B. (1981) "Mitochondrial Genetics and Functions." In *Molecular Biology of the Yeat Saccharomyces: Life Cycle and Inheritance*. Cold Spring Harbor: Cold Spring Harbor Laboratory, pp. 505–635.

Dujon, B., Colson, A. M., and Slonimski, P. P. (1977) "The Mitochondrial Genetic Map of *Saccharomyces cervisae:* A Literature Compilation towards a Unique Map." In W. Bandlow *et al.*, eds., *Mitochondria 1977: Genetics and Biogenesis of Mitochondria*. Berlin: de Gruyter, pp. 579–634.

Dunn, L. C. (1917) "Nucleus and Cytoplasm as Vehicles of Heredity." *American Naturalist* 51:286–300.

Dunn, L. C. (1965) *A Short History of Genetics*. New York: McGraw Hill.

East, E. M. (1923) *Mankind at the Crossroads*. New York: Charles Scribner's Sons.

East, E. M., ed. (1927a) *Biology in Human Affairs*. New York: McGraw-Hill.

East, E. M. (1927b) *Heredity and Human Affairs*. New York: Charles Scribner's Sons.

East, E. M. (1934) "The Nucleus-Plasma Problem." *American Naturalist* 68:289–303;402–439.

East, E. M., and Jones, D. F. (1919) *Inbreeding and Outbreeding: Their Genetic and Sociological Significance*. Philadelphia: Lippincott.

Edwardson, J. R. (1956) "Cytoplasmic Male-Sterility." *Botanical Review* 22:696–738.

Ephrussi, Boris (1925) "Sur le chondriome ovarien des *Drosophila melanogaster*." *Comptes-rendus des séances de la Société de Biologie* 92:778–779.

Ephrussi, Boris (1932) *Croissance et régénération dans les cultures de tissus*. Paris: Masson.

Ephrussi, Boris (1933) "Contribution à l'analyse des premiers stades du développement de l'oeuf. Action de la température." *Archives biologiques* 44:1–40.

Ephrussi, Boris (1938) "Aspects of the Physiology fo Gene Action" *American Naturalist* 72:5–23.

Ephrussi, Boris (1949) "Action de l'acriflavine sur les levures." In *Unités biologiques douées de continuité génétique*. Paris: Edition du Centre National de la Recherche Scientifique, pp. 165–180.

Ephrussi, Boris (1950) "The Interplay of Heredity and Environment in the Synthesis of Respiratory Enzymes in Yeast." *Harvey Lectures*, 45–67.

Ephrussi, Boris (1951) "Remarks on Cell Heredity." In: L. C. Dunn, ed., *Genetics in the 20th Century*. New York: Macmillan, pp. 241–262.

Ephrussi, Boris (1953) *Nucleo-Cytoplasmic Relations in Micro-Organisms*. Oxford: Clarendon Press.

Ephrussi, Boris (1958) "The Cytoplasm and Somatic Cell Variation." *Journal of Cellular and Comparative Physiology* 52:35–53.

Ephrussi, Boris (1972) *Hybridization of Somatic Cells*. Princeton: Princeton University Press.

Ephrussi, Boris, Hottinguer, H., and Chimenes, A. M. (1949a) "Action de l'acriflavine sur les levures. I. La mutation 'petite colonie.'" *Annales de l'Institut Pasteur* 76:351–367.

Ephrussi, Boris, Hottinguer, H., and Tavlitzki, J. (1949b) "Action de l'acriflavine sur les levures. II. Etude génétique du mutant 'petite colonie.'" *Annales de l'Institut Pasteur* 76:419–450.

Ephrussi, Boris, L'Héritier, P., and Hottinguer, H. (1949c) "Action de l'acriflavine sur les levures.

VI. Analyse quantitative de la transformation des populations." *Annales de l'Institut Pasteur* 77:64–83.

Fantini, Bernardino (1985) "The Sea Urchin and the Fruit Fly; Cell Biology and Heredity, 1900–1910." *Biological Bulletin* 168:99–106.

Farrall, Lindsay (1979) "The History of Eugenics; A Bibliographic Review." *Annals of Science* 36:111–123.

Fauré -Fremiet, E. (1908) "Evolution de l'appareil mitochondrial dans l'oeuf de *Julus terrestris.*" *Comptes rendus des séances de la Société de Biologie* 64:1057–1059.

Fauré-Fremiet, E., and Mugard, H. (1948) "Ségrégation d'un materiel cortical au cours de la segmentation chez l'oeuf de *Teredo norvegica.*" *Comptes rendus de l'Academie des Sciences* 227:1409–1411.

Fischer, E. (1901) "Experimentelle Untersuchungen über die Vererbung erworbener Eigenschaften." *Zeitschrift für Entomologie* 6:49,363,377.

Fischer, J. L. (1979) "Yves Delage l'epigenèse neo-lamarckienne contre la prédétermination weismanniene." *Revue de synthese* 3:443–458.

Fisher, R. A. (1930) *The Genetical Theory of Natural Selection.* Oxford: Clarendon Press.

Foucault, Michel (1980) *Power/Knowledge,* edited by Colin Gordon, translated by Colin Gordon, Leo Marshall, John Mepham, and Kate Soper. New York: Pantheon Books.

Frankel, Joseph (1983) "What are the Developmental Underpinnings of Evolutionary Changes in Protozoan Morphology?" In B. C. Goodwin, N. Holder, and C. C. Wylie, eds., *Development and Evolution.* Cambridge: Cambridge University Press, pp. 279–314.

Friedman, Bernard (1950) "Heredity is Subject to Control Through Environmental Means." New York *Daily Compass,* April 18.

Friedman, Bernard (1950) "Oh that Capitalist Science." *New York Post,* February 13, p. 35.

Fuerst, J. A. (1984) "The Definition of Molecular Biology and the Definition of Policy: The Role of the Rockefeller Foundation's Policy for Molecular Biology." *Social Studies of Science* 14:225–237.

Gairdner, A. E. (1929) "Male-Sterility in Flax, II. A Case of Reciprocal Crosses Differing in F_2." *Journal of Genetics* 21:117–124.

Gilbert, G. N., and Mulkay, Michael (1984) *Opening Pandora's Box.* Cambridge: Cambridge University Press.

Gilbert, S. F. (1978) "The Embryological Origins of the Gene Theory." *Journal of the History of Biology* 11:307–351.

Gillham, N. W. (1978) *Organelle Heredity.* New York: Raven Press.

Gillham, Nicholas (1965) "Linkage and Recombination between Nonchromosomal Mutations in *Chlamydomonas reinhardi.*" *Proceedings of the National Academy of Sciences* 54:1560–1573.

Gillham, N. W., and Levine, R. P. (1962) "Studies on the Origin of Streptomycin Resistant Mutants in *Chlamydomonas reinhardi.*" *Genetics* 47:1463–1474.

Glass, Bentley (1949) "'The Science of Biology Today,' by Trofim Lysenko. A Review." *Science* 109:404–405.

Goldschmidt, Richard (1916) "Theodor Boveri." *Science* 63:263–270.

Goldschmidt, Richard (1932) "Genetics and Development." *The Biological Bulletin* 63:337–356.

Goldschmidt, Richard (1933) "Some Aspects of Evolution." *Science* 78:539–547.

Goldschmidt, Richard (1934) "The Influence of the Cytoplasm upon Gene-Controlled Heredity." *American Naturalist* 68:5–23.

Goldschmidt, Richard (1938) *Physiological Genetics.* London: McGraw-Hill.

Goldschmidt, Richard (1940) *The Material Basis of Evolution.* New Haven: Yale University Press.

Goldschmidt, Richard (1958) *Theoretical Genetics.* Berkeley: University of California Press.

Graham, L. R. (1972) *Science and Philosophy in the Soviet Union.* New York: Alfred A. Knopf.

Grégoire, Victor (1925) "Les limites du mendelisme et le rôle des chromosomes dans l'hérédité." *Revue des questions scientifiques* 4:117–155.

Grégoire, Victor (1927) "Génétique et Cytologie." *Bulletins de la Classe des Sciences, Académie Royale de Belgique* 13:856–874.

Gregory, W. K. (1917) "Genetics versus Paleontology." *American Naturalist* 51:622–635.

Griffith, Frederick (1928) "The Significance of Pneumococcal Types." *Journal of Hygiene* 27:113–159.

Grmek, M. D., and Fantini, Bernardino (1982) "Le rôle du hasard dans la naissance du modèle de l'opéron." *Review of the History of Science* 35:193–215.

Guilliermond, A. (1913) "Recherche cytologique sur la mode de formation de l'amidon et sur les plasts végétaux." *Archives de l'anatomie microscopique* 14:309–428.

Guyénot, E. (1924) *L'hérédité*. Paris: Doin.

Guyer, M. F. (1907) "Do Offspring Inherit Equally from Each Parent?" *Science* 25:1006–1010.

Guyer, M. F. (1909) "Deficiencies of the Chromosome Theory of Heredity." *University Studies* 5:3–19. Cincinnati: University of Cincinnati Press.

Guyer, M. F. (1911) "Nucleus and Cytoplasm in Heredity." *American Naturalist* 45:284–305.

Guyer, M. F. (1924) "Further Studies on Inheritance of Eye Defects Induced in Rabbits." *Journal of Experimental Zoology* 38:449–475.

Hadorn, E. (1937) "Die entwicklungsphysiologische Auswirkung der disharmonischen Kern-Plasma Kombination bein Bastardinerogon', *Triton palinatus* (♀) *X Triton crintahs* (♂)." *Archives für Entwicklungsmechanik* 136:97–104.

Haldane, J. B. S. (1938) "The Nature of Interspecific Differences." In G. R. de Beer, ed., *Evolution*. Oxford: Clarendon Press, pp. 19–94.

Haldane, J. B. S. (1954) *The Biochemistry of Genetics*. London: George Allen and Unwin Ltd.

Hamburger, Viktor (1980a) "Embryology and the Modern Synthesis in Evolutionary Theory." In Ernst Mayr and W. Provine, eds., *The Evolutionary Synthesis*. Cambridge: Harvard University Press, pp. 97–111.

Hamburger, Viktor (1980b) "Evolutionary Theory in Germany: A Comment." In Ernst Mayr and W. Provine, eds., *The Evolutionary Synthesis*. Cambridge: Harvard University Press, pp. 303–308.

Hämmerling, J. (1929) "Dauermodifikationen." In *Handbuch der Vererbungswissenschaft,* Bol. 1, Lieferung II.

Haraway, Donna (1976) *Crystals, Fabrics and Fields. Metaphors of Organicism in Twentieth Century Biology*. New Haven: Yale University Press.

Harder, Richard (1927) "Zur Frage nach der Rolle von Kern und Protoplasma im Zellgeschehen und bei der 'Übertragung von Eigenschaften.'" *Zeitschrift für Botanik* 19:337–407.

Harrison, R. G. (1921) "On Relations of Symmetry in Transplanted Limbs." *Journal of Experimental Zoology* 32:1–136.

Harrison, R. G. (1936) "Relations of Symmetry in the Developing Embryo." *Collecting Net* 11:217–226.

Harrison, R. G. (1937) "Embryology and Its Relations." *Science* 85:369–374.

Harrison, R. G. (1940) "Cellular Differentiation and Internal Environment." In G. R. Moulton, ed., *The Cell and Protoplasm*. Washington: The Science Press, pp. 77–97.

Harvey, E. B. (1932) "The Developmment of Half and Quarter Eggs of *Arbacia punctulata* and of Strongly Centrifuged Whole Eggs." *Biological Bulletin* 62:155–167.

Harvey, E. B. (1935) "Cleavage without Nuclei." *Science* 82:277.

Harvey, E. B. (1942) "Maternal Inheritance in Echinoderm Hybrids." *Journal of Experimental Zoology* 91:213–226.

Harwood, Jonathan (1984) The Reception of Morgan's Chromosome Theory in Germany: Interwar Debate Over Cytoplasmic Inheritance." *Medizinhistorisches Journal* 19:3–32.

Harwood, Jonathan (1985) "Geneticists and the Evolutionary Synthesis in Interwar Germany." *Annals of Science* 42:279–301.

Haurowitz, F. (1950) *Chemistry and Biology of Proteins*. New York: Academic Press.

"Heredity: Creative Energy." *American Breeder's Magazine* 1 (1910):79.

Hershey, A. D. (1970) "Genes and Hereditary Characteristics." *Nature* 226:697–700.

Hopkins, F. G. (1932) "Some Aspects of Biochemistry: The Organising Capacities of Specific Catalysts" (Second Purser Memorial Lecture). *Irish Journal of Medical Science,* pp. 333–357.

Horder, T. J. (1983) "Embryological Bases of Evolution." In G. C. Goodwin, N. Holder, and C. C. Wylie, eds., *Development and Evolution*. Cambridge: Cambridge University Press, pp. 315–352.

Horowitz, N. H., and Leupold, U. (1951) "Some Recent Studies Bearing on the One Gene-One Enzyme Hypothesis." *Cold Spring Harbor Symposia on Quantitative Biology* 16:65–74.

Irwin, M. R. (1951) "Genetics and Immunology." In L. C. Dunn, ed., *Genetics in the 20th Century*. New York: Macmillan, pp. 173–219.

Jacob, F. (1976) *The Logic of Life. A History of Heredity,* translated by B. E. Spillmann. New York: Vintage Books.

Jacob, François, and Monod, Jacques (1961) "Genetic Regulatory Mechanisms in the Synthesis of Proteins." *Journal of Molecular Biology* 3:318–356.

Jenkinson, J. W. (1909) *Experimental Embryology*. Oxford: Clarendon Press.

Jenkinson, J. W. (1913) *Vertebrate Embryology*. Oxford: Clarendon Press.

Jennings, H. S. (1927) "Diverse Doctrines of Evolution, Their Relation to the Practice of Science and Life." *Science* 65:19–25.

Jennings, H. S. (1930) "Heredity and Mutation in Relation to Environment in Protozoa." *Collecting Net* 5:81–87.

Jennings, H. S. (1931) "The Cell in Relation to Its Environment." *Journal of the Maryland Academy of Sciences* 2:25–32.

Jennings, H. S. (1937) "Formation, Inheritance and Variation of the Teeth in *Difflugia corona*. A Study of the Morphogenic Activities of Rhizopod Protoplasm." *Journal of Experimental Zoology* 77:287–336.

Jennings, H. S. (1940) "Chromosomes and Cytoplasm in Protozoa." In F. R. Moulton, ed., *The Cell and Protoplasm*. Washington: The Science Press, pp. 44–55.

Jinks, J. L. (1963) "Cytoplasmic Inheritance in Fungi." In W. J. Burdette, ed., *Methodology in Basic Genetics*. San Francisco: Holden Day.

Jinks, J. L. (1964) *Extrachromosomal Inheritance*. Englewood Cliffs, New Jersey: Prentice-Hall.

Johannsen, Wilhelm (1911) "The Genotype Conception of Heredity." *American Naturalist* 45:129–159.

Johannsen, Wilhelm (1923) "Some Remarks about Units in Heredity." *Hereditas* 4:133–141.

Jollos, Victor (1934a) "Dauermodifikationen und Mutationen bei Protozoen." *Archiv für Protistenkunde* 83:197–234.

Jollos, Victor (1934b) "Induction of Mutations by High Temperature in *Drosophila*." *Genetics* 20:42–69.

Jollos, Victor (1934c) "Inherited Changes Produced by Heat Treatment in *Drosophila melanogaster*." *Genetics* 16:476–494.

Jones, D. F. (1944) "Edward Murray East." *National Academy Biographical Memoirs* 23:217–242.

Joravsky, D. (1970) *The Lysenko Affair*. Cambridge: Harvard University Press.

Judson, H. F. (1979) *The Eighth Day of Creation. The Makers of the Revolution in Biology*. New York: Simon and Schuster.

Just, E. E. (1932) "On the Origin of Mutations." *American Naturalist* 66:61–74.

Just, E. E. (1939) *The Biology of the Cell Surface*. Philadelphia: P. Blakiston's Son and Co.

Keller, Evelyn Fox (1983) *A Feeling for the Organism. The Life and Work of Barbara McClintock*. San Francisco: Freeman.

Kevles, D. J. (1980) "Genetics in the United States and Great Britain, 1890–1930: A Review with Speculations." *Isis* 71:441–455.

Kevles, D. J. (1985) *In the Name of Eugenics: Genetics and the Uses of Human Heredity*. New York: Alfred A. Knopf.

Kimmelman, B. A. (1983) "The American Breeder's Association: Genetics and Eugenics in an Agricultural Context, 1903–13. *Social Studies of Science* 13:163–204.

King, T. J., and Briggs, R. (1956) "Serial Transplantation of Embryonic Nuclei." *Cold Spring Harbor Symposia on Quantitative Biology* 21:271–290.

Kohler, R. E. (1976) "The Management of Science: The Experience of Warren Weaver and the Rockefeller Foundation Program in Molecular Biology." *Minerva* 14:279–306.

Kohler, R. E. (1978) "A Policy for the Advancement of Science: The Rockefeller Foundation, 1924–29." *Minerva* 16:480–515.

Kuhn, T. S. (1970) *The Structure of Scientific Revolutions,* 2nd ed. Chicago: University of Chicago Press.

Lecourt, D. (1976) *Proletarian Science? The Case of Lysenko*. introduction by Louis Althusser. New Left Books.

Lederberg, Joshua (1951) "Genetic Studies in Bacteria." In L. C. Dunn, ed., *Genetics in the 20th Century*. New York: Macmillan, pp. 263–289.

Lederberg, Joshua (1952) "Cell Genetics and Hereditary Symbiosis." *Physiological Reviews* 32:403–430.

Lederberg, J., and Lederberg, E. M. (1956) "Infection and Heredity." In P. Rudnick, ed., *Cellular Mechanisms in Differentiation and Growth*. Princeton: Princeton University Press.

Lehmann, E. (1928) "Reziprok verschiedene Bastarde in irhrer Bedeutung für das Kern-Plasma-Problem." *Tübinger Naturwissenschaft Abhandlung* 11:1–39.

Lewontin, Richard, and Levins, Richard (1976) "The Problem of Lysenkoism." In Hilary Rose and Steven Rose, eds., *The Radicalisation of Science*. London: Macmillan, pp. 32–65.

L'Héritier, P. (1948) "Sensitivity to CO_2 in *Drosophila*—A Review." *Heredity* 2:325–348.

L'Héritier, P. (1949) "Genoide sensibilisant la Drosophile à l'anhydrique carbonique." In *Unités biologiques douées de continuité génétique*. Paris: Edition du Centre National de la Recherche Scientifique, pp. 113–122.

L'Héritier, P. (1955) "Les virus integrés et l'unité cellulaire." *L'Anné Biologique* 31:481–496.

L'Héritier, P. (1964) "Qu-est-ce que l'hérédité." In: *Hérédité et Génétique*. Paris: Librairie Artheme Fayard, pp. 11–21.

L'Héritier, P. (1970) "*Drosophila* Viruses and Their Role as Evolutionary Factors." *Evolutionary Biology* 4:185–209.

L'Héritier, P., and Hugon de Sceoux, F. (1947) "Transmission par greffe et injection de la sensibilité héréditaire au gaz carbonique chez la Drosophile." *Bulletin Biologique de la France et de la Belgique* 81:70–91.

Lillie, F. R. (1906) "Observations and Experiments Concerning the Elementary Phenomena of Embryonic Development in Chaetopterus." *Journal of Experimental Zoology* 3:153–267.

Lillie, F. R. (1909) "The Theory of Individual Development." *Popular Science Monthly* 75:239–252.

Lillie, F. R. (1927) "The Gene and the Ontogenetic Process." *Science* 64:361–368.

Lillie, F. R. (1942) "Ernest Everette Just." *Science* 95:10–11.

Limoges, Camille (1976) "Natural Selection, Phagocytosis, and Predaptation: Lucien Cuénot, 1887–1914." *Journal of the History of Medicine and Allied Sciences* 31:176–214.

Limoges, Camille (1980) "A Second Glance at Evolutionary Biology in France." In Ernst Mayr and W. B. Provine, eds., *The Evolutionary Synthesis*. Cambridge: Harvard University Press, pp. 309–320.

Lindegren, C. C. (1946) "A New Gene Theory and an Explanation of the Phenomenon of Dominance to Mendelian Segregation of the Cytogene." *Proceedings of the National Academy of Sciences* 32:68–69.

Lindegren, C. C. (1949) *The Yeast Cell, Its Genetics and Cytology*. St. Louis: Educational Publishers Inc.

Lindegren, C. C. (1966) *The Cold War in Biology*. Ann Arbor, Michigan: Planarian Press.

Lindegren, C. C., and Lindegren, G. (1946) "The Cytogene Theory." *Cold Spring Harbor Symposia on Quantitative Biology* 11:115–129.

Loeb, Jacques (1916) *The Organism as a Whole*. New York: Putnam's Sons.

Loeb, Jacques (1917) "Is Species-Specificity a Mendelian Character?" *Science* 45:191–193.

Ludmerer, K. M. (1972) *Genetics and American Society: A Historical Appraisal*. Baltimore: Johns Hopkins University Press.

Luria, S. E. (1966) "Macromolecular Metabolism." *Journal of General Physiology* (*Supplement*), 49:330.

Luria, S. E., and Delbrück, M. (1943) "Mutations of Bacteria from Virus Sensitivity to Virus Resistance." *Genetics* 28:491–511.

Lwoff, A. (1944) *L'évolution physiologique. Etudes des pertes de fonctions chez les microorganismes*. Paris: Hermann.

Lwoff, A. (1949) "Les organites doués de continuité génétique chez les Protistes." In *Unités biol-*

ogiques doueés de continuité génétique. Paris: Edition du Centre National de la Recherche Scientifique, pp. 7–23.

Lwoff, A. (1950) *Problems of Morphogenesis in Ciliates: The Kinetosomes in Development, Reproduction and Evolution*. New York: John Wiley and Sons.

Lwoff, A. (1981) "Souvenirs," *Le courrier du CNRS* no 41, juillet, pp. 4–7.

Lwoff, A., and Dusi, H. (1935) "La suppréssion éxperimentale des chloroplaste chez *Euglena mesnili*." *Comptes rendus des séances de la société de biologie* 119:1092.

Lysenko, T. D. (1946) *Heredity and Its Variability*, translated by T. Dobzhansky. New York: King's Crown Press.

Lysenko, T. D. (1948) *The Science of Biology Today*. New York: International Publishers.

Maienschein, Jane (1978) "Cell Lineage, Ancestral Reminiscence, and the Biogenetic Law." *Journal of the History of Biology* 11:129–158.

Manning, K. R. (1983) *Black Apollo of Science: The Life of Ernest Everett Just*. New York: Oxford University Press.

Margulis, L. (1981) *Symbiosis in Cell Evolution*. San Francisco: W.H. Freeman and Company.

Mathews, A. P. (1915) *Physiological Chemistry*, New York: William Wood and Company.

Maynard Smith, J. (1983) "Evolution and Development." In B. C. Goodwin, N. Holder, and C. C. Wylie, eds., *Development and Evolution*. Cambridge: Cambridge University Press, pp. 33–47.

Mayr, Ernst (1980) "Prologue: Some Thoughts on the History of the Evolutionary Synthesis." In Ernst Mayr and W. Provine, eds., *The Evolutionary Synthesis*. Cambridge: Harvard University Press.

Mayr, Ernst (1982) *The Growth of Biological Thought*. Cambridge: Harvard University Press.

Mayr, Ernst, and Provine, W. B., eds. (1980) *The Evolutionary Synthesis: Perspectives on the Unification of Biology*, Cambridge: Harvard University Press.

McCarty, M. (1985) *The Transforming Principle*. New York: Norton.

McClelland, C. E. (1980) *State, Society and University in Germany, 1700–1914*. Cambridge: Cambridge University Press.

McClintock, B. (1956) "Controlling Elements and the Gene." *Cold Spring Harbor Symposia on Quantitative Biology* 21:197–216.

Medawar, P. B. (1947) "Cellular Inheritance and Transformation." *Biological Reviews* 22:360–389.

Medvedev, Z. A. (1971) *The Rise and Fall of T. D. Lysenko*, translated by I. Michael Lerner. New York: Anchor Books.

Melchers, G. (1961a) "Abteilung v. Wettstein." *Jahrbuch der Max-Planck-Gesellschaft zur Förderung der Wissenschaften* 2:130–134.

Melchers, G. (1961b) "Max-Planck-Institut für Biologie in Tübingen." *Jahrbuch der Max-Planck-Gesellschaft zur Förderung der Wissenschaften* 5:111–145.

Meves, F. (1908) "Die Chondriosomen als Träger erblicher Anlagen." *Archiv für Mikroskopische Anatomie* 72:816–867.

Michaelis, P. (1929) "Über den Einfluss von Kern und Plasma auf die Vererbung." *Biologisch Zentralblatt* 49:302–316.

Michaelis, P. (1954) "Cytoplasmic Inheritance in *Epilobium* and its Theoretical Significance." *Advances in Genetics* 6:287–401.

Michaelis, Peter (1965a) "Cytoplasmic Inheritance in *Epilobium* (A Survey)." *Nucleus* 8:83–92.

Michaelis, P. (1965b) "II. The Occurrence of Plasmon-Differences in the Genus *Epilobium* and the Interactions between Cytoplasm and Nuclear Genes (A Historical Survey)." *Nucleus* 8:93–108.

Michaelis, P. (1966) "I. The Proof of Cytoplasmic Inheritance in *Epilobium* (A Historical Survey as Example for the Necessary Proceeding)." *Nucleus* 9:1–16.

Mitchell, M. B., and Mitchell, H. K. (1952) "A Case of 'Maternal' Inheritance in *Neurospora crassa*." *Proceedings of the National Academy of Sciences* 38:442–449.

Moewus, Franz (1940a) "Uber Mutationen der Sexual-Gene bei Chlamydomonas." *Biologisches Zentralblatt* 60:597–626.

Moewus, Franz (1940b) "Die Analyse von 42 erblichen Eigenschaften der *Chlamydomonas euga-*

metos—Gruppe. I. Teil: Zellform, Membran, Geisseln, Chloroplasts, Pyrenopid, Augenfleck, Zellteilung." *Zeitschrift für induktive Abstammungs und Vererbungslehre* 78:418–462.

Moewus, Franz (1940c) "Die Analyse von 42 erblichen Eigenschaften der *Chlamydomonas eugametos*—Gruppe. II. Teil: Zellresistenz, Sexualität, Zygote, Besprechung der Ergebnisse." *Zeitschrift für induktive Abstammungs und Vererbungslehre* 78:463–500.

Moewus, Franz (1940d) "Die Analyse von 42 erblichen Eigenschaften der *Chlamydomonas eugametos*—Gruppe. II. Teil: Die 10 Koppelungsgruppen." *Zeitschrift für induktive Abstammungs und Vererbungslehre* 78:501–522.

Monod, Jacques (1947) "The Phenomenon of Enzymatic Adaptation and Its Bearings on Problems of Genetics and Cellular Differentiation." *Growth,* 2:223–289.

Monod, Jacques (1956) "Remarks on the Mechanism on Enzyme Induction." In *Enzymes: Units of Biological Structure and Function,* Henry Ford Hospital International Symposium. New York: Academic Press, pp. 7–28.

Monod, Jacques (1972) *Chance and Necessity,* translated from the French by A. Wainhouse. New York: Vintage Books.

Monod, Jacques, and Cohn, M. (1952) "La Biosynthèse Induite des Enzymes (Adaptation Enzymatique)." *Advances in Enzymology* 13:67–119.

Monod, Jacques, and François, J. (1961) "General Conclusions: Teleonomic Mechanisms in Cellular Metabolism, Growth and Differentiation." *Cold Spring Harbor Symposia on Quantitative Biology* 26:389–401.

Morgan, T. H. (1893) "An Organism Produced Sexually without Characteristics of the Mother." *American Naturalist* 27:222–232.

Morgan, T. H. (1895a) "Studies of 'Partial' Larvae of Sphaerechinus." *Archiv für Entwicklungsmechanik* 2:81–125.

Morgan, T. H. (1895b) "The Fertilization of Non-Nucleated Fragments of Echinoderm-Eggs." *Archiv für Entwicklungsmechanik* 2:267–280.

Morgan, T. H. (1910) "Chromosomes and Heredity." *American Naturalist* 65:449–496.

Morgan, T. H. (1917) "The Theory of the Gene." *American Naturalist* 51:513–544.

Morgan, T. H. (1919) *The Physical Basis of Heredity*. Philadelphia: J.B. Lippincott Co.

Morgan, T. H. (1926a) "Genetics and the Physiology of Development." *Am. Nat*. 60:489–515.

Morgan, T. H. (1926b) *The Theory of the Gene*. New Haven: Yale University Press.

Morgan, T. H. (1932) "The Rise of Genetics. II." *Science* 76:261–267.

Morgan, T. H. (1934) *Embryology and Genetics*. New York: Columbia University Press.

Morgan, T. H., Sturtevant, A. H., Muller, H. J., and Bridges, C. B. (1915) *The Mechanism of Mendelian Heredity*. London: Constable and Company Ltd.

Mulkay, Michael, and Gilbert, G. N. (1982) "Accounting for Error: How Scientists Construct Their Social World When They Account for Correct and Incorrect Belief." *Sociology* 16:165–183.

Muller, H. J. (1929) "The Gene as the Basis of Life." *Proceedings of the International Congress of Plant Sciences,* (Ithaca, 1926), 1:897–921.

Muller, H. J. (1930) "Radiation and Genetics." *American Naturalist* 64:220–251.

Muller, H. J. (1948a) "The Destruction of Science in the U.S.S.R." *Saturday Review of Literature* 31 (December 4):13–15.

Muller, H. J. (1948b) "Back to Barbarism—Scientifically." *Saturday Review of Literature* 31 (December 11):8–10.

Muller, H. J. (1949a) "Genetics in the Scheme of Things." *Proceedings of the 8th International Congress of Genetics: Hereditas* (*Suppl. Vol.*), pp. 96–127.

Muller, H. J. (1949b) "It Still Isn't a Science. A Reply to George Bernard Shaw." *Saturday Review of Literature* 32:11–12.

Muller, H. J. (1949c) "The Russian Counterrevolution Against Biological Science. A Review of Conway Zirkle's 'The Death of a Science in Russia.'" *New York Herald Tribune,* Dec. 11.

Muller, H. J. (1951) "The Development of the Gene Theory." In L. C. Dunn, ed., *Genetics in the 20th Century*. New York: Macmillan, pp. 77–100.

Mullins, N. C. (1971) "The Development of a Scientific Specialty: The Phage Group and the Origins of Molecular Biology." *Minerva* 10:51–82.

Nanney, D. L. (1953) "Mating Type Determination in *Paramecium aurelia,* a Model of Nucleo-Cytoplasmic Interaction." *Proceedings of the National Academy of Sciences* 39:113–119.

Nanney, D. L. (1954) "Mating Type Determination in *Paramecium aurelia.* A Study in Cellular Heredity." In D. H. Wenrich, ed., *Sex in Microorganisms.* Washington, D.C.: American Association for the Advancement of Science, pp. 266–283.

Nanney, D. L. (1957) "The Role of the Cytoplasm in Heredity." In W. D. McElroy and H. B. Glass, eds., *The Chemical Basis of Heredity.* Baltimore: Johns Hopkins Press, pp. 134–164.

Nanney, D. L. (1958a) "Epigenetic Control Systems." *Proceedings of the National Academy of Sciences* 44:712–717.

Nanney, D. L. (1958b) "Epigenetic Factors Affecting Mating Type Expression in Certain Ciliates." *Cold Spring Harbor Symposia on Quantitative Biology* 23:327–335.

Nanney, D. L. (1968) "Cortical Patterns in Cellular Morphogenesis." *Science* 160:496–502.

Nanney, D. L. (1980) *Experimental Ciliatology.* New York: John Wiley and Sons.

Nanney, D. L. (1982a) "T. M. Sonneborn: An Interpretation." *Annual Review of Genetics* 15:1–9.

Nanney, D. L. (1982b) "T. M. Sonneborn: An Appreciation." *Genetics* 102:1–7.

Nanney, D. L. (1983) "The Cytoplasm and the Ciliates." *Journal of Heredity* 74:163–170.

Nanney, D. L. (1984) "Microbial Precursors of Developmental Processes." *Verh. Dtsch. Zool. Ges.* 77:24–29.

Nanney, D. L., and Caughey, P. A. (1953) "Mating Type Determination in *Tetrahymena pyriformis.*" *Proceedings of the National Academy of Sciences* 39:1057–1063.

Needham, J. G. (1919) "Methods of Securing Better Cooperation between Government and Laboratory Zoologists in the Solution of Problems of General or National Importance." *Science* 49:455–458.

Noack, K. L. (1931) "Ueber Hypericum-Kreuzungen I. Die Panaschure der Bastarde Zwischen *Hypericum acutum* Moench und *Hypericum monitanun* L." *Zeitschrift für induktive Abstammungs und Vererbungslehre* 59:77–101.

Olby, Robert (1974) *The Path to the Double Helix.* London: Macmillan.

Olby, Robert (1984) "The Sheriff and the Cowboys: Or Weaver's Support of Astbury and Pauling." *Social Studies of Science* 14:244–247.

Oppenheimer, J. M. (1965) "Questions Posed by Classical Descriptive and Experimental Embryology." In J. A. Moore, ed., *Ideas in Modern Biology.* New York: The Natural History Press, pp. 205–227.

"Party Line Genetics." *Newsweek,* September 6, 1948, p. 45.

Pauly, August (1905) *Darwinismus und Lamarckismus.* Munich: E. Reinhardt Verlag.

Pearson, K. (1898) "Mathematical Contributions to the Theory of Evolution. On the Law of Ancestral Heredity." *Proceedings of the Royal Society* 62:386–412.

Pearson, Karl (1900) *The Grammar of Science,* 2nd ed. London: Macmillan.

Pellew, Caroline (1929) "The Genetics of Unlike Reciprocal Hybrids." *Cambridge Philosophical Society Biological Reviews and Proceedings* 14:209–217.

Pfeifer, E. J. (1965) "The Genesis of American Neo-Lamarckism." *Isis* 56:156–167.

Pfetsch, Frank (1974) *Zur Entwicklung der Wissenschaftspolitik in Deutschland 1750–1914.* Berlin: Duncker and Humblot.

Plate, L. (1913) *Selectionsprinzip und Probleme der Artbildung,* 4th ed., Leipzig and Berlin: W. Engelmann.

Plate, L. (1925) *Die Abstammungslehre.* Jena: G. Fischer.

Plough, Harold (1954) "Edwin Grant Conklin." *Genetics* 39:1–3.

Plough, H. H., and Ives, P. T. (1932) "New Evidence of the Production of Mutations by High Temperature, with a Critique of the Concept of Directed Mutations." *Proceedings of the Sixth International Congress of Genetics* 2:156–157.

Plough, H. H., and Ives, P. T. (1934) "Heat Induced Mutations in *Drosophila.*" *Proceedings of the National Academy of Sciences* 20:268–273.

Plough, H. H., and Ives, P. T. (1935) "Induction of Mutations by High Temperature in *Drosophila.*" *Genetics* 20:42–69.

Portier, P. (1918) *Les symbiotes.* Paris: Masson et Cie.

Preer, J. R. (1946) "Some Properties of the Genetic Cytoplasmic Factor in *Paramecium*." *Proceedings of the National Academy of Sciences* 32:247–253.

Preer, J. R. (1948a) "A Study of Some Properties of the Cytoplasmic Factor, "Kappa," in *Paramecium aurelia,* Variety 2." *Genetics* 33:349–404.

Preer, J. R. (1948b) "The Killer Cytoplasmic Factor Kappa, Its Rate of Reproduction, the Number of Particles per Cell, and Its Size." *American Naturalist* 82:35–42.

Preer, J. R. (1950) "Microscopically Visible Bodies in the Cytoplasm of the "Killer" Strains of *Paramecium aurelia.*" *Genetics* 35:344–362.

Preer, J. R. (1963) Discussion following the paper of D. L. Nanney, "Cytoplasmic Inheritance in Protozoa." In W. J. Burdette, ed., *Methodology in Basic Genetics.* San Francisco: Holden-Day, pp. 353–378.

Preer, J. R. (1969) "Genetics of Protozoa." In *Research in Protozoology, Vol. 3.* New York: Pergamon Press, pp. 133–278.

Prenant, Marcel (1943) *Biology and Marxism,* translated by C. Desmond Greaves with a foreword by Joseph Needham. New York: International Publishers.

Prezent, I. I. (1947) *Agrobiologia,* Translated by Eugenia Artschuager. Moscow.

Prosser, C. L. (1965) "Levels of Biological Organization and Their Physiological Significance." In John A. Moore, ed., *Ideas in Modern Biology.* New York: The Natural History Press, pp. 357–390.

Provine, W. B. (1971) *The Origins of Theoretical Population Genetics.* Chicago: The University of Chicago Press.

Provine, W. B. (1979) "Francis B. Sumner and the Evolutionary Synthesis." In W. Coleman and C. Limoges, eds., *Studies in the History of Biology.* Baltimore: The Johns Hopkins University Press, pp. 211–240.

Renner, Otto (1929) "Die Arbastarde bei Pflanzen." *Handbuch der Vererbungswissenschaft* 7:1–161.

Renner, Otto (1934) "Die pflanzlichen Plastiden als selbständige Elemente der genetischen Konstitution." *Berichter der Mathematisch-Physischen Klasse der Sächsischen Akademie der Wissenenschaften zu Leipzig* 86:241–266.

Renner, Otto (1936) "Zur Kenntnis der nichtmendelnden Buntheit der Laubblätter." *Flora* 30:218–290.

Renner, Otto (1961) "William Bateson und Carl Correns." In: *Sitzungs Berichte der Heidelberger Akademie der Wissenschaften,* pp. 159–181.

Rensch, Bernard (1980) "Neo-Darwinism in Germany." In Ernst Mayr and W. B. Provine, eds., *The Evolutionary Synthesis.* Cambridge: Harvard University Press, pp. 284–302.

Rhoades, M. M. (1933) "The Cytoplasmic Inheritance of Male Sterility in *Zea Mays.*" *Journal of Genetics* 27:71–93.

Rhoades, M. M. (1943) "Genic Induction of an Inherited Cytoplasmic Difference." *Proceedings of the National Academy of Sciences* 29:327–329.

Rhoades, M. M. (1946) "Plastid Mutations." *Cold Spring Harbor Symposia on Quantitative Biology* 11:202–207.

Rhoades, M. M. (1955) "Interaction of Genic and Non-Genic Hereditary Factors and the Physiology of Non-Genic Inheritance." *Encyclopedia of Plant Physiology* 1:19–57.

Ringer, Fritz (1969) *The Decline of the German Mandarins.* Cambridge: Harvard University Press.

Rizet, G. (1942) "Sur l'hérédité d'un caractère de croissance et du pouvoir germinatif des spores dans une lignée de l'Ascomysete *Podospora anserina.*" *Bulletin de la société Linneénne de Normandie* 2:131–136.

Rizet, G., Marcou, D., and Schecroun, J. (1958) "Deux phénomènes d'hérédité cytoplasmique chez l'ascomycete *P. anserina.*" *Bull. Soc. Fr. Physiol. veg.* 4:136–159.

Rolls-Hansen, Nils (1985) "A New Perspective on Lysenko?" *Annals of Science* 42:261–278.

Rosenberg, C. E. (1976) *No Other Gods.* Baltimore: Johns Hopkins University Press.

Ross, H. (1941) "Über die Verschiedenheiten des dissimilatorischen Stoffwechsels in reziproken *Epilobium*-Bastarden und die physiologisch-genetische Ursache der reziproken Unterschiede. I. Die Aktivität der Peroxydase in reziproken *Epilobium*-Bastarden mit der Sippe Jena." *Zeitschrift für induktive Abstammungs und Vererbungslehre* 79:503–529.

Ross, H. (1948) "Über die Verschiedenheiten des dissimilatorischen Stoffwechsels in reziproken *Epilobium*-Bastarden und die physiologisch-genetische Ursache der reziproken Unterschiede. V. Über die Peroxydaseaktivität in gehemmten und enthemmten Wuchsformen reziproker *Epilobium*-Bastarde mit der *hirsutum*-Sippe Jena." *Zeitschrift für induktive Abstammungs und Vererbungslehre* 82:187–196.

Rostand, J. (1928) *Les chromosomes artisans de l'hérédité et du sexe*. Paris: Hachette.

Sager, Ruth (1954) "Mendelian and Non-Mendelian Inheritance of Streptomycin Resistance in *Chlamydomonas reinhardi*." *Proceedings of the National Academy of Sciences* 40:356–363.

Sager, Ruth (1955) "Inheritance in the Green Alga *Chlamydomonas reinhardi*." *Genetics* 40:476–489.

Sager, Ruth (1960) "Genetic Systems in *Chlamydomonas*." *Science* 132:1459–1465.

Sager, Ruth (1962) "Streptomycin as a Mutagen for Nonchromosomal Genes." *Proceedings of the National Academy of Sciences* 48:2018–2026.

Sager, Ruth (1963) "The Particulate Nature of Nonchromosomal Genes in *Chlamydomonas*." *Proceedings of the National Academy of Sciences* 50:260–268.

Sager, Ruth (1964) "Nonchromosomal Heredity." *New England Journal of Medicine* 271:352–357.

Sager, Ruth (1966) "Mendelian and Non-Mendelian Heredity: A Reappraisal." *Proceedings of the Royal Society* 164:290–297.

Sager, Ruth (1972) *Cytoplasmic Genes and Organelles*. New York and London: Academic Press.

Sager, Ruth, and Ramanis, Zenta (1967) "Biparental Inheritance of Non-Chromosomal Genes Induced by Ultraviolet Irradiation." *Proceedings of the National Academy of Sciences* 58:931.

Sager, Ruth, and Ryan, Francis (1961) *Cell Heredity*. New York: John Wiley and Sons.

Saha, Margaret Samosi (1984) Carl Correns and an Alternative Approach to Genetics: The Study of Heredity in Germany Between 1880 and 1930. Ph.D. Dissertation, Michigan State University.

Scheibe, A. (1961) "Max-Planck-Institut für Züchtungsforschung (Erwin-Baur-Institut) in 'Köln-Vogelsang.'" *Jahrbuch der Max-Planck-Gesellschaft zur Förderung der Wissenschaften* e.V., pp. 823–859.

Schiemann, E. (1934) "Erwin Baur." *Bericht der Deutschen Botanisch Gesellschaft* 52:51–114.

Schmidt, W. (1931) "Richard Wettstein." *XL Jahresbericht des Sonnblick-Vereines*, pp. 3–4.

Schultz, J. (1950) "The Question of Plasmagenes." *Science* 3:403–407.

Schuster, J. A., and Yeo, R. R., eds. (1986) *The Politics and Rhetoric of Scientific Method*. Dordrecht, Holland: D. Reidel Publishing Company.

Schwanitz, F. (1932) "Expérimentelle Analyse der Genom- und Plasmonwirkung bei Moosen V. Protonemaregeneration aus Blättchen, Chloroplastengrösse, Chloroplastenzahl, assimilatorische Relation." *Zeitschrift für induktive Abstammungs und Vererbungslehre* 62:232–248.

Shull, A. F. (1916) "Cytoplasm and Heredity." *Ohio Journal of Science* 17:1–8.

Shull, G. H. (1946) "Hybrid Seed Corn." *Science* 103:547–550.

Sinnott, E. W., and Dunn, L. C. (1939) *Principles of Genetics*, 3rd ed. New York: McGraw-Hill.

Sinnott, E. W., and Dunn, L. C. and T. Dobzhansky (1950) *Principles of Genetics*, 4th ed. New York: McGraw-Hill.

Sirks, M. J. (1938) "Plasmatic Inheritance." *Botanical Review* 4:113–131.

Slonimski, P. (1952) *Recherche sur la formation des enzymes respiratoires chez la levure*. Paris: Thèse, Faculté des Sciences.

Slonimski, P., and Ephrussi, Boris (1949) "Action de l'acriflavine sur les levures V. Les systèmes des cytochromes des mutants 'petite colonie.'" *Annales de l'Institut Pasteur* 77:47–63.

Sonneborn, T. M. (1930a) "Cause, Inheritance, and Effects of the Chain Forming Tendency in the Cilate Protozoan, *Colpidium Campylum*." *Collecting Net* 5:232–234.

Sonneborn, T. M. (1930b) "Genetic Studies on *Stenostomum incaudatum* (nov. spec.) I. The Nature and Origin of Differences among Individuals Formed During Vegetative Reproduction." *Journal of Experimental Zoology* 57:57–108.

Sonneborn, T. M. (1930c) "Genetic Studies on *Stenostomum incaudatum*. II. The Effects of Lead Acetate on the Hereditary Constitution." *Journal of Experimental Zoology* 57:409–439.

Sonneborn, T. M. (1931) "McDougall's Lamarckian Experiment." *American Naturalist* 65:541–550.

Sonneborn, T. M. (1932) "Experimental Production of Chains and its Genetic Consequences in the Ciliate Protozoan, *Colpidium campylum*." *Biological Bulletin* 62:258–293.

Sonneborn, T. M. (1933) "Mendelian Methods Applied to the Ciliate Protozoan, *Paramecium aurelia*." *American Naturalist* 67:72.

Sonneborn, T. M. (1936) "Factors Determining Conjugation in *Paramecium aurelia*. I. The Cyclical Factor. The Recency of Nuclear Reorganization. II. Genetic Diversities between Stocks or Races." *Genetics* 21:503–518.

Sonneborn, T. M. (1937) "Sex, Sex Inheritance and Sex Determination in *Paramecium aurelia*." *Proceedings of the National Academy of Sciences* 23:378–395.

Sonneborn, T. M. (1938) "Mating Types in *Parmecium aurelia:* Diverse Conditions for Mating in Different Stocks; Occurrence, Number and Interrelations of the Types." *Proceedings of the American Philosophical Society* 79:411–434.

Sonneborn, T. M. (1939) "Sexuality and Related Problems in *Paramecium*." *Collecting Net* 14:7–84.

Sonneborn, T. M. (1942) "Inheritance in Ciliate Protozoa." *American Naturalist* 76:46–62.

Sonneborn, T. M. (1943) "Gene and Cytoplasm. I. The Determination and Inheritance of the Killer Character in Variety 4 of *Paramecium aurelia*. II. The Bearing of the Determination and Inheritance of Characters in *Paramecium aurelia* on the Problems of Cytoplasmic Inheritance, *Pneumonococcus* Transformations, Mutations and Development." *Proceedings of the National Academy of Sciences* 29:329–343.

Sonneborn, T. M. (1946) "Experiment Control of the Concentration of Cytoplasmic Genetic Factors in *Paramecium*." *Cold Spring Harbor Symposia on Quantitative Biology* 11:236–255.

Sonneborn, T. M. (1947a) "A New Genetic Mechanism and Its Relation to Certain Types of Cancer." *Quarterly Bulletin of the Indiana University Medical Center* 9:51–55.

Sonneborn, T. M. (1947b) "Developmental Mechanisms in *Paramecium*." *Growth Symposium* 11:291–307.

Sonneborn, T. M. (1948a) "Genes, Cytoplasm and Environment in *Paramecium*." *Scientific Monthly* 67:154–166.

Sonneborn, T. M. (1948b) "Herbert Spencer Jennings." *Genetics* 33:1–4.

Sonneborn, T. M. (1948c) "Symposium on Plasmagenes, Genes and Characters in *Paramecium aurelia*." *American Naturalist* 82:26–34.

Sonneborn, T. M. (1948d) "The Determination of Hereditary Antigenic Differences in Genetically Identical *Paramecium* Cells." *Proceedings of the National Academy of Sciences* 34:413–418.

Sonneborn, T. M. (1949) "Beyond the Gene." *American Scientist* 37:33–59.

Sonneborn, T. M. (1950a) "Heredity, Environment, and Politics." *Science* 111:529–539.

Sonneborn, T. M. (1950b) "Methods in the General Biology and Genetics of *Paramecium aurelia*." *Journal of Experimental Zoology* 113:87–148.

Sonneborn, T. M. (1950c) "*Paramecium* in Modern Biology." *Bios* 21:31–43.

Sonneborn, T. M. (1950d) "Partner of the Genes." *Scientific American* November: 30–39.

Sonneborn, T. M. (1950e) "The Cytoplasm in Heredity." *Heredity* 4:11–36.

Sonneborn, T. M. (1950f) "The Role of Cytoplasm in Heredity." *Centennial Volume of the AAAS* 60:243–247.

Sonneborn, T. M. (1951a) "Beyond the Gene—Two Years Later." In George A. Baitsell, ed., *Science in Progress, 7th Series*. New Haven: Yale University Press, pp. 167–203.

Sonneborn, T. M. (1951b) "The Role of the Genes in Cytoplasmic Inheritance." In L. C. Dunn, ed., *Genetics in the 20th Century*. New York: Macmillan, pp. 291–314.

Sonneborn, T. M. (1955) "Protozoa in the General Biology or Zoology Course." *American Biology Teacher* 17:187–190.

Sonneborn, T. M. (1959) "Kappa and Related Particles in Paramecium." *Advances in Virus Research* 6:229–356.

Sonneborn, T. M. (1960) "The Gene and Cell Differentiation." *Proceedings of the National Academy of Sciences* 46:149–165.

Sonneborn, T. M. (1963a) "Bearing of Protozoan Studies on Current Theory of Genic and Cytoplasmic Actions." *Proceedings of the XVI International Zoology Congress* 3:197–202.

Sonneborn, T. M. (1963b) "Does Preformed Cell Structure Play an Essential Role in Cell Heredity?"

In J. M. Allen, ed., *The Nature of Biological Diversity*. New York: McGraw-Hill, pp. 165–221.

Sonneborn, T. M. (1964) "The Differentiation of Cells." *Proceedings of the National Academy of Sciences* 51:915–929.

Sonneborn, T. M. (1968) "H. J. Muller, Crusader for Human Betterment." *Science* 162:772–776.

Sonneborn, T. M. (1975) "Herbert Spencer Jennings 1868–1947." *National Academy of Sciences Biographical Memoirs* 47:143–223.

Sonneborn, T. M., and Beale, G. H. (1949a) "Genes, Plasmagenes and Environment in the Control of Antigenic Traits in *Paramecium aurelia* (variety 4)." *Hereditas* (*Suppl. Vol.*), pp. 451–460.

Sonneborn, T. M., and Beale, G. H. (1949b) "Influence des gènes, des plasmagènes et du milieu dans le déterminisme des caractères antigèniques chez *Paramecium aurelia* (variété 4)." In *Unités biologiques douées de continuité génétique*. Paris: Edition du Centre National de la Recherche Scientifique, pp. 25–36.

Spemann, Hans (1914) "Über verzögerte Kernversorgung von Keimteilen." *Verhandl. Deut. Zool Ges.*, pp. 16–221.

Spemann, Hans (1938) *Embryonic Development and Induction*, reprinted 1967. New York: Hafner Publishing Company.

Spiegelman, S. (1946) "Nuclear and Cytoplasmic Factors Controlling Enzymatic Constitution." *Cold Spring Harbor Symposia on Quantitative Biology* 11:256–277.

Spiegelman, S., and Kamen, M. D. (1946) "Genes and Nucleoproteins in the Synthesis of Enzymes." *Science* 104:581–584.

Spiegelman, S., Lindegren, C. C., and Lindegren, G. (1945) "Maintenance and Increase of a Genetic Character by a Substrate-Cytoplasmic Interaction in the Absence of the Specific Gene." *Proceedings of the National Academy of Sciences* 31:95–102.

Stent, G. S. (1971) *Molecular Genetics: An Introductory Narrative*. San Francisco: W.H. Freeman and Company.

Stubbe, H. (1950–51) "Nachruf auf Fritz von Wettstein." *Jahrbuch der Deutschen Akademie der Wissenschaften zu Berlin*, pp. 1–12.

Sturtevant, A. H. (1923) "Inheritance of Direction of Coiling in *Limnaea*." *Science* 58:269–270.

Sturtevant, A. H. (1959) "Thomas Hunt Morgan." *National Academy of Sciences Biographical Memoirs* 33:282–325.

Sturtevant, A. H. (1965) *A History of Genetics*. New York: Harper & Row.

Sturtevant, A. H., and Beadle, G. W. (1939) *An Introduction to Genetics*. New York: Dover Publications.

Sumner, F. B. (1928) "Observations on the Inheritance of a Multifactor Color Variation in White-footed Mice (*Peromyscus*)." *American Naturalist* 62:193–206.

Tartar, Vance (1961) *The Biology of Stentor*. New York: Pergamon Press.

Ternitz, Charlotte (1912) "Beiträge zur Morphologie und Physiologie der *Euglena gracilis* Klebs." *Jahrb. Wiss. Botan.* 51:435–514.

Tobler, M. (1932) "Experimentelle Analyse der Genom- und Plasmonwirkung bei Moosen IV. Zur Variabilität des Zellvolumens einer Sippenkreuzung von *Funaria hydrometrica* und deren bivalenten Rassen." *Zeitschrift für induktive Abstammungs und Vererbungslehre* 60:39–62.

Toyama, K. (1912–1913) "Maternal Inheritance and Mendelism." *Journal of Genetics* 2:351–405.

Uda, Hajime (1923) "On 'Maternal Inheritance.'" *Genetics* 8:322–335.

Uexküll, J. von (1913) *Bausteine zu einer biologischen Weltanschauung*. München.

Waddington, C. H. (1939) *An Introduction to Genetics*. London:Allen and Unwin.

Waddington, C. H. (1940) *Organisers and Genes*. Cambridge: Cambridge University Press.

Waddington, C. H. (1953) "Role of Plasmagenes." *Nature* 172:784–785.

Wallin, J. E. (1927) *Symbionticism and the Origin of the Species*. Baltimore: Williams and Wilkins.

Watson, J. D., and Crick, F. H. C. (1953) "Genetical Implications of the Structure of Deoxyribonucleic Acid." *Nature* 171:964–967.

Webber, J. (1912) "The Effect of Research in Genetics on the Art of Breeding." *American Breeder's Magazine* 3:29–36.

Weindling, P. J. (1981) "Theories of the Cell State in Imperial Germany." In C. Webster, ed., *Biology, Medicine and Society 1840–1940*. Cambridge: Cambridge University Press.

Weindling, P. J. (1982) *Cell Biology and Darwinism in Imperial Germany: The Contribution of Oscar Hertwig (1849–1922)*. Ph.D. Dissertation, University of London.

Weismann, August (1883) "On Heredity." Republished in *Essays Upon Heredity and Kindred Biological Problems,* translated by E. B. Poulton *et al.*, 2 vols., 2nd ed. Oxford: Clarendon Press, 1891–1892, vol. 1, pp. 67–70.

Wiesmann, August (1885) "Continuity of the Germ-Plasm as the Foundation of a Theory of Heredity." Republished in *Essays Upon Heredity and Kindred Biological Problems,* translated by E. B. Poulton, *et al.*, 2 vols., 2nd ed. Oxford: Clarendon Press, 1891–1892, vol. 1, pp. 163–255.

Weismann, August (1893) *The Germ Plasm: A Theory of Heredity,* translated by W. N. Parker and H. Rofeldt. London: Walter Scott Ltd.

Weiss, Paul (1939) *Principles of Development*. New York: Henry Holt and Co.

Weiss, Paul (1947) "The Problem of Specificity in Growth and Development." *Yale Journal of Biology and Medicine* 19:235–278.

Werskey, Gary (1978) *The Visible College*. New York: Holt, Rinehart and Winston.

Wettstein, Fritz von (1939) "Botanik Paläobotanik, Vererbungsforschung und Abstammungslehre," *Paläobiologie* 7:154–168.

Wettstein, Fritz von (1924) "Morphologie und Physiologie des Formwechsels der Moose auf genetischer Grundlage I." *Zeitschrift für induktive Abstammungs und Vererbungslehre* 33:1–236.

Wettstein, Fritz von (1926) "Über plasmatische Vererbung, sowie Plasma- und Genwirkung." *Nachrichten von Gesellschaft der Wissenschaft zu Göttingen, Math-Phys. Kl.*, pp. 250–281.

Wettstein, Fritz von (1928) "Über plasmatische Vererbung und über das Zusammenwirken von Genen und Plasma." *Berichten der Deutschen Botanischen Gesellschaft* 46:32–49.

Wettstein, Fritz von (1934) "Carl E. Correns." *Naturwissenschaften Jahrgang* 22:1–8.

Wettstein, Fritz von (1937) "Die genetische und entwicklungsphysiologische Bedeutung des Cytoplasmas." *Zeitschrift für induktive Abstammungs und Vererbungslehre* 73:345–366.

Wettstein, Fritz von (1939) "Carl E. Correns." *Berichten der Deutschen Botanischen Gesellschaft* 56:140–160.

Wettstein, Richard (1935) *Handbuch der Systematischen Botanik*. Leipzig and Vienna: Franz Dehtiche. Reprinted, Amsterdam: Asher & Co.

Wheeler, W. M. (1923) "The Dry-Rot of Our Academic Biology." *Science* 57:61–71.

Whitman, C. O. (1893) "The Inadequacy of the Cell-Theory of Development." *Journal of Morphology* 8:639–658.

Wigglesworth, V. B. (1945) "Growth and Form in an Insect." In *Essays on Growth and Form*. Oxford: Oxford University Press, pp. 24–40.

Wilkie, D. (1969) *The Cytoplasm in Heredity*. New York: John Wiley & Sons.

Wilson, E. B. (1896) *The Cell in Development and Inheritance*. New York: Macmillan.

Wilson, E. B. (1900) *The Cell in Development and Inheritance*, 2nd ed. New York: Macmillan.

Wilson, E. B. (1914) "The Bearing of Cytological Research on Heredity." *Science* 88:333–352.

Wilson, E. B. (1918) "Thedor Boveri." In W. C. Roentgen, ed., *Erinnerungen an Theodor Boveri*. Tübingen: J.C.B. Mohr, pp. 67–89.

Wilson, E. B. (1923) "The Physical Basis of Life." *Science* 57:277–278.

Wilson, E. B. (1925) *The Cell in Development and Heredity,* 3rd ed. New York: Macmillan.

Winge, O. (1935) "On *Haplophase* and *Diplophase* in Some *Saccaromycetes*." *Comptes-rendus des travaux du Laboratoire Carlsberg, série physiologie* 21:77–111.

Winge, O., and Lausten, O. (1938) "Artificial Species Hybridization in Yeast." *Comptes-rendus des travaux du Laboratoire Carlsberg, série physiologie* 22:235–244.

Winge, O., and Lausten, O. (1940) "On a Cytoplasmic Effect of Inbreeding in Homozygous Yeast." *Comptes-rendus des travaux du Laboratoire Carlsberg série physiologie* 23:17–39.

Winkler, Hans (1901) "Ueber Merogonie und Befruchtung." *Jahrbucher für Wissenschaftlichen Botanik* 36:753–775.

Winkler, Hans (1924) "Über die Rolle von Kern und Protoplasma bei der Vererbung." *Zeitschrift für induktive Abstammungs und Vererbungslehre* 33:238–253.

Woolf, Virginia (1928) *A Room of One's Own.* Republished 1975, Middlesex, England: Penguin Books.

Wright, Sewall (1941) "The Physiology of the Gene." *Physiological Reviews* 21:487–527.

Wright, Sewall (1945) "Genes as Physiological Agents." *American Naturalist* 79:289–303.

Wright, Sewall (1958) "Genetics, the Gene, and the Hierarchy of Biological Sciences." *Tenth International Congress of Genetics,* pp. 475–489.

Young, Robert (1978) "Getting Started on Lysenkoism." *Radical Science Journal* 6/7:81–106.

Yoxen, E. J. (1984) "Scepticism about the Centrality of Technology Transfer in the Rockefeller Foundation Programme in Molecular Biology." *Social Studies of Science* 14:248–252.

Zirkle, Conway (1949) "Death of a Science in Russia." Reprinted 1970 in Philip Appleman, ed., *Darwin.* New York: W.W. Norton & Company, pp. 551–564.

Index